AF595552

Reptile Evolution and Genetics - Special Issue Dedicated to the Memory of Prof. Teresa Capriglione

Reptile Evolution and Genetics - Special Issue Dedicated to the Memory of Prof. Teresa Capriglione

Editor

Ettore Olmo

MDPI • Basel • Beijing • Wuhan • Barcelona • Belgrade • Manchester • Tokyo • Cluj • Tianjin

Editor
Ettore Olmo
Università Politecnica delle
Marche
Ancona, Italy

Editorial Office
MDPI
St. Alban-Anlage 66
4052 Basel, Switzerland

This is a reprint of articles from the Special Issue published online in the open access journal *Animals* (ISSN 2076-2615) (available at: https://www.mdpi.com/journal/animals/special_issues/Reptile_Evolution_and_Genetics_Teresa_Capriglione).

For citation purposes, cite each article independently as indicated on the article page online and as indicated below:

LastName, A.A.; LastName, B.B.; LastName, C.C. Article Title. *Journal Name* **Year**, *Volume Number*, Page Range.

ISBN 978-3-0365-8170-5 (Hbk)
ISBN 978-3-0365-8171-2 (PDF)

Contents

About the Editor

Ettore Olmo

Ettore Olmo is Full Professor (Emeritus) of Cytology and Histology in the Department of Life and Environmental Sciences, Marche Polytechnic University in Ancona (Italy). He previously was Full Professor of Cytology and Histology in the Faculty of Sciences at the University of Naples "Federico II" (Italy). He was President of the Italian Zoological Union and at present is Editor-in-Chief of *The European Zoological Journal*. Ettore Olmo has published 150 scientific papers on cytogenetics and the molecular evolution of repetitive DNA of reptiles, amphibians, fishes and mollusks. He is a member of various international scientific societies; in particular, he was a member of the International Society of Molecular Evolution and the Standing Committee of the International Chromosomes Conference.

Editorial

Reptile Evolution and Genetics: An Overview

Ettore Olmo

Department of Life and Environmental Sciences, Università Politecnica delle Marche via Brecce Bianche Ancona, 60121 Ancona, Italy; e.olmo@univpm.it

The study of evolution has been indissolubly linked to the study of heredity since its inception [1]. Therefore, genetic and genomic research is essential to understand the phylogenetic relationship at different taxonomic levels and to outline the main evolutionary trends.

The works in this Special Issue contribute to developing more efficient molecular methods, such as the most recent DNA sequencing techniques; isolating single chromosomes or part of them via flow sorting or microdissection; amplifying specific DNA sequences; and identifying the exact localization of these probes on mitotic chromosomes or interphase nuclei.

These methodological approaches particularly demonstrated that the genome of eukaryotes, besides functional genes, contains different types of DNA like the repetitive sequences. The most noteworthy applications included highly repetitive satellite DNAs and mobile elements (transposons and retrotransposons), whose role is not still completely understood but could play a non-negligible role in evolutionary processes [2].

Since classical karyological research evidenced an extensive variation in chromosome number and structure among species, further information gleaned from molecular biology and genome sequencing regarding chromosome structure and molecular composition confirmed that these differences are widespread and might have evolutionary consequences [3].

In 1994, Gauthier [4] proposed a cladistic definition of reptiles as a monophyletic crown group containing the so-called non-avian reptiles (turtles, lizards, snakes, and crocodiles) and birds, their common ancestors, and all their descendants. Non-avian reptiles are a very interesting group for genetic and genomic studies. They occupy a critical position in the evolution of amniotes, and their evolutionary history is well known thanks to the existing meticulous fossil records. They show a wide morphological and ecological variability with shapes adapted to broadly diverse environments. Large amounts of collected data define their biogeography, biology, and physiology and evidence certain peculiarities in their reproductive and developmental biology, such as viviparity [5] and parthenogenetic reproduction [6] in some species and the transition from strictly genetic to environmental sex determination [7]. Specific characteristics are also evident from the cytogenetic point of view, such as the different gene distribution between macro- and microchromosomes and a wide interspecific and intraspecific variability in chromosome number and morphology [8].

This Special Issue contains 12 articles advancing studies on the composition and evolution of the nuclear and mitochondrial DNAs of non-avian reptiles and of the genetic mechanisms linked to biological and evolutionary peculiarities, such as sex determination, differences in DNA methylation between various tissues, and the influence of incubation temperature on gene expression.

One relevant paper concerns genome evolution and phylogenomic [9] and evidences the advancement of whole genome sequencing in the general framework of the karyology and composition of non-avian reptiles. This study shows that genomic resources in non-avian reptiles have now accumulated more slowly than in other amniotic groups despite the extraordinary diversity of phenotypic and genomic traits.

A survey of phylogenomic investigation shows a prevalence of whole genome sequencing, especially regarding the analysis of ultraconserved elements (UCEs). However,

Citation: Olmo, E. Reptile Evolution and Genetics: An Overview. *Animals* **2023**, *13*, 1924. https://doi.org/10.3390/ani13121924

Received: 6 June 2023
Accepted: 6 June 2023
Published: 8 June 2023

many other types of markers exist and are increasingly well represented, being extracted from genome assembly in silico, including some with more significant information potential than UCEs for specific investigations.

Genome sequencing research collectively identified 139 reptilian species, providing a rich resource for in silico harvesting of information-rich markers for phylogenomics and a platform for finding the connection between genomes and phenotypic evolution. These breakthroughs could open a new era of integration of non-avian reptile comparative biology, natural history, cytogenetics, and genomics.

Karyological and genomic studies have evidenced that reptiles are a karyological heterogeneous group in which some orders and suborders exhibit characters similar to those of anamniotes, and others show similarities with homeothems. The class also presents different evolutionary trends in genome and chromosome size and composition [8]. The karyological influence on evolution can occur at different levels: chromosomal, genomic, and molecular.

In non-avian reptiles, notable differences in chromosome number and composition variability can be found between and within families. These variations seem to have had different effects on evolutionary mechanisms such as speciation [10]. Examples of these differences are evidenced in the following two research works.

A study on karyotype diversification in species of Malagasy leaf-toed geckos *Uroplatus* [11] shows that these species of gecckonid have 38 to 34 uniarmed chromosomes and tend to progressively reduce the chromosome number via the translocation of microchromosomes (especially those carrying NOR) to the telomere of macrochromosomes without changing the general morphology of the chromosomes. This evolutionary tendency towards chromosome reduction is largely shared by the gecko clade. Although the translocation to the telomeres of macrochromosomes in some species occurs, the translocation to the centromere also occurs with changes in morphology from uniarmed to biarmed.

The content of heterochromatin in these reptiles is very limited, and in some species, putative heteromorphic sex chromosomes were found.

The situation is very different in the iguanid *Liolaemus monticola* [12]. In this species, several chromosome races differ in the number and morphology of chromosomes due to centric fission and pericentric inversions on various pairs of chromosomes. These different races are arranged in a latitudinal sequence of increasing karyotype complexity from south to north.

The existence of the different chromosome races in this lizard suggests a complex evolutionary history of chromosomal rearrangements, population isolation by barriers, and hybridization. The results of this study evidence that chromosome variation could have a relevant role in reptile evolution, especially for rich-species groups such as *Liolaemus* lizards.

Investigations at the molecular level are also of significant interest, including an extensive study on the satellite DNAs in snake heterochromatin [13], which provides information on general mechanisms of molecular evolution.

This study reports the isolation of four DNA satellite families in snake species of different families: Colubridae and Viperidae. Three of these satellites are common to species of both families, while one is typical of viperids.

Analysis using FISH and BLAST methods shows that one DNA is mainly localized at the centromere of species belonging to both families, whereas the others form clusters specific on specific chromosomes or subsets of chromosomes.

Overall, the above-mentioned results, especially those on the localization of satellite DNAs, demonstrate the conservation of these repetitive elements in snakes. These results contrast the commonly shared opinion that satellite DNAs evolve extremely quickly and are usually species or genus specific [2,14]. The situation in snakes corroborates the "library" model according to which different satellite DNA families coexist in the genome of different species, and the appearance or disappearance of some of these sequences depends on changes in copy number.

Meiosis plays an essential role in controlling variability at the chromosome and gene levels [15].

Two articles studied meiosis in squamate species. Spangenberg et al. [16] analyzed mitosis and meiosis in the common adder *Vipera berus*, which has a bimodal karyotype with 16 macrochromosomes and 22 microchromosomes, using antibodies against meiotic components such as synaptinemal complex and DNA mismatch repair proteins MLH1.

Results of this research using the high-resolution SC karyotyping technique revealed the morphology of microchromosomes and differences in the dynamics of bivalent assembly between macro- and microchromosomes during meiotic prophase for the first time. Immunostaining of MLH1 showed that crossing over sites at pachytene is 49.5, and the number of MLH1 sites per bivalent reached 11, similar to that found in several species of Agamids. These MLH1 sites are higher in microbivalents than macrobivalents. This finding can be related to the enrichment of genes found generally in snake microchromosomes [17,18].

A second paper on meiosis is a review of meiotic chromosomes in the viviparous lizard *Zootoca vivipara* [19], which possesses female heterogamety and multiple sex chromosomes with variable W sex chromosome morphology and composition.

Multiple sex chromosomes and their change may influence meiosis and female meiotic drive, and they may play a role in reproductive isolation [20,21]. In two cryptic taxa of *Z. vivipara* with different W chromosomes, meiosis in spermatogenesis and oogenesis proceeded without disturbance. No variability in the chromosome pairing at the early stages of prophase I and no significant disturbance in chromosome segregation at the anaphase-telophase have been discovered. This suggests that there should be a factor maintaining multiple sex chromosomes, their equal transmission, and the course of meiosis in these cryptic forms of *Z. vivipara*. In this regard, it is interesting that the presence of interspersed elements and transposons in this lizard species is preferentially localized at centromeric, pericentromeric, and telomeric regions that are often of key importance for the spatial orientation of chromosomes in the nucleus and segregation during meiosis [22]. Therefore, we may assume that the specific cytogenetic and genomic composition of the W chromosomes and the SINE-Zv sequences in the peritelomeric–telomeric regions might play a role in the meiotic process and the behavior of sex chromosomes of *Z. vivipara*.

Another field of interest is the molecular mechanism of sex differentiation during embryonic development in TSD species. A study of the influence of temperature on gene expression in leopard geckos shows that temperature exposition during development modifies the expression of genes related to gonadal differentiation and those involved in different developmental pathways [23].

A different situation was found on the methylation level in the turtle *Chrysemys picta*, where gonads exhibit differential DNA methylation between males and females. However, no sexual differences can be recorded in somatic tissues. The results of this research highlight that differential DNA methylation is tissue specific and plays a role in gonadal formation, sex development, and maintenance post hatching, but not in the somatic tissues [24].

Besides studies on chromosomes and the nuclear genome, two papers investigate mitochondrial genome evolution.

One of them studied the DNA barcodes of terrapins and showed that this method is an excellent way to measure the diversity of a population. An analysis of the CO1 DNA of several Malaysian terrapins (eight geoemididae, three emididae, and one pelomedusid) provides new insight into the classification of terrapins and reveals the existence of potential cryptic species [25].

Another research article examined the evolutionary potential and phylogenetic utilities of duplicated CO1 control regions in some species of varanids [26]. Sequence analysis and phylogenetic relationship revealed that divergence between orthologous copies from different individuals was lower than the paralogous copies from the same individuals, indicating an independent evolution of the CRs. These results suggest that CO1 copies seem to have acquired concerted evolution across different species.

Promising perspectives for future studies of the evolution and genetics of reptiles derive from an extensive review of antimicrobials in snake venoms and a tentative hypothesis of the karyotype of dinosaurs.

Snakes have the relevant ability to live in different environments, resist different pathogens, and eat different prey; this could be linked to an immunity similar to that of mammals [27].

One of the major problems facing public health is the growing resistance of microbes to antibiotics, so multiple scientific approaches have been employed to find new antimicrobials with high therapeutic indexes. As a result, several natural secretions, including snake venoms, have been considered sources of bioactive compounds [28,29].

The review by Oguiura et al. [30] shows that snake venoms are rich in biomolecules that can be explored as biological tools for potential anti-inflammatory, analgesic, antitumor, and antimicrobial agents. This work also describes new beta-defensin sequences of *Sistrurus miliaris*. Another significant result obtained by Oguiura and colleagues is the advantage of using multidisciplinary approaches, including sequence phylogeny, with traditional techniques for searching for new molecules with therapeutic potential.

The paper from Griffin et al. [31] is an intriguing review on the state of the art of tentative reconstruction of dinosaurs' karyotypes research.

The divergence between the main lineage of crown reptiles, Lepidosauromorpha (tuatara, lizards, and snakes) and Archosauromorpha (turtles, crocodiles, dinosaurs, and birds), dates back to about 250 million years ago [32]. Despite the ancient divergence time, all crown reptiles' chromosomes and genome variability are low. In particular, most species, except crocodiles, of the two lineages have a karyotype characterized by the presence of macrochromosomes and microchromosomes [10], and recent studies suggest that this pattern was probably established about 255 years ago before the first divergence of main lineages of crown reptiles.

As no intact DNA is available from fossil dinosaurs, information about extinct dinosaurs' karyotypes can be inferred via comparative analysis of chromosomes and genomes of several species of birds and non-avian reptiles.

One approach, based on aligning chromosome-level assemblies from extant birds, determined the most likely ancestral karyotype of all birds [33]. A similar approach was used to reproduce the diapsid ancestral karyotype [34].

Another approach using chicken chromosome painting on chromosome sets of various turtle species evidenced the synteny in macrochromosomes of birds and turtles [35,36].

All of these results show that the avian chromosome pattern remained unchanged not only in most birds but also many extinct dinosaurs with a high degree of certainty [37].

The papers from this Special Issue summarize the state of genetic and genomic studies in reptiles and highlight that reptiles are a good model for studying the genetic and molecular basis of some key moments in vertebrate evolution. However, it is clear that the information collected so far is not sufficient to delineate a complete picture and that it would be important to increase the number of the whole genome sequencings and to deepen the knowledge of the molecular bases underlying some important cytological mechanisms such as meiotic pairing and segregation, the role of repetitive DNAs on the structure of chromosomes and its variations, sex determination, and the interaction between genetic and morphological level.

Conflicts of Interest: The author declares no conflict of interest.

References

1. Futuyma, D.J. *Evolutionary Biology*; Sinauer Associates Inc. Publishers: Sunderland, MA, USA, 1966.
2. Biscotti, M.A.; Heslop-Harrison, J.S.; Olmo, E. Repetitive DNA in eukaryotic genome. *Chromosome Res.* **2015**, *23*, 415–420. [CrossRef] [PubMed]
3. Peichel, C. Chromosome Evolution: Molecular Mechanisms and Evolutionary Consequences. *J. Hered.* **2017**, *108*, 1–2. [CrossRef] [PubMed]

4. Gauthier, J.A. The diversifiction of the amniotes. In *Major Features of Vertebrate Evolution*; Prothero, D.R., Scoch, R.M., Eds.; The Paleontological Society: Knoxville, TN, USA, 1994; pp. 129–159.
5. Pyron, R.A.; Burbrink, F.T. Early Origin of Viviparity and Multiple Reversion in Ovparity in Squamate Reptiles. *Ecol. Lett.* **2014**, *17*, 13–21. [CrossRef] [PubMed]
6. Darevsky, I.; Kupriyanova, L.; Uzzell, T. Parthenogenesis in reptiles. In *Biology of the Reptilia*; Gans, C., Ed.; Wiley: New York, NY, USA, 1985; pp. 412–516.
7. Bull, J.J. *Evolution of Sex Determining Mechanisms*; Benjamin/Cummings Publishing Company Inc.: Menlo Park, CA, USA, 1983; 316p.
8. Olmo, E. Trends in the Evolution of Reptilian Chromosomes. *Integr. Comp. Biol.* **2008**, *48*, 486–493. [CrossRef]
9. Card, D.C.; Jennings, B.W.; Edwards, S.V. Genome Evolution and the Future of Phylogenomics of Non-Avian Reptiles. *Animals* **2023**, *13*, 471. [CrossRef]
10. Olmo, E. Rate of Chromosomes Change and Speciation in Reptiles. *Genetica* **2005**, *125*, 185–203. [CrossRef]
11. Mezzasalma, M.; Brunelli, E.; Odierna, G.; Guarino, F.M. First Insight on the Karyotype Diversification of the Endemic Malagasy Leaf Toed Geckos (Squamata: Gekkonidae: *Uroplatus*). *Animals* **2022**, *12*, 2054. [CrossRef]
12. Lamborot, M.; Osea, C.G.; Aravena-Munoz, N.; Véliz, D. Chromosomal Evolution of the *Liolaemus monticola* (Liolaemidae) Complex: Chromosomal and Molecular Aspects. *Animals* **2022**, *12*, 3372. [CrossRef]
13. Lisachov, A.; Rumyantsev, A.; Prokopov, D.; Ferguson-Smith, M.; Trifonov, V. Conservation of Major Satellite DNAs in Snake Heterochromatin. *Animals* **2023**, *13*, 334. [CrossRef]
14. Ruiz-Ruano, F.J.; Lopez-Leòn, M.D.; Cabrero, J.; Camacho, J.P.M. High throughput Analysis of the Satellitome illuminates satellite DNA evolution. *Sci. Rep.* **2016**, *6*, 28333. [CrossRef]
15. Protacio, R.U.; Davidson, M.K.; Wahls, W.P. Adaptive Control of the Meiotic Recombinaton Landscape by DNA Site-Dependent Hotspots with Implication for Evolution. *Front. Genet.* **2022**, *13*, 947572. [CrossRef]
16. Spangenberg, V.; Redekop, H.; Simanovsky, S.A.; Kolomiets, O. Cytogenetic Analysis of the Bimodal Kartiotype of the Common European Adder *Vipera berus* (Viperidae). *Animals* **2022**, *12*, 3563. [CrossRef]
17. Axelson, F.; Webster, M.T.; Smith, N.G.; Burt, D.W.; Ellegren, H. Comparison of the Chicken and Turkey Genomes reveals a Higher Rate of Nucleotide Divergence on Microchromosomes than Macrochromosomes. *Genome Res.* **2005**, *15*, 120–125. [CrossRef]
18. Srikulnath, K.; Ahmad, S.F.; Sangchat, W.; Panthum, Y. Why Do Some Vertebrates have Microchromosomes? *Cell* **2021**, *10*, 2182. [CrossRef]
19. Kupriyanova, L.; Safronova, L.A. Brief Review of Meiotic Chromosomes in Early Spermatogenesis and Oogenesis and Mitotic Chromosomes in the Viviparous Lizard *Zootoca vivipara* (Squamata: Lacertidae) with Multiple Sex Chromosomes. *Animals* **2023**, *13*, 19. [CrossRef]
20. King, M. The Evolution of sex chromosomes in lizards. In *Evolution and Reproduction*; Australian Academy of Sciences: Canberra, Australia, 1977; pp. 55–60.
21. King, M. Species evolution. In *The Rate of Chromosome Change*; Cambridge University Press: Cambridge, UK, 1993; 336p.
22. Axelson, F.; van Albrechtsen, A.P.; Li, J.; Megens, H.J.; Vereijken, A.L.J.; Crooijmans, R.P.M.A.; Groenen, M.A.M.; Ellegren, H.; Wierslev, E.; Nielsen, R. Segregation Distorsion in Chicken and the Evolutionary Consequences of Female Meiotic Drive in Birds. *Heredity* **2010**, *105*, 290–298. [CrossRef]
23. Pallotta, M.M.; Fogliano, C.; Carotenuto, R. Temperature Incubation Influences Gonadal Gene Expression during Leopard Gecko Development. *Animals* **2022**, *12*, 3186. [CrossRef]
24. Mizoguchi, B.A.; Valenzuela, N. A Cautionay Tale of Sexing by Methylation: Hybrid Bisulfite-Conversion Sequencing of Immunoprecipritated Methylated DNA in *Chrysemis picta* Turtles with Temperature-Dependent Sex Determination Reveals Contrasting Patterns of Somatic and Gonadal Methylaton but No Unobtrusive Sex Diagnostic. *Animals* **2023**, *13*, 117. [CrossRef]
25. Mohd Salleh, M.H.; Esa, Y.; Mohammed, R. Global Terrapin Character-based DNA Barcodes: Assessment of the Mitochondrial CO! Gene and Conservation Status Revealed a Putative Crypic Species. *Animals* **2022**, *12*, 1720. [CrossRef]
26. Thapana, W.; Ariyaraphong, N.; Wongtienchai, P.; Laopichienpong, N.; Singchat, W.; Panthum, T.; Ahmad, S.F.; Kraichak, E.; Muangmai, N.; Duengkae, P.; et al. Concerted and Independent Evolution of Control Region 1 and 2 of Water Monitor Lizard (*Varanus salvator macromaculatus*) and Different Phylogenetic Informative Markers. *Animals* **2022**, *12*, 148. [CrossRef]
27. Rios, F.M.; Zimmermann, L.M. *Immunology of Reptiles*; John Wiley & Sons Ltd.: Chichester, UK, 2015.
28. Kumar, P.; Kizhakkedathu, J.N.; Straus, S.K. Antimicrobial Peptides Diversity Mechanism of Action and Strategies to Improve the Activity and Biocompatibility in Vivo. *Biomolecules* **2018**, *8*, 4. [CrossRef] [PubMed]
29. Stiles, B.G.; Sexton, F.W.; Weinstein, S.A. Antibacterial Effects of Different Snake Venoms Purification and Charcterization of Antibacterial Proteins from *Pseudechis australis* (Australian King Brown or Muga Snake) Venom. *Toxicon* **1991**, *29*, 1129–1141. [CrossRef] [PubMed]
30. Oguiura, N.; Sanches, L.; Duarte, P.V.; Sulca-Lopez, M.A.; Machini, T. Past, Present and Future of Naturally Occurring Antimicrobials Related to Snake Venom. *Animals* **2023**, *13*, 744. [CrossRef] [PubMed]
31. Griffin, D.K.; Larkin, D.M.; O'Connor, R.E.; Romanov, M.N. Dinosaurs Comparative Cytogenomics of Their Reptile Cousins and Avian Descendants. *Animals* **2023**, *13*, 106. [CrossRef] [PubMed]
32. Pritchard, A.C.; Nesbit, S.J. A Bird-like Skull in a Triassic Diapsid Reptile Increases Heterogeneity of Morphologival and Phylogenetic Radiation of Diapsida. *R. Soc. Open Sci.* **2017**, *4*, 170499. [CrossRef]

33. Romanov, M.N.; Farrè, M.; Lightgow, P.E.; Fowler, K.E.; Skinner, M.D.; O'Connor, R.; Fonseka, G.; Backstrom, N.; Matsuda, Y.; Nishida, C.; et al. Reconstruction of Gross Avian Genome Structure Organization and Evolution Suggests that the Chicken Lineage most Closely Resembles the Dinosaur Avian Ancestor. *BMC Genome* **2014**, *15*, 1060. [CrossRef]
34. Lightgow, P.E.; O'Connor, R.; Smith, D.; Fonseka, G.; Rathje, C.; Frodsham, R.; O'Brien, P.C.; Ferguson-Smith, M.A.; Skinner, B.M.; Griffin, D.K.; et al. Novel Tools for Characterising Inter-and Intra Chromosomal Rearrangements in Avian Microchromosomes. In Proceedings of the 2014 Meeting on Avian Model Systems, Cold Spring Harbor, NY, USA, 5–8 March 2014; Cold Spring Laboratory: Cold Spring Harbor, NY, USA, 2014; p. 56.
35. Matsuda, Y.; Nishida-Umehara, C.; Tarui, H.; Kurowa, A.; Yamada, K.; Isobe, T.; Ando, J.; Fujiwara, A.; Hirano, Y.; Nishimura, O.; et al. Highy Conserved Linkage Homology Between Bird and Turtle Chromosomes Are Precise Counterparts of Each Other. *Chromosomes Res.* **2005**, *13*, 601–615. [CrossRef]
36. Badenorst, D.; Hillier, L.W.; Literman, R.; Montiel, E.E.; Radhakrishnan, S.; Shen, Y.; Minx, P.; Janes, D.E.; Warren, W.C.; Edwards, S.V.; et al. Physical Mapping and Refinement of the Painted Turtle Genome (*Chrysemys picta*) Inform Amniote Genome Evolution and Challenge Turtle-Bird Chromosomal Consevation. *Genome Biol. Evol.* **2015**, *7*, 2038–2050. [CrossRef]
37. O'Connor, R.E.; Kiazim, L.; Skinner, B.; Fonseka, G.; Joseph, S.; Jennings, R.; Larkin, D.M.; Grlffin, D.K. Patterns of Microchromosome Organisation Remain Highly Conserved Throughout Avian Evolution. *Chromosoma* **2019**, *128*, 21–29. [CrossRef]

Review

Genome Evolution and the Future of Phylogenomics of Non-Avian Reptiles

Daren C. Card [1,2,*], W. Bryan Jennings [3,4,5] and Scott V. Edwards [1,2]

1 Department of Organismic & Evolutionary Biology, Harvard University, Cambridge, MA 02138, USA
2 Museum of Comparative Zoology, Harvard University, Cambridge, MA 02138, USA
3 Department of Evolution, Ecology & Organismal Biology, University of California Riverside, Riverside, CA 92521, USA
4 Departamento de Ecologia, Programa de Pós-Graduação em Ecologia e Evolução, Universidade do Estado do Rio de Janeiro, Rio de Janeiro 20550-013, Brazil
5 Departamento de Vertebrados, Museu Nacional, Universidade Federal do Rio de Janeiro, Rio de Janeiro 20490-040, Brazil
* Correspondence: dcard@fas.harvard.edu

Simple Summary: As a group of organisms, non-avian reptiles, most of which are the ~11,000 species of lizards and snakes, are an extraordinarily diverse group, displaying a greater diversity of genetic, genomic, and phenotypic traits than mammals or birds. Yet the number of genomes available for non-avian reptiles lags behind that for other major vertebrate groups. Here we review the diversity of genome structures and reproductive and genetic traits of non-avian reptiles and discuss how this diversity can fuel the next generation of whole-genome phylogenomic analyses. Whereas most higher-level phylogenies of non-avian reptile groups have been driven by a group of markers known as ultraconserved elements (UCEs), many other types of markers, some with likely greater information content than UCEs, exist and are easily mined bioinformatically from whole-genomes. We review methods for bioinformatically harvesting diverse marker sets from whole genomes and urge the community of herpetologists to band together to begin collaboratively constructing a large-scale, whole-genome tree of life for reptiles, a process that has already begun for birds and mammals. Such a resource would provide a much-needed high-level view of the phylogenetic relationships and patterns of genome evolution in this most diverse clade of amniotes.

Abstract: Non-avian reptiles comprise a large proportion of amniote vertebrate diversity, with squamate reptiles—lizards and snakes—recently overtaking birds as the most species-rich tetrapod radiation. Despite displaying an extraordinary diversity of phenotypic and genomic traits, genomic resources in non-avian reptiles have accumulated more slowly than they have in mammals and birds, the remaining amniotes. Here we review the remarkable natural history of non-avian reptiles, with a focus on the physical traits, genomic characteristics, and sequence compositional patterns that comprise key axes of variation across amniotes. We argue that the high evolutionary diversity of non-avian reptiles can fuel a new generation of whole-genome phylogenomic analyses. A survey of phylogenetic investigations in non-avian reptiles shows that sequence capture-based approaches are the most commonly used, with studies of markers known as ultraconserved elements (UCEs) especially well represented. However, many other types of markers exist and are increasingly being mined from genome assemblies in silico, including some with greater information potential than UCEs for certain investigations. We discuss the importance of high-quality genomic resources and methods for bioinformatically extracting a range of marker sets from genome assemblies. Finally, we encourage herpetologists working in genomics, genetics, evolutionary biology, and other fields to work collectively towards building genomic resources for non-avian reptiles, especially squamates, that rival those already in place for mammals and birds. Overall, the development of this cross-amniote phylogenomic tree of life will contribute to illuminate interesting dimensions of biodiversity across non-avian reptiles and broader amniotes.

Citation: Card, D.C.; Jennings, W.B.; Edwards, S.V. Genome Evolution and the Future of Phylogenomics of Non-Avian Reptiles. *Animals* **2023**, *13*, 471. https://doi.org/10.3390/ani13030471

Academic Editor: Ettore Olmo

Received: 16 December 2022
Revised: 13 January 2023
Accepted: 15 January 2023
Published: 29 January 2023

Keywords: anonymous loci; GC content; genome size; isochores; karyotype; natural history; reduced representation; repetitive elements; sex determination and chromosomes; target capture; ultraconserved elements

1. Introduction

Amniote vertebrates are an important clade encompassing humans, model organisms such as mouse and chicken, and many other non-model taxa, which has collectively become the most well-studied radiation of eukaryotes [1,2]. Among amniotes, there are two major evolutionary lineages—mammals and reptiles—that vary in several major natural history characteristics, such as the presence of hair versus scales, the production of milk for nourishing young, and the features of the skeletal system, especially skull structure and jaw articulation. Significant variation also exists among reptiles, resulting in four major groups that are often studied in isolation from mammals and one another: (1) birds (Class Aves), (2) crocodylians (Class Reptilia, Order Crocodylia), (3) turtles (Class Reptilia, Order Testudines), and (4) squamate reptiles (Class Reptilia, Order Squamata) [3]. Dinosaurs (including birds), crocodylians, and turtles form one major clade of reptiles, Archosauromorpha, whereas squamates and the unique taxon tuatara (Class Reptilia, Order Sphenodontia, *Sphenodon punctatus*) form the other major reptilian clade, Lepidosauromorpha. Archosauromorpha and Lepidosauromorpha diverged approximately 281 million years ago (MYA) and most of the major reptilian lineages had emerged by approximately 250 MYA [4,5]. Based on numbers of extant species, there are large differences in the diversity of each major reptilian lineage (Figure 1). Tuatara (1 species), crocodylians (27 species) and turtles (356 species) have relatively few species [6,7] whereas mammals, birds, and squamates comprise the vast majority of amniotes. In contrast to mammals and birds, whose species counts have been relatively stable (current counts of 6495 [https://www.mammaldiversity.org/ (accessed on 1 December 2022)] and 10,906 species [https://birdsoftheworld.org/bow/home (accessed on 1 December 2022)], respectively), new squamates continue to be described at a high rate, resulting in thousands of new species having been recognized in the last 10 years and a total species count (11,349 species of squamates as of March 2022; [6,7]) that now surpasses birds, which had long been regarded as the most species-rich group of tetrapods (Figure 1).

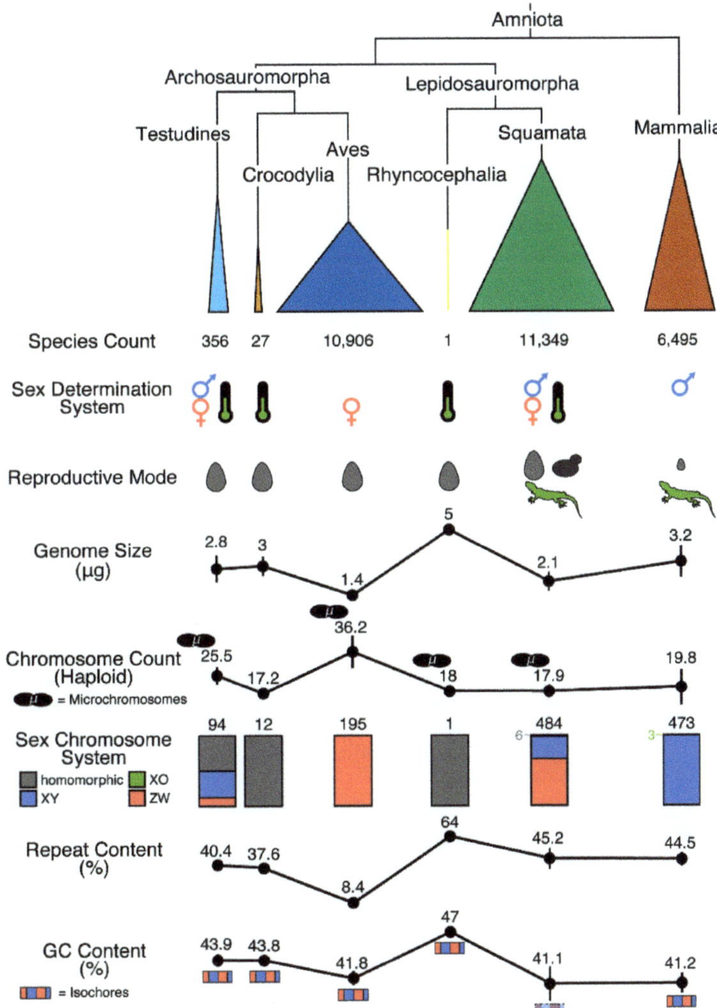

Figure 1. Overview of the natural history of amniotes, including non-avian reptiles, in a phylogenetic context. The width of clades on the phylogeny is proportional to species diversity, which are noted for each clade. For sex determination, GSD is denoted by the male and female symbols for male and female heterogamety, respectively, and TSD is denoted by the thermometer symbol [8,9]. Reproductive mode is indicated with an egg (oviparity), a lizard (viviparity), and a budding yeast symbol (parthenogenesis) [8,10–12]. Note the small egg for mammals that reflects the oviparous Monotremata (5 extant species). For genome size (C-value), data from the Animal Genome Size Database [13] were averaged per species and the clade-wise average was calculated as the mean of these species estimates. Karyotype is reported as the mean number of haploid chromosome counts per clade based on the ACC database (https://cromanpa94.github.io/ACC/ (accessed on 1 December 2022)) and lineages with microchromosomes present are indicated with a symbol near the mean chromosome count. Sex chromosome data were gathered from the Tree of Sex database [14]: the proportions of homomorphic, XY, XO, and ZW sex chromosome systems for each clade are indicated with the total species sample size per clade. The small number of squamates with homomorphic sex chromosomes (N = 6) and mammals with XO sex chromosomes (N = 3) are noted, and for counting purposes, complex XY and ZW systems were set to XY and ZW systems, respectively. For repeat

content (reported as percentage of the total genome), data from the literature (see [15–21] and references therein) were averaged per clade. For GC content (reported as percentage of the total genome), data retrieved from the NCBI Genome Assembly database [22] were averaged per species and the clade-wise average was calculated as the mean of these species estimates. Clades with isochore structure are indicated with symbols below the GC estimate [23–31] and the isochore symbol for Squamata has a broken border and faded color to indicate the partial loss of isochores in some proportion of species in that lineage. Bars behind the data points are standard deviation. Data gathered from databases were retrieved on 1 December 2022. This figure was inspired by Janes et al. [32].

The recent rapid taxonomic growth of non-avian reptiles, especially squamates, has paralleled early growth and development of genomic resources in these clades, which is beginning to enable a range of investigations in the established but rapidly evolving field of phylogenomics. Phylogenomics is the field of study concerned with using genome-wide data to infer the evolution of genes, genomes, and the tree of life [33]. Phylogenomics datasets are a product of complex patterns of evolution evident across genomic loci, many of which are influenced by various natural history characteristics of the focal taxa, and the imperfect process of producing and extracting meaningful information from genomics data. Therefore, phylogenomics investigations are both motivated and confounded by the natural histories of the taxa of interest and the underlying characteristics of the genomics data [34–36]. Moreover, the importance of reference genomes in phylogenomic investigations is growing, and interest is also increasing in using phylogenomics approaches to study the evolutionary history and unique natural histories of non-avian reptiles [15,37]. In anticipation of these developments, here, we review and discuss the rich natural histories and available reference genomes of non-avian reptiles and considerations for future phylogenomics investigations based on genomic resources in these lineages.

2. Non-Avian Reptiles Are Highly Variable in Physical Traits with Strong Links to the Genome

Non-avian reptiles—turtles, crocodylians, squamates, and tuatara—exhibit many interesting natural history characteristics ranging from physical traits to the composition and structure of the genome [32]. Physical traits that are normally invariant in well-studied amniotes such as mammals and birds are often variable across non-avian reptiles and even within certain reptile clades, making these lineages interesting and important for many biological investigations. Differences in sex determination are evident among non-avian reptiles and two major forms of sex determination have evolved in amniotes: (1) genetic sex determination (GSD), in which biological sex is determined genetically by the presence, absence, or dosage of a particular locus or allele during development, and (2) environmental sex determination (ESD), in which environmental conditions during development, often temperature, controls sex, normally resulting in clutches that are largely or exclusively one sex due to incubation conditions [8]. Whereas mammals and birds are well known examples of clades with only GSD, crocodylians and the tuatara are clades where ESD apparently functions exclusively (Figure 1; see caption for the details of the datasets and their summarization) [38–40]. In contrast to this pattern, turtles and squamates each are characterized by species or clades with either GSD or ESD (Figure 1) [8,41–44] and some interesting examples in which environmental temperature can override known GSD [45–51]. Overall, squamate reptile sex determination remains poorly understood relative to other amniote clades due to the complexity of sex determination and large numbers of apparent transitions between sex determination mechanisms across squamate species studied so far, although such patterns in squamates also offer an unparalleled opportunity to understand the genetics and evolution of all forms of amniote sex determination.

The complex interplay between environment and organism development that characterizes aspects of sex determination in non-avian reptiles also functions to drive interesting and complex patterns in the evolution of reproductive mode in these lineages. As is the case for most amniotes, sexual reproduction dominates among non-avian reptiles, and like birds and unlike most mammals, all turtles, crocodylians, and the tuatara are oviparous (Figure 1) [10,52,53]. Squamate reptiles, on the other hand, exhibit all major modes of sexual reproduction known from amniotes—oviparity, viviparity, and ovivivparity—and also reproduce asexually via various modes of parthenogenesis (Figure 1) [11,53,54]. Sexual reproductive mode can turnover rapidly in squamate reptiles [55] and numerous squamate species are capable of reproduction via both oviparity and viviparity (e.g., *Zootoca vivipara*, *Lerista bougainvillii*, and *Saiphos equalis*; see [56–61]), a situation that has driven the hypothesis that uterine retention is selectively advantageous in cooler environments [62,63]. Squamates can also reproduce via obligate parthenogenesis (Figure 1), resulting in species or populations composed entirely of females, including in certain geckos (*Lepidodactylus lugubris* [64] and *Hemidactylus garnotii* [65]), the well-known 'flowerpot snake' (*Indotyphlops braminus* [66]), and several hybrid species from the genera *Cnemidophorus*/*Aspidoscelis* [67] and *Darevskia* [68]. Numerous examples of facultative parthenogenesis have recently been documented in captive squamates, including the Komodo dragon [69] and various snakes [70–76], and wild populations of pit vipers [77]. As was the case with sex determination, squamate reptiles are an ideal group for investigating the genetics and evolution of reproductive mode and unappreciated examples of unique reproductive modes likely remain to be discovered. Overall, non-avian reptiles possess unparalleled variation in two major natural history traits, sex determination and reproductive mode, which each drive complex evolutionary patterns genome-wide.

3. Substantial Variation in Genome Size and Karyotype among Non-Avian Reptiles

At the cellular level, genome size and karyotype comprise an important aspect of biology that can impact other aspects of natural history, including physical traits and patterns of genetic variation [44]. Genome size, in particular, has strong links with the activity of repetitive elements, organism longevity, metabolic rate, and the rate of development [78,79], and has a known impact on cellular physiology, nuclear volume, and overall cell size [80]. Genome size can be measured by mass or by the combined length of all chromosomes and these measures generally correspond 1:1, such that 1 picogram (pg) of DNA corresponds to a 1 gigabasepair (Gbp) genome. Genome size varies greatly among amniotes, generally ranging from a mean of 1.4 pg in birds to 3.2 pg in mammals, and the range of genome sizes in reptiles alone is similarly broad (Figure 1). Both turtles and crocodylians have mean genome sizes that are similar to mammals at 2.8 and 3.0 pg, respectively, and the genome of the tuatara is the largest of any amniote studied to date at 5 pg (Figure 1). Squamate genome sizes are more tightly distributed around an intermediate genome size between birds and mammals at 2.1 pg (Figure 1). Previous investigations of the evolution of genome size in reptiles has yielded nuanced conclusions about the rates of genome size evolution, which have been inferred to be gradual overall [81], but potentially faster in taxa with larger genomes [82].

Karyotypic variation in reptiles is also high relative to what is observed in mammals due mainly to the presence of microchromosomes in several reptile clades [83]. Microchromosomes are approximately half the size of macrochromosomes on average [84] and have higher GC content [85], gene densities [86], and recombination rates [87] and lower densities of repetitive elements [85]. Recent studies of microchromosomes, first in snakes [37] and since more broadly [88,89], indicate that they may have unique functional characteristics relative to macrochromosomes, such as higher rates of interchromosomal contacts between loci of chromosomes in the nucleus of cells. Aside from birds, microchromosomes are present in squamates, tuatara, and turtles, but absent in crocodylians (Figure 1) [90–92]. Mammal and bird karyotypes have been particularly well-studied but karyotypical variation for all amniote lineages is well known [32,44]. A mean haploid chromosome count ranging from approximately 17 to 20 describes most lineages of amniotes, including Crocodylia, Rhyncocephalia, Squamata, and Mammalia, although Mammalia has a far greater variance in haploid chromosome count than the other lineages (Figure 1). Birds have a similarly broad distribution in haploid chromosome count but a substantially larger number of chromosomes on average (36.2), whereas turtles have an intermediate mean haploid chromosome count of 25.5 (Figure 1). The breadth of karyotypic diversity in reptiles far exceeds what is observed in mammals, especially when including birds, making this clade ideal for investigating karyotype evolution in amniotes.

The mechanism of sex determination impacts the evolution of sex chromosome systems in the case of GSD, which can result in homomorphic sex chromosomes—sex chromosomes that are superficially similar and hard to identify as linked to sex—and two forms of heteromorphic sex chromosomes—sex chromosomes that are able to be distinguished in males or females due to the evolution of a degenerated chromosome. Heteromorphic chromosomes were originally observed using cytogenetic methodologies and largely continue to be identified in this manner, although detecting less obvious homomorphic chromosomes is also possible, but has been rarely pursued due to increased difficulty. In heteromorphic sex chromosome systems, the evolution of a visually degenerated chromosome can result from sexual conflict, which is overcome through the suppression of recombination via inversions. The degenerated sex chromosome can be inherited paternally, resulting in a XY sex chromosome system, or maternally, resulting in a ZW sex chromosome system [93–95]. Again, whereas mammals and birds are all characterized by a common sex chromosome system (XY and ZW, respectively), non-avian reptiles have far more nuanced variation in the form of sex chromosomes across different lineages. In the tuatara and crocodylians, ESD putatively results in lower genetic sexual conflict, resulting in (visually) homomorphic sex chromosomes [96–98]. Homomorphic sex chromosomes have also been observed in turtles and squamates, especially those species where ESD is known, but both of these lineages also exhibit species with XY and ZW sex chromosomes [43,44]. The greatest known variation in sex determination and sex chromosome systems is evident in geckos [99]. However, our knowledge of sex determination/chromosomes is still relatively incomplete in non-avian reptiles, especially squamates, and new, interesting phylogenetic patterns of sex chromosomes are regularly being discovered, including the recent discovery of largely homomorphic XY sex chromosomes that evolved independently in Henophidian snakes [100] that violated long-held assumptions that all snakes possessed ZW sex chromosomes [101,102]. Altogether, non-avian reptiles possess the most complex evolutionary patterns of sex chromosome systems of all amniotes, with squamates emerging as a relatively powerful system for interrogating the evolution of sex chromosomes. Finally, we focus here on phylogenomics using the nuclear genome, and do not discuss mitogenomics. However, we note that non-avian reptiles have interesting patterns of evolution of the mitochondrion that should be considered in phylogenomic investigations (e.g., see [103,104]), such as a snake-specific duplicate control region and high rates of adaptive evolution of snake mitochondrial metabolic proteins [105,106].

4. Dynamic Features of Sequence Composition in Non-Avian Reptiles

At the sequence level, non-avian reptile genomes are characterized by several unique characteristics that may impact downstream phylogenomics investigations. Non-avian reptile genomes contain a diverse repertoire of repetitive elements that is only beginning to be explored. Most knowledge of amniote repeat element landscapes is based on early genomic investigations in mammals and birds [85,107–109], where there is a dichotomy in genomic architecture. Mammals have a relatively rich diversity of repeat elements that form a substantial portion of the genomes of these organisms (e.g., at least 50% of the human genome [107], with other mammals having similar patterns), correlating with larger genomes. In contrast, bird genomes are generally relatively streamlined, containing much less repeat diversity dominated largely by chicken repeat (CR1) long interspersed nuclear elements (LINEs) that form a much smaller portion of the already much smaller genomes of most bird species (typically < 20% of ~1 Gbp genomes; [85,110,111]). Early studies of non-avian reptile repeat landscapes using BAC-end sequences revealed high diversity in tuatara and various squamate lineages [112,113]. More recent studies based on whole-genome sequences have only emerged since 2011 and have established an additional dichotomy in the evolution of repeat landscapes between the non-avian Archosauromorpha (crocodylians and turtles) and Lepidosauromorpha (squamates and tuatara). Crocodylians and turtles have relatively homogeneous repeat landscapes that superficially resemble birds and a reduced rate of new TE family invasion/evolution [16], although larger proportions of the genomes of these species are comprised of repeat elements (>35% [16,17]). In contrast, squamates and tuatara have an extremely rich diversity of repeat elements that exceeds the diversity of mammals. As could be hypothesized given the unique evolutionary history of the tuatara, a large proportion of the genome of this species consists of repeats, and this diverse repeat landscape is unique from any other amniote [18]. Squamates, despite showing a fairly even distribution of genome sizes, have large amounts of variation in the proportions of their genomes composed of repetitive elements, ranging from lower proportions in most lizards of ~30–40% to higher proportions in many colubroid snakes of 50% or greater. Squamate genome repeat landscapes are dominated by three types of LINE families (CR1, BovB, and L2) and have high proportions of DNA transposons, in contrast with other amniote genomes where one LINE family typically dominates [15,114,115]. Moreover, while other amniotes have fairly inactive repeat landscapes, where only one or a few repeats has continued to proliferate in the genomes of these organisms (e.g., L1 LINEs or Alu elements in humans [107]), several repeat types, subtypes, and families appear simultaneously active in squamate genomes [15]. Indeed, the patterns of repeat evolution observed in squamates [15] challenge the accordion model of repeat element evolution in vertebrate genomes that was based on data from mammals and birds [19]. Moreover, the striking phylogenetic pattern of higher repeat element proportions characterizing the genomes of snakes and especially the venomous lineages of snakes has led to speculation that the seeding of repetitive loci, especially microsatellites, which may drive the rapid evolution of tandemly duplicated gene families that are important in evolutionary novelties in these clades: *Hox* genes and the serpentine body plan and various toxin gene families that function in venom [116–118]. Finally, although most repeat element activity is limited to the nucleus, there are many documented cases of horizontal transfer of repetitive elements between divergent lineages, including non-avian reptiles, apparently mediated by viruses or blood-sucking ectoparasites [15,114,119–128].

Even at the level of the nucleotide, studies of GC content indicate non-avian reptile genomes have unique features. Overall GC content varies greatly across amniotes. Mammals and birds have similar GC content with a mean across species of ~41–42%, although mammalian genomes have greater variation in GC content (Figure 1). Squamate reptile genomes have a similar mean GC content, but far greater variation than even mammalian genomes, whereas turtle and crocodylian genomes have elevated GC content (43.9% and 43.8%, respectively) and the tuatara has by far the highest GC content known from any amniote (47%; Figure 1). The composition of bases across the genome is not uniform and large genomic tracts (>100 kb) with relatively homogenous, biased base composition—often called isochores—can form. Isochores are well defined in mammalian and avian genomes but generally absent in fish and amphibians [129,130]. GC-rich isochores, in particular, correlate with several other genomic features, such as recombination rate [131], gene density [132], epigenetic modifications [133], intron length [134], and replication timing [135]. The correlation between GC content and recombination rate is particularly profound for genome biology and evolution, especially in light of documented GC-based repair biases, and mechanistic links between GC content and recombination rate may be a strong driver of variation in recombination rate across the genome [136–141].

Early investigations of isochores in reptiles first used CsCl fractionation [142] and, later, GC values at third-codon positions as a proxy for isochores (GC3 [23,24,143,144])—a practice that has since come into question because GC3 only explains a small proportion of variation in the GC content in the regions flanking genes [145]—although ideally such investigations are based on high-quality genome assemblies (e.g., [25,26,146]). Analysis of the first non-avian reptile genome, that of the green anole (*Anolis carolinensis* [147]), established that there was little evidence for isochore structure in this species relative to what is observed in mammals and chicken [27]. However, a subsequent investigation questioned this result [28] and an additional study established that snake genomes have a higher degree of GC-isochore structure than seen in *Anolis* [115], indicating more complex evolutionary patterns of isochore structure in squamate reptiles that remain to be thoroughly investigated. Reference genomes suggest that turtles and crocodylians also show patterns consistent with isochore structure [29,30]. Despite retaining (or perhaps secondarily evolving) isochores, snakes have lower GC content than *Anolis* and the evolution of GC at third codon positions (GC3) trends towards AT richness, in contrast to the GC bias notable in mammalian genomes [115]. Beyond GC content, investigations of nucleotide substitution patterns indicates that squamates generally have higher substitution rates that are similar to those from mammals [16,115], with interesting bursts in the rates of evolution associated with root branches of snakes and colubroid snakes [115]. In contrast, birds have modest substitution rates and analyses of turtle and crocodylian genomes show these lineages have extremely slow substitution rates [16]. Overall, evolution in the composition of non-avian reptile genomes has resulted in remarkably different genomic environments in these lineages, which will need to be taken into account during downstream phylogenomics investigations.

5. Summary of Available Reference Genomes for Non-Avian Reptiles

Despite possessing a range of interesting natural histories, genomics resources, and therefore phylogenomics investigations, in non-avian reptiles have only emerged since the publication of the first non-avian reptile genome in 2011, that of the green anole [147]. Since this release, genomic resources for increasing numbers of non-avian reptiles have emerged. These genomic resources are most often available from National Center for Biotechnology Information (NCBI), but other data repositories can be used (e.g., DNAZoo and GenomeArk) and a growing number of sequencing initiatives are targeting non-avian reptiles for reference genomes, making it difficult to collate all resources available for these clades. While genomes are available from relatively large proportions of smaller reptilian clades (i.e., crocodylians, turtles, and tuatara), genomic resources in the species-rich squamate reptiles have been slower in developing than similar resources in mammals and birds, where reference genomes have been constructed for approximately 9% and 6% of species, respectively (Figure 2). However, in recent years the pace of sequencing has increased in non-avian reptiles, especially squamates, due to technological advances and improving economics. An exhaustive accounting indicates there are 165 publicly available and 23 announced (i.e., expected in the future) non-avian reptile reference genomes (Figure 3; see caption for the details of the datasets and their summarization). These genomes collectively represent 139 reptilian species, with redundancy in the form of multiple assemblies of varying quality from the same source material and multiple assemblies sourced from different animals, sometimes representing distinct populations (Figure 3). Reference genomes are available for 31 (9% of known species) turtle species, 4 (15%) crocodylian species, 1 (100%) rhyncocephalian, and 84 (<1%) squamate species. Most non-avian reptile genomes have been released since 2020 (Figures 2 and 3). Moreover, there are significant differences in assembly characteristics (i.e., length and GC content) and quality (i.e., N50s and BUSCO scores) between genomes due to technical differences in assembly production and evolutionary differences in the genomic characteristics of amniotes (Figure 3).

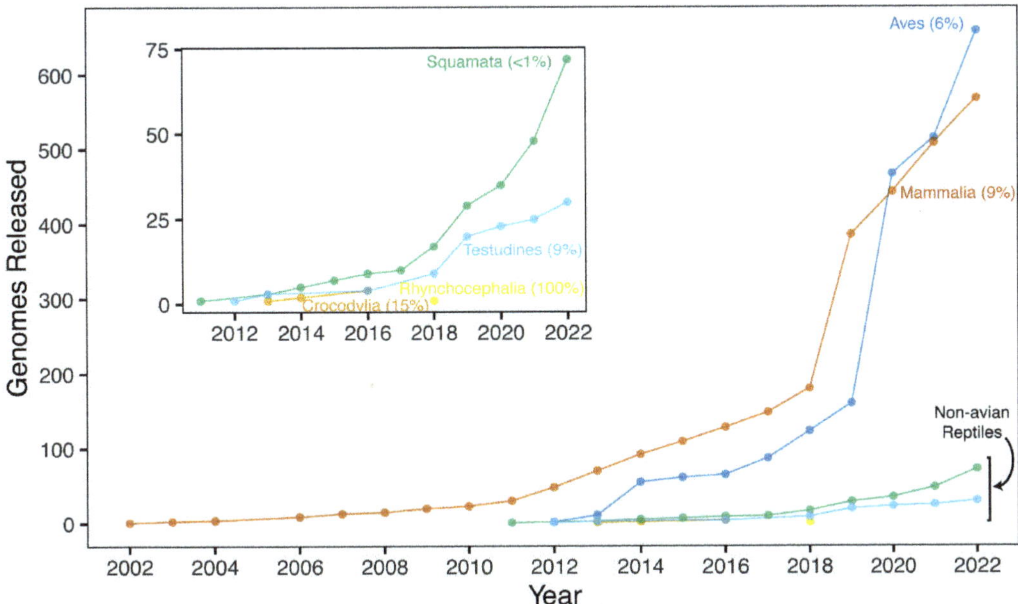

Figure 2. Temporal accumulation of genomes available on NCBI for major amniote clades (data retrieved 1 December 2022). Inset: Details of the growth in the number of available genomes for non-avian reptiles. Note: The counts from this dataset represent a subset of the full non-avian reptile genomes dataset presented in Figure 3, as many genomes are available from sources other than NCBI. This figure was inspired by Bravo et al. [148].

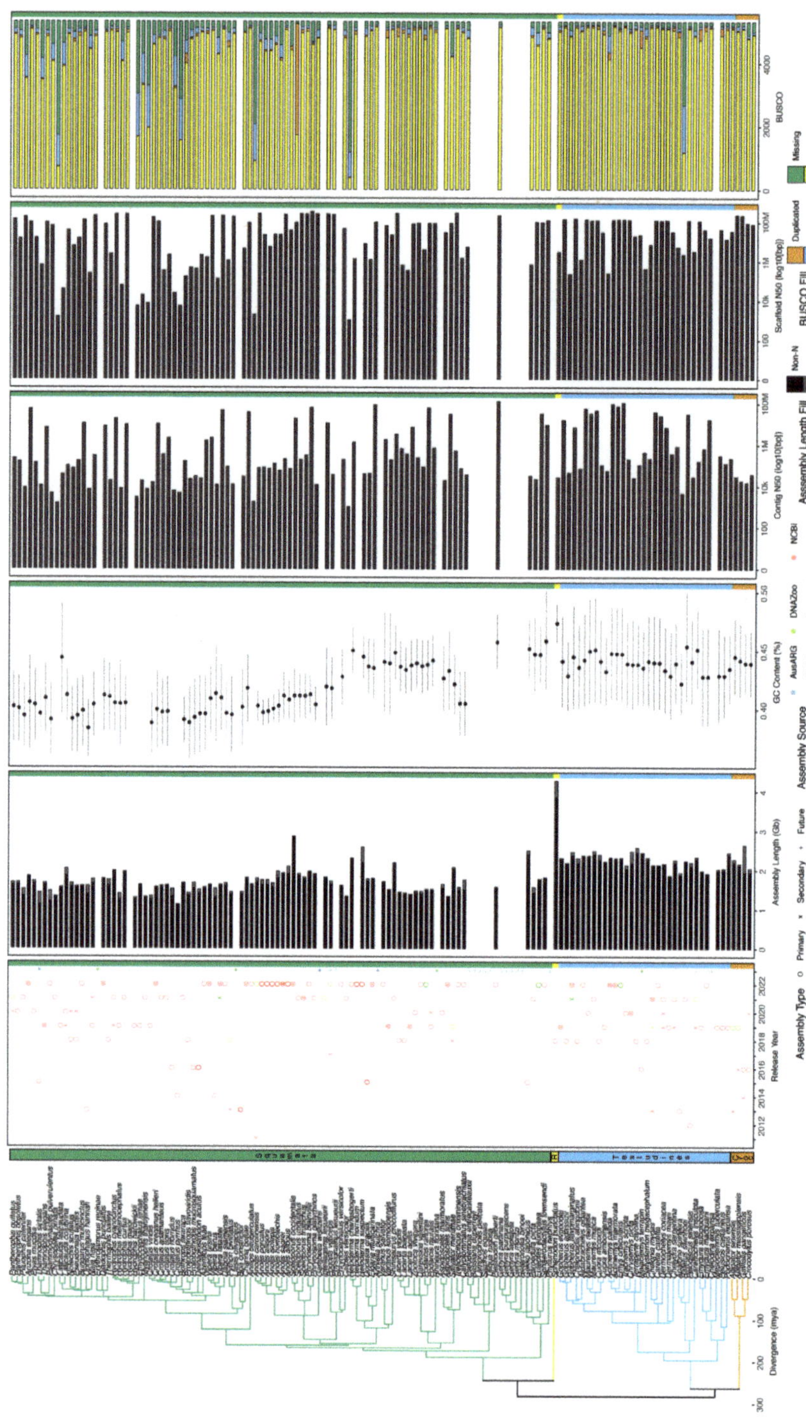

Figure 3. Phylogenetic summary of available reference genomes for non-avian reptiles. The topology and divergence times were gathered

from the TimeTree database (accessed 1 December 2022) [4,5]. For taxa that were not already included in TimeTree, we used existing studies of *Gehyra* [149–152], *Heloderma* [153], *Physignathus* [154], *Gopherus* [155,156], *Actinemys* [157], *Cuora* [158,159], and *Myanophis* [160,161] to place taxa and determine the approximate divergence time. Horizontal bars delineate the major clades: Squamata, Rhyncocephalia ("R"), Testudines, and Crocodylia ("Croc"). The colored bars to the right of each panel indicate each clade and aid in visualization. Publicly-available and announced genomes were collated from NCBI, the Genome10K/VGP/EBGP GenomeArk website (https://genomeark.github.io/genomeark-all/ (accessed on 1 December 2022)), the DNAZoo website (https://www.dnazoo.org/assemblies (accessed on 1 December 2022)), the Australian Amphibian and Reptile Genomics (AusARG) initiative website (https://ausargenomics.com/ (accessed on 1 December 2022)), the California Conservation Genomics Project (CCGP) website (https://www.ccgproject.org/reptiles (accessed on 1 December 2022)), and other locations noted in the literature. For each assembly, we gathered the release date, total assembly length and number of ambiguous (N) bases, and calculated scaffold N50 and contig N50 after breaking scaffolds at runs of >25 Ns. We also ran BUSCO v. 5.4.2 [162] in 'genome' mode with the tetrapoda_odb10 dataset to assess the completeness of genomes based on 5310 generally conserved, single-copy tetrapod genes and used bedtools v. 2.29.0 [163] and seqtk v. 1.3-r106 (https://github.com/lh3/seqtk (accessed on 1 December 2022)) to calculate GC content in 500 kb genomic windows (where a minimum of 250 kb of non-N bases were present). Some genomes were not contiguous enough for GC content distributions to be estimated. Where multiple assemblies were available for a species, we plotted the release date and source of each assembly but only quantify genomic characteristic and quality metrics for the primary assembly with the highest-quality assembly based on contiguity and BUSCO results, most of which were designated as the primary assembly on NCBI. Secondary assemblies are those additional assemblies for a given species and future assemblies reflect forthcoming genomes for species that were publicly announced where data are not yet available.

6. Why Are There So Few Genomes of Non-Avian Reptiles?

As we have seen, genome sequencing in non-avian reptiles has lagged behind progress in birds, where there are now hundreds of genomes and an increasing number based on long-read sequencing [148]. Additionally, there is a paucity of long-read, high-quality or even chromosome-scale genomes from non-avian reptiles. This paucity likely stems from the academic orientation of the many biologists interested in herpetology. Few if any reptiles can claim the exalted status of a 'model organism', especially in the fields of genetics, developmental biology, or cell biology. This is not to say that non-avian reptiles cannot serve as important models for many fields, such as ecology and adaptive radiation: the large number of studies on *Anolis* lizards is clear evidence of this [164]. Nevertheless, the fields for which non-avian reptiles are models tend not as yet to be the fields that require genomes. Of course, there have been several studies that have effectively linked genome variation and ecology in non-avian reptiles, especially for *Anolis* [165,166]. In many ways, the availability of a high-quality genome from *Anolis carolinensis* in 2011 has attracted investigators to that species in diverse contexts; however, by the metrics we use in Figure 3, such activity would not increase the number of non-avian reptile genomes. The increased number of genomes from the genus *Anolis* is beginning to reveal the potential of comparative genomics in non-avian reptiles [167–169]. However, as has been evident in ornithology, it takes a strong, ambitious, and sustained focus by the research community on comparative biology, as well as the availability of multiple models for other fields, such as molecular and cellular biology, to drive the accumulation of genomes from multiple species. For example, birds are models in neuroscience (zebra finch, *Taeniopygia guttata*) and developmental biology (chicken, *Gallus gallus*) that are useful enough to compete for priority for particular scientific problems with well-funded models of genetics and genomics, such as mice.

In the era before whole-genome sequencing became routine, several researchers put in place important genomic resources for non-avian reptiles, such as BAC libraries, cell lines, and short-read sequence archives that helped move the field forward [170–172]. Several BAC libraries, such as those for the tuatara (*Sphenodon punctatus*) and western painted turtle (*Chrysemys picta*) helped fuel subsequent genome projects of these species, or refinement of assemblies via mapping [18,171,173], and provided useful resources for early phylogenomic analyses [174]. Additionally, some of the first glimpses of the structure of non-avian reptile genomes—including larger-scale observations of GC content [113,175], transposable element abundances [81,112], and non-coding conserved elements [172,176]—came from such resources. Many of these early observations of genome structure in non-avian reptiles were only indirect, and have been vastly improved upon with better tools and direct genome-scale analyses [15,37,100]. Furthermore, low availability of high-quality molecular specimens continues to hamper efforts to build genomic resources for non-avian reptiles, especially when using emerging technologies and approaches capable of constructing highly contiguous genome assemblies (e.g., long read sequencing). Nonetheless, the continued paucity of non-avian reptile genomes may have a more practical source. For example, the small number of non-avian reptile genomes may simply be a consequence of the smaller number of researchers studying reptiles and, ultimately, the smaller sector of society that is engaged in reptile-related activities and community science. Finally, because reptile genomes are on average about twice as big as avian genomes, the sheer cost and labor required to assemble a high-quality reptile genome may be prohibitive. With the advent of increasingly inexpensive long-read sequencing, the production of high-quality reptile genomes may finally ramp up and achieve cruising speed.

7. First-Generation Phylogenomic Data Acquisition: Reduced Representation Approaches

Less than a decade after the first draft of the human genome was published [107,177], the melding of two new technological innovations would enable researchers to amass datasets of hundreds to thousands of loci—one to two orders of magnitude more loci than was hitherto possible to acquire using PCR-based approaches [8]. A major advance was the development of massively parallel or "short-read" DNA sequencing, a genomic sequencing platform that far surpassed the data output of the classical Sanger sequencing platform [178,179]. Despite the sudden appearance of several types of short-read genome sequencers during the mid-2000s, only one of them—the Illumina sequencer [180,181]—would come to dominate the genome sequencing and phylogenomics scenes [8], a situation that has remained largely unchanged to the present.

Since this technological breakthrough, phylogenomic studies have routinely utilized genome-wide data consisting of hundreds to thousands of loci ("loci" defined here as DNA segments of at least ~200 base pairs [bp] in length) to estimate species trees and associated historical demographic parameters [8]. A key advantage to using such "big data" in analyses assuming a multispecies coalescent model [182] compared to the one- to several-locus datasets of early molecular phylogenetic and phylogeographic studies is that confidence intervals around parameter estimates are expected to be far more precise (i.e., narrower) than estimates obtained from smaller numbers of loci. This is because the gene tree for each locus is thought to approximate an independent realization of the coalescent process [183–186]. Accordingly, statistical precision surrounding parameter estimates can be improved simply by increasing the numbers of independent loci [8,187–189]. Indeed, several empirical studies have corroborated this basic tenet of multilocus population genetics [190–193].

Another innovation especially important in phylogenomics was the development of a broad array of molecular techniques for constructing sequencing libraries. Many of these techniques enrich for particular regions of the genome in various ways, which enables a reduced representation of the genome to be preferentially generated for targeted loci [194]. Most early phylogenomics investigations have favored reduced representation approaches for economic reasons, as the cost of sequencing remains a major financial bottleneck for most research groups and sequencing a small percentage of the genome (typically < 5%) results in proportional savings in sequencing cost. The innovation most used for phylogenomics investigations is in-solution hybrid selection—a methodological spinoff of the microarray technology from the early 2000s [195]. In-solution hybrid selection or "target capture" allowed researchers to selectively sequence only targeted DNA sequence loci using Illumina sequencing. By hybridizing 60–120 bp biotinylated RNA oligonucleotide probes to complementary target genomic fragments in the reaction mixture for each sample (individual), DNA sequence data obtained from the probe-annealing and flanking regions could be obtained for hundreds to thousands of genome-wide loci (see reviews in Jennings [8] and Andermann et al. [196]). Therefore, the probe set, often called a "bait set," can effectively "fish out" the DNA fragments containing the sequences that are complementary to the probes—and, importantly, the accompanying flanking sequences—from a solution of random genomic DNA fragments (i.e., the shotgun library). Once these fragments containing the target sequences have been isolated, they can be sequenced with adequate coverage per locus and analyzed accordingly (Figure 4). Consequently, the standard Illumina sequencing-target capture workflow can regularly churn out immense multilocus datasets in phylogenomics studies in a cost-effective manner.

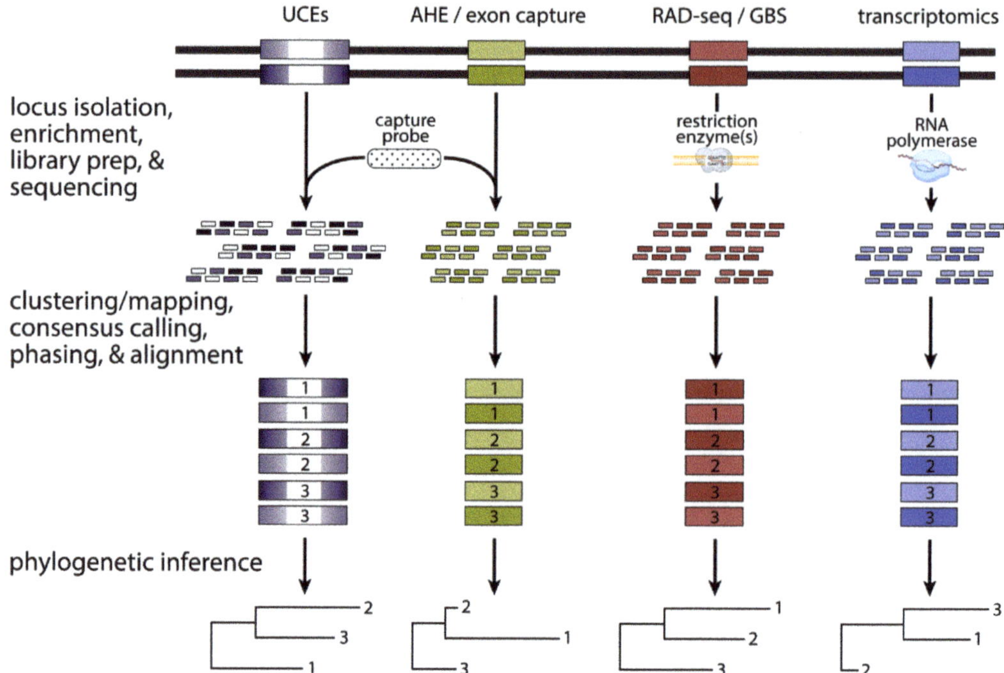

Figure 4. Graphical overview of various reduced representation approaches used in phylogenomics investigations. Alternative depictions are presented for different methods of enriching for particular loci in the genome: two kinds of target capture (targeting UCEs and AHEs or exons), RAD-seq (also known as GBS), and transcriptomics. In each case, the color indicates the location of phylogenetically informative signal in the locus, which typically comprises the whole extent of the target locus, except in the case of UCEs, where this signal is found in the regions flanking the locus. These classes of loci, or markers, are depicted along a diploid genome for a single sample, with heterozygous variation in the form of two alleles at each locus indicated with alternative shading. Although only a single sample is indicated, these approaches would be applied to all samples of interest in parallel, ultimately resulting in sequencing for all samples (e.g., N = 3 samples depicted below). For target capture, the genome is fragmented, and oligonucleotide probes are used to enrich for the target loci. For RAD-seq and transcriptomics, regions of interest are isolated and enriched simultaneously by restriction enzymes and cellular RNA polymerase transcription activity followed by in vitro reverse transcription, respectively. Importantly, of the three general methods, only target capture requires a priori sequencing data and knowledge to construct oligonucleotide probes. After this isolation and enrichment step, all methods proceed generally the same way with standard library preparation and sequencing steps. The resulting sequencing data are also generally analyzed similarly by bioinformatically parsing data to recover sample-specific sequences (three samples are indicated) and clustering sequences by similarity to enable consensus calling (not shown), although a reference genome can aid in this process. Variation across loci is ideally phased to recover the original heterozygous state—two phased alleles per sample are depicted. Phased sequence data for each sample and locus can then be aligned and used for phylogenetic inference.

The first two probe sets for obtaining hundreds to thousands of phylogenomic loci in animals were the "ultraconserved elements loci" (so-called "UCEs" [197]) and "Anchored Hybrid Enrichment loci" (so-called "AHEs" [198]) probe sets (Figure 4). Both probe sets have since propelled studies involving many groups of non-avian reptiles (e.g., [199–201]).

Although all these probes were designed to hybridize to highly conserved genomic sequences in the genomes of tetrapods and vertebrates, respectively, the template sequences used to design the probes fundamentally differed between UCEs and AHEs. The UCE loci probes anneal to non-coding elements called "ultraconserved elements" or "UCEs," which have remained virtually unchanged for hundreds of millions of years [202] and may function as regulatory elements that control gene expression of nearby genes. Although their highly conserved nature makes UCEs ideal probe targets in species, these sequences contain insufficient numbers of variable sites to be useful in phylogenomic studies. Accordingly, the less-conserved flanking sequences surrounding the actual UCEs are used in phylogenomic studies (Figure 4). Moreover, because these flanking sequences contain a spectrum of sites ranging from completely non-conserved to highly conserved sites, UCE loci have been informative at both "shallow" (<ca. 10 million years ago [MYA]) and "deep" (>ca. 10 MYA) timescales [197]. The genomic targets of AHE probes, on the other hand, are highly conserved exons whose flanking sequences are useful for shallow and deep timescales [198].

An early example of a reptile-specific, mixed-marker probe set emerged when Singhal et al. [203] designed an all-encompassing probe set specifically for squamate reptiles termed "SqCL set", which contains probes for harvesting 5052 UCEs, 372 AHEs, and ~50 other "legacy loci" that have been useful for reconstructing the squamate tree of life [204,205]. A survey of the literature over the past five years shows the tremendous impact that UCE and AHE loci have had on phylogenomic studies of non-avian reptiles, as 25 of the 35 target capture studies (71%) employed either or both probe sets (Table 1). Moreover, both sets of loci have been successfully applied to non-avian reptile taxa on both shallow and deep timescales (see [203]; Table 1). Consequently, in one sense the UCE and AHE probe sets are analogous to the first "universal" PCR primers [206], which, together with PCR, launched the field of molecular phylogenetics [207] and modernized phylogeography [208].

Table 1. Phylogenomic studies of non-avian reptile clades published between 2017 and 2022 that used at least 100 DNA sequence loci.

Study	Taxon	Type of Loci	# of Loci	# Samples	Depth of Divergences
Ashman et al., 2018 [209]	lizards	exons	547	64	shallow
Blair et al., 2022 [210]	lizards	UCEs	3157	34	shallow
Blom et al., 2017 [211]	lizards	exons	2840	28	shallow
Blom et al., 2019 [212]	lizards	exons	2457	135	shallow
Bragg et al., 2018 [213]	lizards	exons	2364	123	shallow-deep
Brennan et al., 2021 [214]	lizards	AHEs	388	103	shallow-deep
Bryson et al., 2017 [215]	lizards	UCEs	3282	58	shallow
Domingos et al., 2017 [216]	lizards	AHEs	422	30	shallow
Freitas et al., 2022 [217]	lizards	exons	625	69	shallow
Garcia-Porta et al., 2019 [218]	lizards	AHEs + other	6593 (324 AHEs + 6269 other)	262	shallow-deep
Grummer et al., 2018 [219]	lizards	UCEs + exons	589 (541 UCEs + 44 exons)	29	shallow
Morando et al., 2020 [220]	lizards	UCEs + exons	588 (540 UCEs + 44 exons)	26	deep
Moritz et al., 2018 [149]	lizards	exons	1636	56	shallow
Panzera et al., 2017 [221]	lizards	UCEs	581 (538 UCEs + 43 exons)	16	higher
Ramírez-Reyes et al., 2020 [222]	lizards	RAD-seq	78,970–549,193	90	shallow
Reilly et al., 2022a [223]	lizards	exons	709	99	shallow-deep

Table 1. *Cont.*

Study	Taxon	Type of Loci	# of Loci	# Samples	Depth of Divergences
Reilly et al., 2022b [224]	lizards	exons	1154	104	shallow
Reynolds et al., 2022 [225]	lizards	UCEs	4055	82	shallow-deep
Rodriguez et al., 2018 [226]	lizards	UCEs	2690	119	shallow-deep
Schools et al., 2022 [227]	lizards	UCEs	5060	30	higher
Singhal et al., 2018 [228]	lizards	exons	2668	25	shallow
Skipwith et al., 2019 [229]	lizards	UCEs	4268	290	shallow-deep
Tucker et al., 2017 [230]	lizards	AHEs	316	16	shallow
Wood et al., 2020 [231]	lizards	UCEs	772–4715	42	deep
Zozaya et al., 2022 [232]	lizards	exons	1429	33	shallow
Bernstein and Ruane 2022 [233]	snakes	AHEs + UCEs + other	1–642 UCEs, 1–39 AHEs, 2–11 other	156	shallow-deep
Blair et al., 2019 [234]	snakes	UCEs	3384	54	shallow
Chen et al., 2017 [235]	snakes	AHEs	304	88	shallow
Esquerré et al., 2020 [236]	snakes	AHEs	376	50	deep
Hallas et al., 2022 [237]	snakes	RAD-seq	22,289–48,867	49	shallow
Li et al., 2022 [238]	snakes	exons + other	3023 (1948 exons + 1948 other)	24	deep
Myers et al., 2022 [239]	lizards	RAD-seq	2950	74	shallow
Natusch et al., 2021 [240]	snakes	AHEs	421	19	shallow
Nikolakis et al., 2022 [241]	snakes	UCEs	3383–4146	43	shallow
Ruane and Austin 2017 [242]	snakes	UCEs	2318	10	deep
Burbrink et al., 2020 [243]	squamates	AHEs	394	289	deep
Singhal et al., 2021 [244]	squamates	AHEs + UCEs + other	5462 (372 AHEs + 5052 UCEs + 38 other)	92	deep
Streicher and Wiens 2017 [245]	squamates	UCEs	2738	24	deep
Shaffer et al., 2017 [246]	turtles	UCEs + other	539	24	deep

Although target capture has been the most commonly used approach in non-avian reptile phylogenomics, other reduced representation methodologies have been applied as well. In an approach commonly referred to as RAD-seq (restriction-site association DNA sequencing; also commonly called genotyping by sequencing [GBS]), one [247–249] or more [250] restriction enzymes are used to enrich homologous regions of the genome in a flexible and economical way (Figure 4). The RAD-seq approaches have been commonly applied for population genomics investigations and are occasionally used in phylogenomics studies, although only at shallower phylogenetic scales in three studies of non-avian reptiles (Table 1). Known issues with stochastic locus fallout due to accumulating variation in restriction sites restrict RAD-seq investigations to relatively shallow phylogenetic scales [251,252]. In theory, transcriptomics could be used to generate homologous sequencing data from protein-coding regions of the genome via RNA-seq or similar techniques (Figure 4) [253,254]. However, except for the purposes of constructing probes for sequence capture studies, transcriptomics approaches have not been widely applied in non-avian reptiles due to the difficulty in working with RNA and the small number of samples with sufficient quality available for RNA-seq investigations.

The flexibility of target capture allows researchers to mix target loci from different previously constructed capture panels and even include new loci of interest. Indeed, in many of the studies listed in Table 1, researchers opted to develop their own probe sets for obtaining thousands of annotated exon sequences. For example, many of these studies (e.g., [149,223,224]) used a de novo transcriptome approach to develop a custom exon-capture probe set for each study species group (see [254,255] for the protocol). Although this do-it-yourself approach to designing a custom exon-capture probe set increases project costs in terms of laboratory and bioinformatics work and in consumables, this approach enjoys several advantages over the universal bait kits described earlier for shallow-scale phylogenomic studies: (1) a reference genome is not needed to design loci probe sets; (2) multiple sequence alignments are simpler for coding (e.g., exons) than non-coding (e.g., UCE loci) DNA sequences; (3) larger numbers (>1000) of exonic loci can be developed using the de novo transcriptomic approach compared to the 400–500 loci found obtainable from an AHE set; and (4) thousands of exonic loci that exhibit a wide range of evolutionary rates can be harvested, which contrasts with UCEs and AHE loci [254,255]. However, for studies at deep timescales, universal probe kits will likely perform best and enable researchers to cost-effectively outsource library construction and sequencing to service providers. The advent of massively parallel, reduced representation approaches, especially target capture, have unquestionably revolutionized phylogenomics, as researchers have been able to affordably infer species trees and associated historical demographic parameters with unprecedented accuracy and precision.

8. Genome-Scale Phylogenomics: In Silico Investigation of Markers Extracted from Whole Genomes

In a landmark study, Jarvis et al. [256] inferred the higher-level relationships in the avian tree of life using newly generated complete genome sequences for 48 species, ushering in an era of truly genome-scale phylogenomics. Since then, there have been continual advances in short-read sequencing in terms of sequence output and reduction in cost per Gigabase (Gb) of sequences, as well as the development of high-quality, long-read DNA sequencing (e.g., PacBio platform). These improvements in genomic sequencing are now making it practical for researchers working on non-avian reptiles to not only acquire a chromosome-level reference genome assembly for their study organism, but also to obtain large numbers of resequenced genomes for that species or for species in the clade of interest. Indeed, early examples of this approach are already occurring: the California Conservation Genomics Project (CCGP), which is a consortium of 114 principal investigators, is nearly finished with the amassing of high-quality reference genomes and 100–150 resequenced genomes for each of 235 focal species found in marine and terrestrial habitats throughout the state of California [257]. Eight of these species are non-avian reptiles and thus a total of eight high quality reference genomes—three of which were recently published [258–260]—plus associated resequenced genome sequences (total of ~800–1200 datasets) will soon be completed. It therefore appears certain that the numbers of population genomic and phylogenomic studies based solely on full genome sequences will accelerate in the future. However, despite these anticipated developments, reduced representation approaches will continue to play an important role in phylogenomics because of cost effectiveness and the wide availability of genetic specimens in natural history museums [242,255] or elsewhere that are degraded or otherwise not suitable for building high-quality genomic resources.

As is the case with reduced representation approaches, the quality of genomics data can have a large impact on the ability of researchers to perform downstream phylogenomic investigations and must be taken into account. The quality of genome assemblies, the forthcoming foundation for truly phylogenomics-scale research, can also be quite variable due to differences in the genomic characteristics of organisms and the practices used to generate genomics datasets and digitally assemble a representation of the genome of an organism, which have changed significantly over time. Moreover, although there are a growing number of high-quality genomes available for amniotes, a "complete genome" is difficult to construct and carries a high burden of proof that has rarely been met, although recently, a first complete, "telomere-to-telomere" reference genome was constructed for human [261]. Equally important for most downstream biological investigations that use genomic resources is a high-quality annotation of all repetitive elements, protein-coding genes, and other important features of the genome that can form the foundation for phylogenomics studies and contribute to genome biology and evolution. Annotation quality is a function of the underlying genome assembly quality, the quantity and quality of functional genomics data used as biological evidence to guide annotation (especially RNA-seq data, but other types of omics data could also be used), and the bioinformatic approach used for annotation. Moreover, additional steps are necessary to estimate homology of genomic loci, such as annotated protein-coding regions, across genomes, a critical prerequisite for phylogenomics studies. Therefore, when evaluating publicly available reference genomes for use in a phylogenomics investigation, it is important to evaluate the quality of genomic resources and build other quality control considerations into the analysis of these comparative genomics datasets.

One major advantage that the complete genome approach enjoys over reduced representation approaches is that computational, or in silico, acquisition of hundreds to thousands of DNA sequence loci from complete genome sequences is much simpler than the target capture workflow [193,262]. For example, to illustrate the phylogenomic utility of UCE markers, McCormack et al. [263] designed a set of in silico UCE probes using available genome data and then performed in silico target capture of target UCE loci from 29 genome sequences for placental mammals. Although there are simpler in silico-based methods for acquiring a comparable dataset from complete genome sequences (as acknowledged by McCormack et al. [263]), their study nonetheless hinted at the promise of in silico extraction of phylogenomic loci from whole genome sequences. Moreover, Costa et al. [193] later designed a Python-based software pipeline that can, in automatic fashion, extract the target loci sequences from complete genome sequences, perform multiple sequence alignments of each locus, and output ready-to-analyze data files. A test run of this program using the human, chimpanzee, gorilla, and orangutan genomes quickly produced a 242 AHE locus dataset. Although analyses of far larger numbers of complete genome sequences would require more time for the software to finish the analysis, the time needed to generate a phylogenomic dataset will undoubtedly still be less than the one- to several-week time requirement for the target capture workflow.

Perhaps an even more important advantage of the complete genome approach is that it will provide researchers, for the first time, an effective way to obtain orthologous sequences from multiple individuals' genomes for hundreds to thousands of "anonymous loci" [193]. Anonymous loci, which comprise a distinctive marker class first developed by Karl and Avise [264], are ideal DNA sequence markers for phylogeographic and shallow-scale phylogenomic analyses that employ the multi-species coalescent because of their neutral or near-neutral characteristics [8,193,265]—UCEs, AHEs, and other exonic loci violate the neutrality assumption to some degree, making their application to these types of studies uncertain (see [8]). Historically, anonymous loci datasets have been notoriously difficult to obtain, as genomic cloning methods and allele separation methods such as single-stranded conformation polymorphism gels or PCR cloning were the only means by which these types of data could be acquired [190,191,264,266]. Even target capture has done little to help increase the use of anonymous loci in phylogenomic studies because a reference genome is required to generate template sequences for probe kit design—an expensive process that must be iterated for every study because probes for one species or organismal group will likely not perform well for another given the lack of sequence conservation in these markers and their flanking sequences. In silico-based searches for anonymous loci in complete genome sequences are not impacted by these problems, making it straightforward to extract these sequences, align them locus by locus, and output the data in common file formats ready for phylogenomic analyses (Figure 5). As a proof-of-concept illustration of this approach, software called ALFIE (**A**nonymous **L**oci **Fi**nd**e**r; Figure 5) developed by Costa et al. [193] extracted sequences for 292 presumably neutral and genealogically independent anonymous loci (average locus length ~1 kb; total of 292,169 nucleotide sites) from the human, chimpanzee, gorilla, and orangutan genomes. Given that half of the studies listed in Table 1 focused on clades with shallow-time divergences, in silico acquisition of anonymous loci from complete genomes will likely have a large positive impact on phylogeographic, population genomic, and shallow-scale phylogenomic studies in the future.

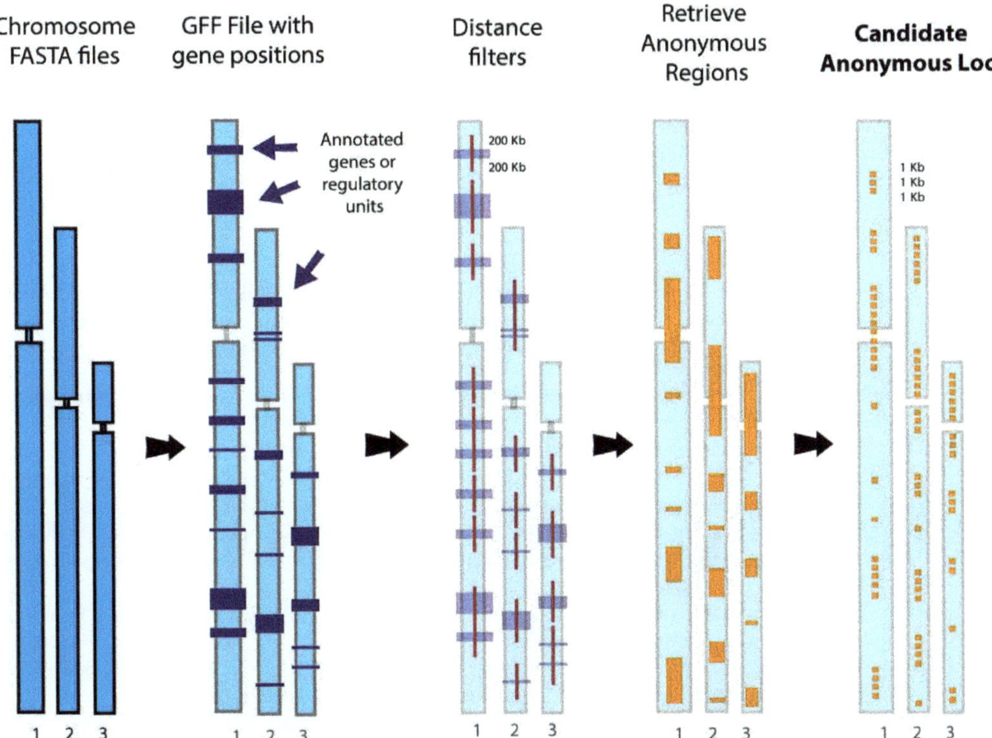

Figure 5. The ALFIE software pipeline for in silico extraction of anonymous loci sequences from complete genome sequences and assembling ready-to-analyze data sets. The user first inputs genome sequences in FASTA format, one of which must be a reference genome with a GFF (general features format) file of genomic annotations, namely protein-coding genes, and regulatory regions. The program then maps the presumably neutral intergenic or "anonymous" regions by applying a user-specified physical distance threshold (in base pairs [bp]). This filter discards all chromosomal regions that contain known functional elements and their flanking sequences (up to the threshold distance), thereby helping to ensure that retained anonymous regions are unaffected by natural selection (e.g., background selection). The anonymous regions are then split into user-specific locus lengths (in bp), which are referred to as "candidate anonymous loci." In the final steps (not shown), the program uses candidate anonymous loci as query sequences to conduct BLAST searches against all input genomes, keeping only single-copy loci in all genomes, before saving them to a FASTA file. Next, the program conducts multiple sequence alignments for all loci before using a second user-defined distance threshold (in bp) to retain loci that are spaced far enough from other sampled loci that they likely meet the independent gene tree assumption. Lastly, the program outputs the dataset in NEXUS, PHYLIP, and FASTA formats, and can use other included modules to find in automated fashion the best DNA substitution model and gene tree for each locus (figure modified after Figure 1 in Costa et al. [193]). See also Jennings [189] for further explanation and extensions of physical distance threshold theory. Reprinted with permission from Costa et al. [193].

9. Allele Phasing Is a Much-Neglected Component of Most Phylogenomic Workflows

Both the hybrid-capture and in silico approaches to isolating loci for phylogenomics routinely miss a key component of the phylogenomic workflow: allele phasing (Figure 4). Allele phasing tries to reconstruct the actual alleles that comprise a locus over a region of the genome in a diploid organism. Phased alleles are the most natural way to represent genetic diversity within and among species, yet most if not all phylogenetic trees for non-avian reptiles, whether using coalescent or concatenation approaches, neglect to attempt to resolve the two alleles that comprise all loci of diploid organisms. We will not review the different types of allele phasing here, except to say that the approach became popular in the early 2000s with software such as PHASE and fastPHASE [267,268]. Most phylogenomic studies, knowingly or unknowingly, analyze loci that do not represent natural alleles because they are unphased and are usually arbitrary amalgamations of the two alleles found at a particular locus. Neglecting to phase loci has been a major, unacknowledged gap in the program of molecular systematics ever since DNA began to be used routinely in the 1980s. Several studies have demonstrated convincingly that allele phasing improves phylogenetic and phylogeographic inference at multiple temporal scales [196,269,270]. The phylogenomics community likely misses many intriguing insights due to the rampant lack of phasing. Reptile phylogenomics and phylogenomics generally should work towards making allele phasing a routine part of phylogenomic workflows.

10. New Reptile Genomes Will Fuel the Future of Reptile Phylogenomics and Genome-Phenotype Discovery via Comparative Genomics

As we have seen, there are critical differences in the production of UCE data in the wet lab, via hybrid capture, and bioinformatically from whole genomes. For example, hybrid capture approaches may result in loci with few flanking regions, especially if the source DNA is degraded, as it often is with historical museum specimens [271]. By contrast, UCEs harvested in silico from whole genomes provide the flexibility to modulate the length of the flanking regions, allowing the researcher to find a balance between maximizing the number of variable sites in the flanking regions with the uncertainty that comes with the inevitable degradation of the alignments of those regions [272]. Consequently, although in silico methods rely on expensive production and assembly of whole genomes, this approach will likely become the norm in phylogenomics of non-avian reptiles.

Another reason why whole genomes will help drive a new generation of reptile phylogenomic studies is that they immediately make available a wealth of marker types that will allow easier comparison of loci of different evolutionary dynamics and phylogenetic information content. A major question in phylogenomics today is what is the optimal marker for a phylogenomic study? This question, in turn, depends somewhat on the method by which phylogenies will be built; concatenation versus coalescent approaches. Regardless of one's predilections towards one or the other method, a recent study [273] showed that, across a wide variety of phylogenomic data sets, there was strong evidence in the sequence data for heterogeneity and lack of concordance among gene trees—sufficient evidence for many researchers that coalescent approaches, which attempt to accommodate such heterogeneity, should be favored. Whereas concatenation approaches need not pay much attention to the information content of individual loci, relying instead on the summed signal across loci, coalescent approaches—especially "two-step" approaches that build gene trees from each locus prior to amalgamating their signal in a species tree—depend critically on well-resolved gene trees [243,274]. Several phylogenetic and phylogeographic models based on the multispecies coalescent model rely on so-called "sequence-based markers" [275]—sets of aligned sites from which gene trees can be built [276,277]. Sequence-based markers, of which UCEs are one type, constitute a major data type for modern phylogenomics, and whole-genome sequences will maximize the ability to choose among various marker types judiciously. A wealth of phylogenomic studies have shown that there is a great variety of information content of different marker types: for example, introns routinely surpass exons in phylogenomic performance and display less evidence for clade-wide or lineage-specific shifts in base composition, which can compromise many methods of phylogenetic inference [110,278,279]. However, there are many maker types that have been unexplored to date: for example, we know nothing about the performance of loci occurring between genes—intergenic regions. Such regions are likely to be highly heterogeneous, consisting of transposable elements, non-coding regulatory regions and other types of genomic regions with diverse evolutionary dynamics. Such regions, however, need to be explored, not only to further resolve the tree for reptiles but also to learn about the relative performance of all regions of the genome, rather than the select few that have risen to high popularity in recent years. New 'pangenome' approaches, such as the ProgressiveCactus genome aligner [280] and the Optimized Dynamic Genome/Graph Implementation [281,282] implemented in the Pangenome Graph Builder—methods that eschew a single reference genome and instead align and compare genomes in an 'all versus all' manner—are able to retain all regions of a genome of every species in a comparative study, and are therefore better able to capture complex but potentially phylogenetically informative 'rare genomic changes' across the tree of life.

Finally, whole genomes will be essential to the nascent field of "PhyloG2P"—using phylogenies to connect genomes and phenotypes across the tree of life [283]. PhyloG2P, also known as PhyloGWAS [284], presents an extremely exciting prospect of mapping genes underlying key phenotypes using comparative genomics. Several papers in recent years have demonstrated the power of comparative genomics for understanding the loci, both coding and noncoding, that appear to drive specific phenotypes in specific lineages [285–288]. Examples from amniotes and other taxa reveal PhyloG2P to be a viable endeavor to understand the genetic basis of convergent and lineage-specific traits, such as loss of flight in birds [289], longevity in mammals and fish [290–293], and limb and digital morphology in mammals and squamates [287,294], and several other traits. Additionally, there is an emerging set of statistical models that allows researchers to study evolutionary associations between candidate regions of the genome and the evolution of specific traits on phylogenies [295–298]. However, to our knowledge, these promising approaches have rarely been attempted in non-avian reptile datasets [294]. This shortcoming is evident despite the unique phenotypes in this clade, including the many novel genomic features reviewed here, diverse modes of reproduction and sex determination (Figure 1), and numerous, often derived morphological and physiological traits with poorly known genomic underpinnings, such as ectothermy, venom, and limb reduction or loss [32,51,53,54,118,299,300].

11. Conclusions

In conclusion, we have reviewed the many novel features of non-avian reptile genomes and the challenges they present for genome assembly, phylogenetic inference, and comparative biology. The relatively large genomes of non-avian reptiles, their sometimes high-density of repetitive elements, and the dearth of researchers straddling the connections between genomic and phenotypic evolution have slowed progress in whole genome sequencing and phylogenomics in non-avian reptiles. Indeed, debate remains about the phylogenetic relationships among squamate reptiles—the most species-rich group of non-avian reptiles. Nevertheless, the wealth of distinctive features of non-avian reptile genomes and phenotypes makes them a prime focus for comparative genomics and phylogenetics. Whole genome sequencing not only provides a rich resource for in silico harvesting of information-rich markers for phylogenomics, but also can provide a platform for finding connections between genomes and phenotypic evolution. We look forward to a new era of integration of non-avian reptile comparative biology, natural history, and genomics, fueled by an increased number of high-quality genomes.

Author Contributions: D.C.C., W.B.J. and S.V.E. outlined the scope of the article; D.C.C. and W.B.J. gathered and analyzed the datasets summarized in this article; D.C.C., W.B.J. and S.V.E. wrote and edited the article. All authors have read and agreed to the published version of the manuscript.

Funding: DCC was supported by a postdoctoral fellowship from the National Science Foundation (DEB-1812310). WBJ was supported by a foreign visiting professorship from the Coordenação de Aperfeiçoamaento de Pessoal de Nível Superior (CAPES-8887.571413/2020-00).

Institutional Review Board Statement: Not applicable.

Informed Consent Statement: Not applicable.

Data Availability Statement: The data presented in this study (Figures 1–3) are openly available at GitHub/Zenodo at https://doi.org/10.5281/zenodo.7517351.

Acknowledgments: The authors wish to thank Ettore Olmo for inviting us to write this review. We thank two anonymous reviewers for helpful feedback on this article. WBJ would like to thank Eugenia Zandonà for hosting him in her laboratory at the Universidade do Estado do Rio de Janeiro. Computational analyses presented in this paper were run on the FASRC Cannon cluster supported by the FAS Division of Science Research Computing Group at Harvard University.

Conflicts of Interest: The authors declare no conflict of interest with the content of this article.

References

1. *Amniote Paleobiology: Perspectives on the Evolution of Mammals, Birds, and Reptiles*; Carrano, M.T., Gaudin, T.J., Blob, R.W., Wible, J.R., Eds.; University of Chicago Press: Chicago, IL, USA, 2006; ISBN 978-0-226-09478-6.
2. Shedlock, A.M.; Edwards, S.V. Amniotes (Amniota). In *The Timetree of Life*; Hedges, S.B., Kumar, S., Eds.; Oxford University Press: New York, NY, USA, 2009; pp. 375–379. ISBN 978-0-19-953503-3.
3. Sues, H.-D. *The Rise of Reptiles: 320 Million Years of Evolution*, Illustrated ed.; Johns Hopkins University Press: Baltimore, MA, USA, 2019; ISBN 978-1-4214-2867-3.
4. Hedges, S.B.; Dudley, J.; Kumar, S. TimeTree: A Public Knowledge-Base of Divergence Times among Organisms. *Bioinformatics* **2006**, *22*, 2971–2972. [CrossRef] [PubMed]
5. Kumar, S.; Stecher, G.; Suleski, M.; Hedges, S.B. TimeTree: A Resource for Timelines, Timetrees, and Divergence Times. *Mol. Biol. Evol.* **2017**, *34*, 1812–1819. [CrossRef] [PubMed]
6. Uetz, P.; Etzold, T. The EMBL/EBI Reptile Database. *Herpetol. Rev.* **1996**, *27*, 174–175.
7. Uetz, P.; Freed, P.; Aguilar, R.; Reyes, F.; Hošek, J. The Reptile Database. 2022. Available online: http://www.reptile-database.org (accessed on 15 December 2022).
8. Stöck, M.; Kratochvíl, L.; Kuhl, H.; Rovatsos, M.; Evans, B.J.; Suh, A.; Valenzuela, N.; Veyrunes, F.; Zhou, Q.; Gamble, T.; et al. A Brief Review of Vertebrate Sex Evolution with a Pledge for Integrative Research: Towards 'Sexomics'. *Philos. Trans. R. Soc. B Biol. Sci.* **2021**, *376*, 20200426. [CrossRef]
9. Capel, B. Vertebrate Sex Determination: Evolutionary Plasticity of a Fundamental Switch. *Nat. Rev. Genet.* **2017**, *18*, 675–689. [CrossRef]
10. Angelini, F.; Ghiara, G. Reproductive Modes and Strategies in Vertebrate Evolution. *Boll. Zool.* **1984**, *51*, 121–203. [CrossRef]
11. Avise, J.C. Evolutionary Perspectives on Clonal Reproduction in Vertebrate Animals. *Proc. Natl. Acad. Sci. USA* **2015**, *112*, 8867–8873. [CrossRef] [PubMed]
12. Neaves, W.B.; Baumann, P. Unisexual Reproduction among Vertebrates. *Trends Genet.* **2011**, *27*, 81–88. [CrossRef] [PubMed]
13. Gregory, T.R.; Nicol, J.A.; Tamm, H.; Kullman, B.; Kullman, K.; Leitch, I.J.; Murray, B.G.; Kapraun, D.F.; Greilhuber, J.; Bennett, M.D. Eukaryotic Genome Size Databases. *Nucleic Acids Res.* **2007**, *35*, D332–D338. [CrossRef]
14. Ashman, T.-L.; Bachtrog, D.; Blackmon, H.; Goldberg, E.E.; Hahn, M.W.; Kirkpatrick, M.; Kitano, J.; Mank, J.E.; Mayrose, I.; Ming, R.; et al. Tree of Sex: A Database of Sexual Systems. *Sci. Data* **2014**, *1*, 140015. [CrossRef]
15. Pasquesi, G.I.M.; Adams, R.H.; Card, D.C.; Schield, D.R.; Corbin, A.B.; Perry, B.W.; Reyes-Velasco, J.; Ruggiero, R.P.; Vandewege, M.W.; Shortt, J.A.; et al. Squamate Reptiles Challenge Paradigms of Genomic Repeat Element Evolution Set by Birds and Mammals. *Nat. Commun.* **2018**, *9*, 2774. [CrossRef]
16. Green, R.E.; Braun, E.L.; Armstrong, J.; Earl, D.; Nguyen, N.; Hickey, G.; Vandewege, M.W.; St. John, J.A.; Capella-Gutiérrez, S.; Castoe, T.A.; et al. Three Crocodilian Genomes Reveal Ancestral Patterns of Evolution among Archosaurs. *Science* **2014**, *346*, 1254449. [CrossRef] [PubMed]
17. Brian Simison, W.; Parham, J.F.; Papenfuss, T.J.; Lam, A.W.; Henderson, J.B. An Annotated Chromosome-Level Reference Genome of the Red-Eared Slider Turtle (*Trachemys scripta elegans*). *Genome Biol. Evol.* **2020**, *12*, 456–462. [CrossRef] [PubMed]
18. Gemmell, N.J.; Rutherford, K.; Prost, S.; Tollis, M.; Winter, D.; Macey, J.R.; Adelson, D.L.; Suh, A.; Bertozzi, T.; Grau, J.H.; et al. The Tuatara Genome Reveals Ancient Features of Amniote Evolution. *Nature* **2020**, *584*, 403–409. [CrossRef] [PubMed]
19. Kapusta, A.; Suh, A.; Feschotte, C. Dynamics of Genome Size Evolution in Birds and Mammals. *Proc. Natl. Acad. Sci. USA* **2017**, *114*, E1460–E1469. [CrossRef]
20. Cao, D.; Wang, M.; Ge, Y.; Gong, S. Draft Genome of the Big-Headed Turtle *Platysternon megacephalum*. *Sci. Data* **2019**, *6*, 60. [CrossRef]

21. Das, D.; Singh, S.K.; Bierstedt, J.; Erickson, A.; Galli, G.L.J.; Crossley, D.A., II; Rhen, T. Draft Genome of the Common Snapping Turtle, *Chelydra serpentina*, a Model for Phenotypic Plasticity in Reptiles. *G3 Genes | Genomes | Genetics* **2020**, *10*, 4299–4314. [CrossRef]
22. Kitts, P.A.; Church, D.M.; Thibaud-Nissen, F.; Choi, J.; Hem, V.; Sapojnikov, V.; Smith, R.G.; Tatusova, T.; Xiang, C.; Zherikov, A.; et al. Assembly: A Resource for Assembled Genomes at NCBI. *Nucleic Acids Res.* **2016**, *44*, D73–D80. [CrossRef]
23. Chojnowski, J.L.; Franklin, J.; Katsu, Y.; Iguchi, T.; Guillette, L.J.; Kimball, R.T.; Braun, E.L. Patterns of Vertebrate Isochore Evolution Revealed by Comparison of Expressed Mammalian, Avian, and Crocodilian Genes. *J. Mol. Evol.* **2007**, *65*, 259–266. [CrossRef]
24. Chojnowski, J.L.; Braun, E.L. Turtle Isochore Structure Is Intermediate between Amphibians and Other Amniotes. *Integr. Comp. Biol.* **2008**, *48*, 454–462. [CrossRef]
25. Costantini, M.; Clay, O.; Auletta, F.; Bernardi, G. An Isochore Map of Human Chromosomes. *Genome Res.* **2006**, *16*, 536–541. [CrossRef] [PubMed]
26. Costantini, M.; Filippo, M.D.; Auletta, F.; Bernardi, G. Isochore Pattern and Gene Distribution in the Chicken Genome. *Gene* **2007**, *400*, 9–15. [CrossRef] [PubMed]
27. Fujita, M.K.; Edwards, S.V.; Ponting, C.P. The *Anolis* Lizard Genome: An Amniote Genome without Isochores. *Genome Biol. Evol.* **2011**, *3*, 974–984. [CrossRef] [PubMed]
28. Costantini, M.; Greif, G.; Alvarez-Valin, F.; Bernardi, G. The *Anolis* Lizard Genome: An Amniote Genome without Isochores? *Genome Biol. Evol.* **2016**, *8*, 1048–1055. [CrossRef]
29. St John, J.A.; Braun, E.L.; Isberg, S.R.; Miles, L.G.; Chong, A.Y.; Gongora, J.; Dalzell, P.; Moran, C.; Bed'Hom, B.; Abzhanov, A.; et al. Sequencing Three Crocodilian Genomes to Illuminate the Evolution of Archosaurs and Amniotes. *Genome Biol.* **2012**, *13*, 415. [CrossRef] [PubMed]
30. Shaffer, H.B.; Minx, P.; Warren, D.E.; Shedlock, A.M.; Thomson, R.C.; Valenzuela, N.; Abramyan, J.; Amemiya, C.T.; Badenhorst, D.; Biggar, K.K.; et al. The Western Painted Turtle Genome, a Model for the Evolution of Extreme Physiological Adaptations in a Slowly Evolving Lineage. *Genome Biol.* **2013**, *14*, R28. [CrossRef]
31. Costantini, M.; Cammarano, R.; Bernardi, G. The Evolution of Isochore Patterns in Vertebrate Genomes. *BMC Genom.* **2009**, *10*, 146. [CrossRef]
32. Janes, D.E.; Organ, C.L.; Fujita, M.K.; Shedlock, A.M.; Edwards, S.V. Genome Evolution in Reptilia, the Sister Group of Mammals. *Annu. Rev. Genom. Hum. Genet.* **2010**, *11*, 239–264. [CrossRef]
33. Jennings, W.B. *Phylogenomic Data Acquisition: Principles and Practice*; CRC Press: Boca Raton, FL, USA, 2016; ISBN 978-1-315-18143-1.
34. Duret, L.; Galtier, N. Biased Gene Conversion and the Evolution of Mammalian Genomic Landscapes. *Annu. Rev. Genom. Hum. Genet.* **2009**, *10*, 285–311. [CrossRef] [PubMed]
35. Romiguier, J.; Ranwez, V.; Douzery, E.J.P.; Galtier, N. Contrasting GC-Content Dynamics across 33 Mammalian Genomes: Relationship with Life-History Traits and Chromosome Sizes. *Genome Res.* **2010**, *20*, 1001–1009. [CrossRef]
36. Figuet, E.; Nabholz, B.; Bonneau, M.; Carrio, E.M.; Nadachowska-Brzyska, K.; Ellegren, H.; Galtier, N. Life History Traits, Protein Evolution, and the Nearly Neutral Theory in Amniotes. *Mol. Biol. Evol.* **2016**, *33*, 1517–1527. [CrossRef]
37. Schield, D.R.; Card, D.C.; Hales, N.R.; Perry, B.W.; Pasquesi, G.M.; Blackmon, H.; Adams, R.H.; Corbin, A.B.; Smith, C.F.; Ramesh, B.; et al. The Origins and Evolution of Chromosomes, Dosage Compensation, and Mechanisms Underlying Venom Regulation in Snakes. *Genome Res.* **2019**, *29*, 590–601. [CrossRef] [PubMed]
38. Cree, A.; Thompson, M.B.; Daugherty, C.H. Tuatara Sex Determination. *Nature* **1995**, *375*, 543. [CrossRef]
39. Mitchell, N.J.; Nelson, N.J.; Cree, A.; Pledger, S.; Keall, S.N.; Daugherty, C.H. Support for a Rare Pattern of Temperature-Dependent Sex Determination in Archaic Reptiles: Evidence from Two Species of Tuatara (*Sphenodon*). *Front. Zool.* **2006**, *3*, 9. [CrossRef]
40. González, E.J.; Martínez-López, M.; Morales-Garduza, M.A.; García-Morales, R.; Charruau, P.; Gallardo-Cruz, J.A. The Sex-Determination Pattern in Crocodilians: A Systematic Review of Three Decades of Research. *J. Anim. Ecol.* **2019**, *88*, 1417–1427. [CrossRef]
41. Ewert, M.A.; Nelson, C.E. Sex Determination in Turtles: Diverse Patterns and Some Possible Adaptive Values. *Copeia* **1991**, *1991*, 50–69. [CrossRef]
42. Viets, B.E.; Ewert, M.A.; Talent, L.G.; Nelson, C.E. Sex-Determining Mechanisms in Squamate Reptiles. *J. Exp. Zool.* **1994**, *270*, 45–56. [CrossRef]
43. Pokorná, M.; Kratochvíl, L. Phylogeny of Sex-Determining Mechanisms in Squamate Reptiles: Are Sex Chromosomes an Evolutionary Trap? *Zool. J. Linn. Soc.* **2009**, *156*, 168–183. [CrossRef]
44. Bista, B.; Valenzuela, N. Turtle Insights into the Evolution of the Reptilian Karyotype and the Genomic Architecture of Sex Determination. *Genes* **2020**, *11*, 416. [CrossRef] [PubMed]
45. Robert, K.A.; Thompson, M.B. Viviparous Lizard Selects Sex of Embryos. *Nature* **2001**, *412*, 698–699. [CrossRef]
46. Shine, R.; Elphick, M.J.; Donnellan, S. Co-Occurrence of Multiple, Supposedly Incompatible Modes of Sex Determination in a Lizard Population. *Ecol. Lett.* **2002**, *5*, 486–489. [CrossRef]
47. Quinn, A.E.; Georges, A.; Sarre, S.D.; Guarino, F.; Ezaz, T.; Graves, J.A.M. Temperature Sex Reversal Implies Sex Gene Dosage in a Reptile. *Science* **2007**, *316*, 411. [CrossRef] [PubMed]
48. Radder, R.S.; Quinn, A.E.; Georges, A.; Sarre, S.D.; Shine, R. Genetic Evidence for Co-Occurrence of Chromosomal and Thermal Sex-Determining Systems in a Lizard. *Biol. Lett.* **2008**, *4*, 176–178. [CrossRef]

49. Holleley, C.E.; O'Meally, D.; Sarre, S.D.; Marshall Graves, J.A.; Ezaz, T.; Matsubara, K.; Azad, B.; Zhang, X.; Georges, A. Sex Reversal Triggers the Rapid Transition from Genetic to Temperature-Dependent Sex. *Nature* **2015**, *523*, 79–82. [CrossRef]
50. Hill, P.L.; Burridge, C.P.; Ezaz, T.; Wapstra, E. Conservation of Sex-Linked Markers among Conspecific Populations of a Viviparous Skink, *Niveoscincus ocellatus*, Exhibiting Genetic and Temperature-Dependent Sex Determination. *Genome Biol. Evol.* **2018**, *10*, 1079–1087. [CrossRef] [PubMed]
51. Cornejo-Páramo, P.; Dissanayake, D.S.B.; Lira-Noriega, A.; Martínez-Pacheco, M.L.; Acosta, A.; Ramírez-Suástegui, C.; Méndez-de-la-Cruz, F.R.; Székely, T.; Urrutia, A.O.; Georges, A.; et al. Viviparous Reptile Regarded to Have Temperature-Dependent Sex Determination Has Old XY Chromosomes. *Genome Biol. Evol.* **2020**, *12*, 924–930. [CrossRef]
52. Blackburn, D.G. Classification of the Reproductive Patterns of Amniotes. *Herpetol. Monogr.* **2000**, *14*, 371–377. [CrossRef]
53. Sites, J.W.; Reeder, T.W.; Wiens, J.J. Phylogenetic Insights on Evolutionary Novelties in Lizards and Snakes: Sex, Birth, Bodies, Niches, and Venom. *Annu. Rev. Ecol. Evol. Syst.* **2011**, *42*, 227–244. [CrossRef]
54. Whittington, C.M.; Van Dyke, J.U.; Liang, S.Q.T.; Edwards, S.V.; Shine, R.; Thompson, M.B.; Grueber, C.E. Understanding the Evolution of Viviparity Using Intraspecific Variation in Reproductive Mode and Transitional Forms of Pregnancy. *Biol. Rev.* **2022**, *97*, 1179–1192. [CrossRef]
55. Pyron, R.A.; Burbrink, F.T. Early Origin of Viviparity and Multiple Reversions to Oviparity in Squamate Reptiles. *Ecol. Lett.* **2014**, *17*, 13–21. [CrossRef]
56. Brana, F.; Bea, A. Biomodalité de La Reproduction Chez *Lacerta vivipara* (Reptilia, Lacertidae). *Bull. Société Herpétologique Fr.* **1987**, *44*, 1–5.
57. Heulin, B. Données Nouvelles Sur Les Populations Ovipares de *Lacerta vivipara*. *C. R. Académie Sci. Sér. 3 Sci. Vie* **1988**, *306*, 63–68.
58. Qualls, C.P.; Shine, R.; Donnellan, S.; Hutchinson, M. The Evolution of Viviparity within the Australian Scincid Lizard *Lerista bougainvillii*. *J. Zool.* **1995**, *237*, 13–26. [CrossRef]
59. Smith, S.A.; Shine, R. Intraspecific Variation in Reproductive Mode within the Scincid Lizard *Saiphos equalis*. *Aust. J. Zool.* **1997**, *45*, 435–445. [CrossRef]
60. Braz, H.B.; Scartozzoni, R.R.; Almeida-Santos, S.M. Reproductive Modes of the South American Water Snakes: A Study System for the Evolution of Viviparity in Squamate Reptiles. *Zool. Anz. J. Comp. Zool.* **2016**, *263*, 33–44. [CrossRef]
61. Recknagel, H.; Carruthers, M.; Yurchenko, A.A.; Nokhbatolfoghahai, M.; Kamenos, N.A.; Bain, M.M.; Elmer, K.R. The Functional Genetic Architecture of Egg-Laying and Live-Bearing Reproduction in Common Lizards. *Nat. Ecol. Evol.* **2021**, *5*, 1546–1556. [CrossRef] [PubMed]
62. Shine, R. A New Hypothesis for the Evolution of Viviparity in Reptiles. *Am. Nat.* **1995**, *145*, 809–823. [CrossRef]
63. Noble, D.W.A.; Stenhouse, V.; Schwanz, L.E. Developmental Temperatures and Phenotypic Plasticity in Reptiles: A Systematic Review and Meta-Analysis. *Biol. Rev.* **2018**, *93*, 72–97. [CrossRef] [PubMed]
64. Cuellar, O.; Kluge, A.G. Natural Parthenogenesis in the Gekkonid Lizard *Lepidodactylus lugubris*. *J. Genet.* **1972**, *61*, 14–26. [CrossRef]
65. Kluge, A.G.; Eckardt, M.J. *Hemidactylus garnotii* Duméril and Bibron, a Triploid All-Female Species of Gekkonid Lizard. *Copeia* **1969**, *1969*, 651–664. [CrossRef]
66. McDowell, S.B. A Catalogue of the Snakes of New Guinea and the Solomons, with Special Reference to Those in the Bernice P. Bishop Museum, Part I. Scolecophidia. *J. Herpetol.* **1974**, *8*, 1–57. [CrossRef]
67. Lowe, C.H.; Wright, J.W. Evolution of Parthenogenetic Species of *Cnemidophorus* (Whiptail Lizards) in Western North America. *J. Ariz. Acad. Sci.* **1966**, *4*, 81–87.
68. Darevsky, I. Natural Parthenogenesis in Certain Subspecies of Rock Lizard, *Lacerta saxicola*. *Dokl. Akad. Nauk SSSR* **1958**, *122*, 730–732.
69. Watts, P.C.; Buley, K.R.; Sanderson, S.; Boardman, W.; Ciofi, C.; Gibson, R. Parthenogenesis in Komodo Dragons. *Nature* **2006**, *444*, 1021–1022. [CrossRef] [PubMed]
70. Magnusson, W.E. Production of an Embryo by an *Acrochordus javanicus* Isolated for Seven Years. *Copeia* **1979**, *1979*, 744–745. [CrossRef]
71. Schuett, G.W.; Fernandez, P.J.; Gergits, W.F.; Casna, N.J.; Chiszar, D.; Smith, H.M.; Mitton, J.B.; Mackessy, S.P.; Odum, R.A.; Demlong, M.J. Production of Offspring in the Absence of Males: Evidence for Facultative Parthenogenesis in Bisexual Snakes. *Herpetol. Nat. Hist.* **1997**, *5*, 1–10.
72. Groot, T.V.M.; Bruins, E.; Breeuwer, J.A.J. Molecular Genetic Evidence for Parthenogenesis in the Burmese Python, *Python molurus bivittatus*. *Heredity* **2003**, *90*, 130–135. [CrossRef]
73. Booth, W.; Johnson, D.H.; Moore, S.; Schal, C.; Vargo, E.L. Evidence for Viable, Non-Clonal but Fatherless Boa Constrictors. *Biol. Lett.* **2011**, *7*, 253–256. [CrossRef] [PubMed]
74. Booth, W.; Million, L.; Reynolds, R.G.; Burghardt, G.M.; Vargo, E.L.; Schal, C.; Tzika, A.C.; Schuett, G.W. Consecutive Virgin Births in the New World Boid Snake, the Colombian Rainbow Boa, *Epicrates maurus*. *J. Hered.* **2011**, *102*, 759–763. [CrossRef]
75. Reynolds, R.G.; Booth, W.; Schuett, G.W.; Fitzpatrick, B.M.; Burghardt, G.M. Successive Virgin Births of Viable Male Progeny in the Checkered Gartersnake, *Thamnophis marcianus*. *Biol. J. Linn. Soc.* **2012**, *107*, 566–572. [CrossRef]
76. Card, D.C.; Vonk, F.J.; Smalbrugge, S.; Casewell, N.R.; Wüster, W.; Castoe, T.A.; Schuett, G.W.; Booth, W. Genome-Wide Data Implicate Terminal Fusion Automixis in King Cobra Facultative Parthenogenesis. *Sci. Rep.* **2021**, *11*, 7271. [CrossRef]

77. Booth, W.; Smith, C.F.; Eskridge, P.H.; Hoss, S.K.; Mendelson, J.R.; Schuett, G.W. Facultative Parthenogenesis Discovered in Wild Vertebrates. *Biol. Lett.* **2012**, *8*, 983–985. [CrossRef]
78. Gregory, T.R. Coincidence, Coevolution, or Causation? DNA Content, Cell size, and the C-Value Enigma. *Biol. Rev.* **2001**, *76*, 65–101. [CrossRef] [PubMed]
79. Waltari, E.; Edwards, S.V. Evolutionary Dynamics of Intron Size, Genome Size, and Physiological Correlates in Archosaurs. *Am. Nat.* **2002**, *160*, 539–552. [CrossRef] [PubMed]
80. Gregory, T.R. Genome Size Evolution in Animals. In *The Evolution of the Genome*; Gregory, T.R., Ed.; Academic Press: Burlington, VT, USA, 2005; pp. 3–87. ISBN 978-0-12-301463-4.
81. Organ, C.L.; Moreno, R.G.; Edwards, S.V. Three Tiers of Genome Evolution in Reptiles. *Integr. Comp. Biol.* **2008**, *48*, 494–504. [CrossRef]
82. Oliver, M.J.; Petrov, D.; Ackerly, D.; Falkowski, P.; Schofield, O.M. The Mode and Tempo of Genome Size Evolution in Eukaryotes. *Genome Res.* **2007**, *17*, 594–601. [CrossRef] [PubMed]
83. Olmo, E. Rate of Chromosome Changes and Speciation in Reptiles. *Genetica* **2005**, *125*, 185–203. [CrossRef] [PubMed]
84. Rodionov, A. Micro versus Macro: A Review of Structure and Functions of Avian Micro-and Macrochromosomes. *Russ. J. Genet.* **1996**, *32*, 517–527.
85. Hillier, L.W.; Miller, W.; Birney, E.; Warren, W.; Hardison, R.C.; Ponting, C.P.; Bork, P.; Burt, D.W.; Groenen, M.A.M.; Delany, M.E.; et al. Sequence and Comparative Analysis of the Chicken Genome Provide Unique Perspectives on Vertebrate Evolution. *Nature* **2004**, *432*, 695–716. [CrossRef]
86. Smith, J.; Bruley, C.K.; Paton, I.R.; Dunn, I.; Jones, C.T.; Windsor, D.; Morrice, D.R.; Law, A.S.; Masabanda, J.; Sazanov, A.; et al. Differences in Gene Density on Chicken Macrochromosomes and Microchromosomes. *Anim. Genet.* **2000**, *31*, 96–103. [CrossRef]
87. Rodionov, A.V.; Chel'sheva, L.A.; Soloveĭ, I.V.; Miakoshina, I.A. Chiasma distribution in the lampbrush chromosomes of the chicken *Gallus gallus domesticus*: Hot spots of recombination and their possible role in proper dysjunction of homologous chromosomes at the first meiotic division. *Genetika* **1992**, *28*, 151–160. [PubMed]
88. Perry, B.W.; Schield, D.R.; Adams, R.H.; Castoe, T.A. Microchromosomes Exhibit Distinct Features of Vertebrate Chromosome Structure and Function with Underappreciated Ramifications for Genome Evolution. *Mol. Biol. Evol.* **2021**, *38*, 904–910. [CrossRef] [PubMed]
89. Waters, P.D.; Patel, H.R.; Ruiz-Herrera, A.; Álvarez-González, L.; Lister, N.C.; Simakov, O.; Ezaz, T.; Kaur, P.; Frere, C.; Grützner, F.; et al. Microchromosomes Are Building Blocks of Bird, Reptile, and Mammal Chromosomes. *Proc. Natl. Acad. Sci. USA* **2021**, *118*, e2112494118. [CrossRef]
90. Fredga, K. Chromosomal Changes in Vertebrate Evolution. *Proc. R. Soc. Lond. B Biol. Sci.* **1977**, *199*, 377–397.
91. Damas, J.; Corbo, M.; Lewin, H.A. Vertebrate Chromosome Evolution. *Annu. Rev. Anim. Biosci.* **2021**, *9*, 1–27. [CrossRef]
92. Srikulnath, K.; Ahmad, S.F.; Singchat, W.; Panthum, T. Why Do Some Vertebrates Have Microchromosomes? *Cells* **2021**, *10*, 2182. [CrossRef] [PubMed]
93. Ellegren, H. Sex-Chromosome Evolution: Recent Progress and the Influence of Male and Female Heterogamety. *Nat. Rev. Genet.* **2011**, *12*, 157–166. [CrossRef] [PubMed]
94. Wright, A.E.; Dean, R.; Zimmer, F.; Mank, J.E. How to Make a Sex Chromosome. *Nat. Commun.* **2016**, *7*, 12087. [CrossRef] [PubMed]
95. Abbott, J.K.; Nordén, A.K.; Hansson, B. Sex Chromosome Evolution: Historical Insights and Future Perspectives. *Proc. R. Soc. B Biol. Sci.* **2017**, *284*, 20162806. [CrossRef] [PubMed]
96. Van Doorn, G.S.; Kirkpatrick, M. Turnover of Sex Chromosomes Induced by Sexual Conflict. *Nature* **2007**, *449*, 909–912. [CrossRef]
97. Wright, A.E.; Darolti, I.; Bloch, N.I.; Oostra, V.; Sandkam, B.; Buechel, S.D.; Kolm, N.; Breden, F.; Vicoso, B.; Mank, J.E. Convergent Recombination Suppression Suggests Role of Sexual Selection in Guppy Sex Chromosome Formation. *Nat. Commun.* **2017**, *8*, 14251. [CrossRef] [PubMed]
98. Vicoso, B. Molecular and Evolutionary Dynamics of Animal Sex-Chromosome Turnover. *Nat. Ecol. Evol.* **2019**, *3*, 1632–1641. [CrossRef] [PubMed]
99. Gamble, T.; Coryell, J.; Ezaz, T.; Lynch, J.; Scantlebury, D.P.; Zarkower, D. Restriction Site-Associated DNA Sequencing (RAD-Seq) Reveals an Extraordinary Number of Transitions among Gecko Sex-Determining Systems. *Mol. Biol. Evol.* **2015**, *32*, 1296–1309. [CrossRef] [PubMed]
100. Gamble, T.; Castoe, T.A.; Nielsen, S.V.; Banks, J.L.; Card, D.C.; Schield, D.R.; Schuett, G.W.; Booth, W. The Discovery of XY Sex Chromosomes in a Boa and Python. *Curr. Biol.* **2017**, *27*, 2148–2153. [CrossRef] [PubMed]
101. Ohno, S. *Sex Chromosomes and Sex-Linked Genes*; Monographs on Endocrinology; Springer: Berlin/Heidelberg, Germany, 1967; ISBN 978-3-642-88178-7.
102. Matsubara, K.; Tarui, H.; Toriba, M.; Yamada, K.; Nishida-Umehara, C.; Agata, K.; Matsuda, Y. Evidence for Different Origin of Sex Chromosomes in Snakes, Birds, and Mammals and Step-Wise Differentiation of Snake Sex Chromosomes. *Proc. Natl. Acad. Sci. USA* **2006**, *103*, 18190–18195. [CrossRef]
103. Castoe, T.A.; de Koning, A.P.J.; Kim, H.-M.; Gu, W.; Noonan, B.P.; Naylor, G.; Jiang, Z.J.; Parkinson, C.L.; Pollock, D.D. Evidence for an Ancient Adaptive Episode of Convergent Molecular Evolution. *Proc. Natl. Acad. Sci. USA* **2009**, *106*, 8986–8991. [CrossRef]

104. Macey, J.R.; Pabinger, S.; Barbieri, C.G.; Buring, E.S.; Gonzalez, V.L.; Mulcahy, D.G.; DeMeo, D.P.; Urban, L.; Hime, P.M.; Prost, S.; et al. Evidence of Two Deeply Divergent Co-Existing Mitochondrial Genomes in the Tuatara Reveals an Extremely Complex Genomic Organization. *Commun. Biol.* **2021**, *4*, 1–10. [CrossRef]
105. Jiang, Z.J.; Castoe, T.A.; Austin, C.C.; Burbrink, F.T.; Herron, M.D.; McGuire, J.A.; Parkinson, C.L.; Pollock, D.D. Comparative Mitochondrial Genomics of Snakes: Extraordinary Substitution Rate Dynamics and Functionality of the Duplicate Control Region. *BMC Evol. Biol.* **2007**, *7*, 123. [CrossRef]
106. Castoe, T.A.; Jiang, Z.J.; Gu, W.; Wang, Z.O.; Pollock, D.D. Adaptive Evolution and Functional Redesign of Core Metabolic Proteins in Snakes. *PLoS ONE* **2008**, *3*, e2201. [CrossRef]
107. Lander, E.S.; Linton, L.M.; Birren, B.; Nusbaum, C.; Zody, M.C.; Baldwin, J.; Devon, K.; Dewar, K.; Doyle, M.; FitzHugh, W.; et al. Initial Sequencing and Analysis of the Human Genome. *Nature* **2001**, *409*, 860–921. [CrossRef]
108. Chinwalla, A.T.; Cook, L.L.; Delehaunty, K.D.; Fewell, G.A.; Fulton, L.A.; Fulton, R.S.; Graves, T.A.; Hillier, L.W.; Mardis, E.R.; McPherson, J.D.; et al. Initial Sequencing and Comparative Analysis of the Mouse Genome. *Nature* **2002**, *420*, 520–562. [CrossRef]
109. Warren, W.C.; Clayton, D.F.; Ellegren, H.; Arnold, A.P.; Hillier, L.W.; Künstner, A.; Searle, S.; White, S.; Vilella, A.J.; Fairley, S.; et al. The Genome of a Songbird. *Nature* **2010**, *464*, 757–762. [CrossRef]
110. Zhang, G.; Li, C.; Li, Q.; Li, B.; Larkin, D.M.; Lee, C.; Storz, J.F.; Antunes, A.; Greenwold, M.J.; Meredith, R.W.; et al. Comparative Genomics Reveals Insights into Avian Genome Evolution and Adaptation. *Science* **2014**, *346*, 1311–1320. [CrossRef] [PubMed]
111. Kapusta, A.; Suh, A. Evolution of Bird Genomes—A Transposon's-Eye View. *Ann. N. Y. Acad. Sci.* **2017**, *1389*, 164–185. [CrossRef] [PubMed]
112. Shedlock, A.M. Phylogenomic Investigation of CR1 LINE Diversity in Reptiles. *Syst. Biol.* **2006**, *55*, 902–911. [CrossRef]
113. Shedlock, A.M.; Botka, C.W.; Zhao, S.; Shetty, J.; Zhang, T.; Liu, J.S.; Deschavanne, P.J.; Edwards, S.V. Phylogenomics of Nonavian Reptiles and the Structure of the Ancestral Amniote Genome. *Proc. Natl. Acad. Sci. USA* **2007**, *104*, 2767–2772. [CrossRef] [PubMed]
114. Castoe, T.A.; Hall, K.T.; Guibotsy Mboulas, M.L.; Gu, W.; de Koning, A.P.J.; Fox, S.E.; Poole, A.W.; Vemulapalli, V.; Daza, J.M.; Mockler, T.; et al. Discovery of Highly Divergent Repeat Landscapes in Snake Genomes Using High-Throughput Sequencing. *Genome Biol. Evol.* **2011**, *3*, 641–653. [CrossRef] [PubMed]
115. Castoe, T.A.; de Koning, A.P.J.; Hall, K.T.; Card, D.C.; Schield, D.R.; Fujita, M.K.; Ruggiero, R.P.; Degner, J.F.; Daza, J.M.; Gu, W.; et al. The Burmese Python Genome Reveals the Molecular Basis for Extreme Adaptation in Snakes. *Proc. Natl. Acad. Sci. USA* **2013**, *110*, 20645–20650. [CrossRef]
116. Di-Poï, N.; Montoya-Burgos, J.I.; Miller, H.; Pourquié, O.; Milinkovitch, M.C.; Duboule, D. Changes in Hox Genes' Structure and Function during the Evolution of the Squamate Body Plan. *Nature* **2010**, *464*, 99–103. [CrossRef]
117. Shaney, K.J.; Card, D.C.; Schield, D.R.; Ruggiero, R.P.; Pollock, D.D.; Mackessy, S.P.; Castoe, T.A. Squamate Reptile Genomics and Evolution. In *Toxinology: Venom Genomics and Proteomics*; Gopalakrishnakone, P., Calvete, J.J., Eds.; Springer Netherlands: Dordrecht, The Netherlands, 2014; pp. 1–18. ISBN 978-94-007-6649-5.
118. Perry, B.W.; Gopalan, S.S.; Pasquesi, G.I.M.; Schield, D.R.; Westfall, A.K.; Smith, C.F.; Koludarov, I.; Chippindale, P.T.; Pellegrino, M.W.; Chuong, E.B.; et al. Snake Venom Gene Expression Is Coordinated by Novel Regulatory Architecture and the Integration of Multiple Co-Opted Vertebrate Pathways. *Genome Res.* **2022**, *32*, 1058–1073. [CrossRef]
119. Kordiš, D.; Gubenšek, F. Bov-B Long Interspersed Repeated DNA (LINE) Sequences Are Present in *Vipera ammodytes* Phospholipase A2 Genes and in Genomes of Viperidae Snakes. *Eur. J. Biochem.* **1997**, *246*, 772–779. [CrossRef] [PubMed]
120. Kordiš, D.; Gubenšek, F. Unusual Horizontal Transfer of a Long Interspersed Nuclear Element between Distant Vertebrate Classes. *Proc. Natl. Acad. Sci. USA* **1998**, *95*, 10704–10709. [CrossRef]
121. Kordiš, D.; Gubenšek, F. Molecular Evolution of Bov-B LINEs in Vertebrates. *Gene* **1999**, *238*, 171–178. [CrossRef] [PubMed]
122. Piskurek, O.; Okada, N. Poxviruses as Possible Vectors for Horizontal Transfer of Retroposons from Reptiles to Mammals. *Proc. Natl. Acad. Sci. USA* **2007**, *104*, 12046–12051. [CrossRef]
123. Pace, J.K.; Gilbert, C.; Clark, M.S.; Feschotte, C. Repeated Horizontal Transfer of a DNA Transposon in Mammals and Other Tetrapods. *Proc. Natl. Acad. Sci. USA* **2008**, *105*, 17023–17028. [CrossRef]
124. Novick, P.A.; Basta, H.; Floumanhaft, M.; McClure, M.A.; Boissinot, S. The Evolutionary Dynamics of Autonomous Non-LTR Retrotransposons in the Lizard *Anolis carolinensis* Shows More Similarity to Fish Than Mammals. *Mol. Biol. Evol.* **2009**, *26*, 1811–1822. [CrossRef] [PubMed]
125. Piskurek, O.; Nishihara, H.; Okada, N. The Evolution of Two Partner LINE/SINE Families and a Full-Length Chromodomain-Containing Ty3/Gypsy LTR Element in the First Reptilian Genome of *Anolis carolinensis*. *Gene* **2009**, *441*, 111–118. [CrossRef]
126. Novick, P.; Smith, J.; Ray, D.; Boissinot, S. Independent and Parallel Lateral Transfer of DNA Transposons in Tetrapod Genomes. *Gene* **2010**, *449*, 85–94. [CrossRef]
127. Gilbert, C.; Hernandez, S.S.; Flores-Benabib, J.; Smith, E.N.; Feschotte, C. Rampant Horizontal Transfer of SPIN Transposons in Squamate Reptiles. *Mol. Biol. Evol.* **2012**, *29*, 503–515. [CrossRef] [PubMed]
128. Walsh, A.M.; Kortschak, R.D.; Gardner, M.G.; Bertozzi, T.; Adelson, D.L. Widespread Horizontal Transfer of Retrotransposons. *Proc. Natl. Acad. Sci. USA* **2013**, *110*, 1012–1016. [CrossRef]
129. Bernardi, G. The Vertebrate Genome: Isochores and Evolution. *Mol. Biol. Evol.* **1993**, *10*, 186–204. [CrossRef] [PubMed]
130. Bernardi, G. Isochores and the Evolutionary Genomics of Vertebrates. *Gene* **2000**, *241*, 3–17. [CrossRef] [PubMed]

131. Fullerton, S.M.; Bernardo Carvalho, A.; Clark, A.G. Local Rates of Recombination Are Positively Correlated with GC Content in the Human Genome. *Mol. Biol. Evol.* **2001**, *18*, 1139–1142. [CrossRef] [PubMed]
132. Mouchiroud, D.; D'Onofrio, G.; Aïssani, B.; Macaya, G.; Gautier, C.; Bernardi, G. The Distribution of Genes in the Human Genome. *Gene* **1991**, *100*, 181–187. [CrossRef]
133. Jabbari, K.; Bernardi, G. CpG Doublets, CpG Islands and Alu Repeats in Long Human DNA Sequences from Different Isochore Families. *Gene* **1998**, *224*, 123–128. [CrossRef]
134. Duret, L.; Mouchiroud, D.; Gautier, C. Statistical Analysis of Vertebrate Sequences Reveals That Long Genes Are Scarce in GC-Rich Isochores. *J. Mol. Evol.* **1995**, *40*, 308–317. [CrossRef] [PubMed]
135. Watanabe, Y.; Fujiyama, A.; Ichiba, Y.; Hattori, M.; Yada, T.; Sakaki, Y.; Ikemura, T. Chromosome-Wide Assessment of Replication Timing for Human Chromosomes 11q and 21q: Disease-Related Genes in Timing-Switch Regions. *Hum. Mol. Genet.* **2002**, *11*, 13–21. [CrossRef] [PubMed]
136. Galtier, N.; Piganeau, G.; Mouchiroud, D.; Duret, L. GC-Content Evolution in Mammalian Genomes: The Biased Gene Conversion Hypothesis. *Genetics* **2001**, *159*, 907–911. [CrossRef]
137. Galtier, N. Gene Conversion Drives GC Content Evolution in Mammalian Histones. *Trends Genet.* **2003**, *19*, 65–68. [CrossRef]
138. Marais, G. Biased Gene Conversion: Implications for Genome and Sex Evolution. *Trends Genet.* **2003**, *19*, 330–338. [CrossRef]
139. Meunier, J.; Duret, L. Recombination Drives the Evolution of GC-Content in the Human Genome. *Mol. Biol. Evol.* **2004**, *21*, 984–990. [CrossRef] [PubMed]
140. Galtier, N.; Duret, L.; Glémin, S.; Ranwez, V. GC-Biased Gene Conversion Promotes the Fixation of Deleterious Amino Acid Changes in Primates. *Trends Genet.* **2009**, *25*, 1–5. [CrossRef] [PubMed]
141. Schield, D.R.; Pasquesi, G.I.M.; Perry, B.W.; Adams, R.H.; Nikolakis, Z.L.; Westfall, A.K.; Orton, R.W.; Meik, J.M.; Mackessy, S.P.; Castoe, T.A. Snake Recombination Landscapes Are Concentrated in Functional Regions despite *PRDM9*. *Mol. Biol. Evol.* **2020**, *37*, 1272–1294. [CrossRef] [PubMed]
142. Hughes, S.; Clay, O.; Bernardi, G. Compositional Patterns in Reptilian Genomes. *Gene* **2002**, *295*, 323–329. [CrossRef] [PubMed]
143. Hughes, S.; Zelus, D.; Mouchiroud, D. Warm-Blooded Isochore Structure in Nile Crocodile and Turtle. *Mol. Biol. Evol.* **1999**, *16*, 1521–1527. [CrossRef] [PubMed]
144. Fortes, G.G.; Bouza, C.; Martínez, P.; Sánchez, L. Diversity in Isochore Structure among Cold-Blooded Vertebrates Based on GC Content of Coding and Non-Coding Sequences. *Genetica* **2007**, *129*, 281–289. [CrossRef] [PubMed]
145. Elhaik, E.; Landan, G.; Graur, D. Can GC Content at Third-Codon Positions Be Used as a Proxy for Isochore Composition? *Mol. Biol. Evol.* **2009**, *26*, 1829–1833. [CrossRef]
146. Zhang, C.-T.; Zhang, R. Isochore Structures in the Mouse Genome. *Genomics* **2004**, *83*, 384–394. [CrossRef]
147. Alföldi, J.; Di Palma, F.; Grabherr, M.; Williams, C.; Kong, L.; Mauceli, E.; Russell, P.; Lowe, C.B.; Glor, R.E.; Jaffe, J.D.; et al. The Genome of the Green Anole Lizard and a Comparative Analysis with Birds and Mammals. *Nature* **2011**, *477*, 587–591. [CrossRef]
148. Bravo, G.A.; Schmitt, C.J.; Edwards, S.V. What Have We Learned from the First 500 Avian Genomes? *Annu. Rev. Ecol. Evol. Syst.* **2021**, *52*, 611–639. [CrossRef]
149. Moritz, C.C.; Pratt, R.C.; Bank, S.; Bourke, G.; Bragg, J.G.; Doughty, P.; Keogh, J.S.; Laver, R.J.; Potter, S.; Teasdale, L.C.; et al. Cryptic Lineage Diversity, Body Size Divergence, and Sympatry in a Species Complex of Australian Lizards (*Gehyra*). *Evolution* **2018**, *72*, 54–66. [CrossRef] [PubMed]
150. Oliver, P.M.; Clegg, J.R.; Fisher, R.N.; Richards, S.J.; Taylor, P.N.; Jocque, M.M.T. A New Biogeographically Disjunct Giant Gecko (*Gehyra*: Gekkonidae: Reptilia) from the East Melanesian Islands. *Zootaxa* **2016**, *4208*, 61–76. [CrossRef] [PubMed]
151. Doughty, P.; Bourke, G.; Tedeschi, L.G.; Pratt, R.C.; Oliver, P.M.; Palmer, R.A.; Moritz, C. Species Delimitation in the *Gehyra nana* (Squamata: Gekkonidae) Complex: Cryptic and Divergent Morphological Evolution in the Australian Monsoonal Tropics, with the Description of Four New Species. *Zootaxa* **2018**, *4403*, 201. [CrossRef]
152. Oliver, P.M.; Prasetya, A.M.; Tedeschi, L.G.; Fenker, J.; Ellis, R.J.; Doughty, P.; Moritz, C. Crypsis and Convergence: Integrative Taxonomic Revision of the *Gehyra australis* Group (Squamata: Gekkonidae) from Northern Australia. *PeerJ* **2020**, *8*, e7971. [CrossRef] [PubMed]
153. Douglas, M.E.; Douglas, M.R.; Schuett, G.W.; Beck, D.D.; Sullivan, B.K. Conservation Phylogenetics of Helodermatid Lizards Using Multiple Molecular Markers and a Supertree Approach. *Mol. Phylogenet. Evol.* **2010**, *55*, 153–167. [CrossRef]
154. Hugall, A.F.; Foster, R.; Hutchinson, M.; Lee, M.S.Y. Phylogeny of Australasian Agamid Lizards Based on Nuclear and Mitochondrial Genes: Implications for Morphological Evolution and Biogeography. *Biol. J. Linn. Soc.* **2008**, *93*, 343–358. [CrossRef]
155. Edwards, T.; Tollis, M.; Hsieh, P.; Gutenkunst, R.N.; Liu, Z.; Kusumi, K.; Culver, M.; Murphy, R.W. Assessing Models of Speciation under Different Biogeographic Scenarios; an Empirical Study Using Multi-Locus and RNA-Seq Analyses. *Ecol. Evol.* **2016**, *6*, 379–396. [CrossRef]
156. Edwards, T.; Karl, A.E.; Vaughn, M.; Rosen, P.C.; Torres, C.M.; Murphy, R.W. The Desert Tortoise Trichotomy: Mexico Hosts a Third, New Sister-Species of Tortoise in the *Gopherus morafkai–G. agassizii* Group. *ZooKeys* **2016**, *562*, 131–158. [CrossRef]
157. Spinks, P.Q.; Thomson, R.C.; McCartney-Melstad, E.; Shaffer, H.B. Phylogeny and Temporal Diversification of the New World Pond Turtles (Emydidae). *Mol. Phylogenet. Evol.* **2016**, *103*, 85–97. [CrossRef]
158. Spinks, P.Q.; Thomson, R.C.; Zhang, Y.; Che, J.; Wu, Y.; Bradley Shaffer, H. Species Boundaries and Phylogenetic Relationships in the Critically Endangered Asian Box Turtle Genus *Cuora*. *Mol. Phylogenet. Evol.* **2012**, *63*, 656–667. [CrossRef]

159. Zhou, H.; Jiang, Y.; Nie, L.; Yin, H.; Li, H.; Dong, X.; Zhao, F.; Zhang, H.; Pu, Y.; Huang, Z.; et al. The Historical Speciation of *Mauremys Sensu Lato*: Ancestral Area Reconstruction and Interspecific Gene Flow Level Assessment Provide New Insights. *PLoS ONE* **2015**, *10*, e0144711. [CrossRef]
160. Figueroa, A.; McKelvy, A.D.; Grismer, L.L.; Bell, C.D.; Lailvaux, S.P. A Species-Level Phylogeny of Extant Snakes with Description of a New Colubrid Subfamily and Genus. *PLoS ONE* **2016**, *11*, e0161070. [CrossRef] [PubMed]
161. Köhler, G.; Khaing, K.P.P.; Than, N.L.; Baranski, D.; Schell, T.; Greve, C.; Janke, A.; Pauls, S.U. A New Genus and Species of Mud Snake from Myanmar (Reptilia, Squamata, Homalopsidae). *Zootaxa* **2021**, *4915*, 301–325. [CrossRef]
162. Manni, M.; Berkeley, M.R.; Seppey, M.; Simão, F.A.; Zdobnov, E.M. BUSCO Update: Novel and Streamlined Workflows along with Broader and Deeper Phylogenetic Coverage for Scoring of Eukaryotic, Prokaryotic, and Viral Genomes. *Mol. Biol. Evol.* **2021**, *38*, 4647–4654. [CrossRef] [PubMed]
163. Quinlan, A.R.; Hall, I.M. BEDTools: A Flexible Suite of Utilities for Comparing Genomic Features. *Bioinformatics* **2010**, *26*, 841–842. [CrossRef] [PubMed]
164. Losos, J. *Lizards in an Evolutionary Tree: Ecology and Adaptive Radiation of Anoles*; University of California Press: Berkeley, CA, USA, 2011; ISBN 978-0-520-26984-2.
165. Stuart, Y.E.; Campbell, T.S.; Hohenlohe, P.A.; Reynolds, R.G.; Revell, L.J.; Losos, J.B. Rapid Evolution of a Native Species Following Invasion by a Congener. *Science* **2014**, *346*, 463–466. [CrossRef] [PubMed]
166. Campbell-Staton, S.C.; Cheviron, Z.A.; Rochette, N.; Catchen, J.; Losos, J.B.; Edwards, S.V. Winter Storms Drive Rapid Phenotypic, Regulatory, and Genomic Shifts in the Green Anole Lizard. *Science* **2017**, *357*, 495–498. [CrossRef]
167. Card, D.C.; Van Camp, A.G.; Santonastaso, T.; Jensen-Seaman, M.I.; Anthony, N.M.; Edwards, S.V. Structure and Evolution of the Squamate Major Histocompatibility Complex as Revealed by Two Anolis Lizard Genomes. *Front. Genet.* **2022**, *13*, 979746. [CrossRef]
168. Geneva, A.J.; Park, S.; Bock, D.G.; de Mello, P.L.H.; Sarigol, F.; Tollis, M.; Donihue, C.M.; Reynolds, R.G.; Feiner, N.; Rasys, A.M.; et al. Chromosome-Scale Genome Assembly of the Brown Anole (*Anolis sagrei*), an Emerging Model Species. *Commun. Biol.* **2022**, *5*, 1–13. [CrossRef]
169. Kanamori, S.; Díaz, L.M.; Cádiz, A.; Yamaguchi, K.; Shigenobu, S.; Kawata, M. Draft Genome of Six Cuban *Anolis* Lizards and Insights into Genetic Changes during Their Diversification. *BMC Ecol. Evol.* **2022**, *22*, 129. [CrossRef]
170. Couzin, J. NSF's Ark Draws Alligators, Algae, and Wasps. *Science* **2002**, *297*, 1638–1639. [CrossRef] [PubMed]
171. Wang, Z.; Miyake, T.; Edwards, S.V.; Amemiya, C.T. Tuatara (*Sphenodon*) Genomics: BAC Library Construction, Sequence Survey, and Application to the DMRT Gene Family. *J. Hered.* **2006**, *97*, 541–548. [CrossRef] [PubMed]
172. Janes, D.E.; Organ, C.; Valenzuela, N. New Resources Inform Study of Genome Size, Content, and Organization in Nonavian Reptiles. *Integr. Comp. Biol.* **2008**, *48*, 447–453. [CrossRef] [PubMed]
173. Badenhorst, D.; Hillier, L.W.; Literman, R.; Montiel, E.E.; Radhakrishnan, S.; Shen, Y.; Minx, P.; Janes, D.E.; Warren, W.C.; Edwards, S.V.; et al. Physical Mapping and Refinement of the Painted Turtle Genome (*Chrysemys picta*) Inform Amniote Genome Evolution and Challenge Turtle-Bird Chromosomal Conservation. *Genome Biol. Evol.* **2015**, *7*, 2038–2050. [CrossRef] [PubMed]
174. Thomson, R.C.; Shedlock, A.M.; Edwards, S.V.; Shaffer, H.B. Developing Markers for Multilocus Phylogenetics in Non-Model Organisms: A Test Case with Turtles. *Mol. Phylogenet. Evol.* **2008**, *49*, 514–525. [CrossRef]
175. Shedlock, A.M.; Janes, D.E.; Edwards, S.V. Amniote Phylogenomics: Testing Evolutionary Hypotheses with BAC Library Scanning and Targeted Clone Analysis of Large-Scale DNA Sequences from Reptiles. In *Phylogenomics*; Murphy, W.J., Ed.; Methods in Molecular Biology; Humana Press: Totowa, NJ, USA, 2008; pp. 91–117. ISBN 978-1-59745-581-7.
176. Janes, D.E.; Chapus, C.; Gondo, Y.; Clayton, D.F.; Sinha, S.; Blatti, C.A.; Organ, C.L.; Fujita, M.K.; Balakrishnan, C.N.; Edwards, S.V. Reptiles and Mammals Have Differentially Retained Long Conserved Noncoding Sequences from the Amniote Ancestor. *Genome Biol. Evol.* **2011**, *3*, 102–113. [CrossRef] [PubMed]
177. Venter, J.C.; Adams, M.D.; Myers, E.W.; Li, P.W.; Mural, R.J.; Sutton, G.G.; Smith, H.O.; Yandell, M.; Evans, C.A.; Holt, R.A.; et al. The Sequence of the Human Genome. *Science* **2001**, *291*, 1304–1351. [CrossRef]
178. Shendure, J.; Mitra, R.D.; Varma, C.; Church, G.M. Advanced Sequencing Technologies: Methods and Goals. *Nat. Rev. Genet.* **2004**, *5*, 335–344. [CrossRef]
179. Mardis, E.R. Next-Generation DNA Sequencing Methods. *Annu. Rev. Genom. Hum. Genet.* **2008**, *9*, 387–402. [CrossRef] [PubMed]
180. Bentley, D.R.; Balasubramanian, S.; Swerdlow, H.P.; Smith, G.P.; Milton, J.; Brown, C.G.; Hall, K.P.; Evers, D.J.; Barnes, C.L.; Bignell, H.R.; et al. Accurate Whole Human Genome Sequencing Using Reversible Terminator Chemistry. *Nature* **2008**, *456*, 53–59. [CrossRef]
181. Cronn, R.; Liston, A.; Parks, M.; Gernandt, D.S.; Shen, R.; Mockler, T. Multiplex Sequencing of Plant Chloroplast Genomes Using Solexa Sequencing-by-Synthesis Technology. *Nucleic Acids Res.* **2008**, *36*, e122. [CrossRef] [PubMed]
182. Degnan, J.H.; Rosenberg, N.A. Gene Tree Discordance, Phylogenetic Inference and the Multispecies Coalescent. *Trends Ecol. Evol.* **2009**, *24*, 332–340. [CrossRef] [PubMed]
183. Edwards, S.V.; Beerli, P. Perspective: Gene Divergence, Population Divergence, and the Variance in Coalescence Time in Phylogeographic Studies. *Evolution* **2000**, *54*, 1839–1854. [CrossRef]
184. Arbogast, B.S.; Edwards, S.V.; Wakeley, J.; Beerli, P.; Slowinski, J.B. Estimating Divergence Times from Molecular Data on Phylogenetic and Population Genetic Timescales. *Annu. Rev. Ecol. Syst.* **2002**, *33*, 707–740. [CrossRef]
185. Wakeley, J. *Coalescent Theory: An Introduction*; W.H. Freeman: New York, NY, USA, 2009; ISBN 978-0-9747077-5-4.

186. Wakeley, J.; King, L.; Low, B.S.; Ramachandran, S. Gene Genealogies within a Fixed Pedigree, and the Robustness of Kingman's Coalescent. *Genetics* **2012**, *190*, 1433–1445. [CrossRef] [PubMed]
187. Pluzhnikov, A.; Donnelly, P. Optimal Sequencing Strategies for Surveying Molecular Genetic Diversity. *Genetics* **1996**, *144*, 1247–1262. [CrossRef]
188. Felsenstein, J. Accuracy of Coalescent Likelihood Estimates: Do We Need More Sites, More Sequences, or More Loci? *Mol. Biol. Evol.* **2006**, *23*, 691–700. [CrossRef] [PubMed]
189. Jennings, W.B. On the Independent Gene Trees Assumption in Phylogenomic Studies. *Mol. Ecol.* **2017**, *26*, 4862–4871. [CrossRef]
190. Jennings, W.B.; Edwards, S.V. Speciational History of Australian Grass Finches (*Poephila*) Inferred from Thirty Gene Trees. *Evolution* **2005**, *59*, 2033–2047. [CrossRef]
191. Lee, J.Y.; Edwards, S.V. Divergence Across Australia's Carpentarian Barrier: Statistical Phylogeography of the Red-Backed Fairy Wren (*Malurus melanocephalus*). *Evolution* **2008**, *62*, 3117–3134. [CrossRef]
192. Smith, B.T.; Harvey, M.G.; Faircloth, B.C.; Glenn, T.C.; Brumfield, R.T. Target Capture and Massively Parallel Sequencing of Ultraconserved Elements for Comparative Studies at Shallow Evolutionary Time Scales. *Syst. Biol.* **2014**, *63*, 83–95. [CrossRef]
193. Costa, I.R.; Prosdocimi, F.; Jennings, W.B. In Silico Phylogenomics Using Complete Genomes: A Case Study on the Evolution of Hominoids. *Genome Res.* **2016**, *26*, 1257–1267. [CrossRef] [PubMed]
194. Good, J.M. Reduced Representation Methods for Subgenomic Enrichment and Next-Generation Sequencing. In *Molecular Methods for Evolutionary Genetics*; Orgogozo, V., Rockman, M.V., Eds.; Methods in Molecular Biology; Humana Press: Totowa, NJ, USA, 2011; pp. 85–103. ISBN 978-1-61779-228-1.
195. Gnirke, A.; Melnikov, A.; Maguire, J.; Rogov, P.; LeProust, E.M.; Brockman, W.; Fennell, T.; Giannoukos, G.; Fisher, S.; Russ, C.; et al. Solution Hybrid Selection with Ultra-Long Oligonucleotides for Massively Parallel Targeted Sequencing. *Nat. Biotechnol.* **2009**, *27*, 182–189. [CrossRef]
196. Andermann, T.; Torres Jiménez, M.F.; Matos-Maraví, P.; Batista, R.; Blanco-Pastor, J.L.; Gustafsson, A.L.S.; Kistler, L.; Liberal, I.M.; Oxelman, B.; Bacon, C.D.; et al. A Guide to Carrying Out a Phylogenomic Target Sequence Capture Project. *Front. Genet.* **2020**, *10*, 1047. [CrossRef]
197. Faircloth, B.C.; McCormack, J.E.; Crawford, N.G.; Harvey, M.G.; Brumfield, R.T.; Glenn, T.C. Ultraconserved Elements Anchor Thousands of Genetic Markers Spanning Multiple Evolutionary Timescales. *Syst. Biol.* **2012**, *61*, 717–726. [CrossRef] [PubMed]
198. Lemmon, A.R.; Emme, S.A.; Lemmon, E.M. Anchored Hybrid Enrichment for Massively High-Throughput Phylogenomics. *Syst. Biol.* **2012**, *61*, 727–744. [CrossRef] [PubMed]
199. Crawford, N.G.; Faircloth, B.C.; McCormack, J.E.; Brumfield, R.T.; Winker, K.; Glenn, T.C. More than 1000 Ultraconserved Elements Provide Evidence That Turtles Are the Sister Group of Archosaurs. *Biol. Lett.* **2012**, *8*, 783–786. [CrossRef] [PubMed]
200. Brandley, M.C.; Bragg, J.G.; Singhal, S.; Chapple, D.G.; Jennings, C.K.; Lemmon, A.R.; Lemmon, E.M.; Thompson, M.B.; Moritz, C. Evaluating the Performance of Anchored Hybrid Enrichment at the Tips of the Tree of Life: A Phylogenetic Analysis of Australian *Eugongylus* Group Scincid Lizards. *BMC Evol. Biol.* **2015**, *15*, 62. [CrossRef] [PubMed]
201. Leaché, A.D.; Chavez, A.S.; Jones, L.N.; Grummer, J.A.; Gottscho, A.D.; Linkem, C.W. Phylogenomics of Phrynosomatid Lizards: Conflicting Signals from Sequence Capture versus Restriction Site Associated DNA Sequencing. *Genome Biol. Evol.* **2015**, *7*, 706–719. [CrossRef]
202. Bejerano, G.; Pheasant, M.; Makunin, I.; Stephen, S.; Kent, W.J.; Mattick, J.S.; Haussler, D. Ultraconserved Elements in the Human Genome. *Science* **2004**, *304*, 1321–1325. [CrossRef]
203. Singhal, S.; Grundler, M.; Colli, G.; Rabosky, D.L. Squamate Conserved Loci (SqCL): A Unified Set of Conserved Loci for Phylogenomics and Population Genetics of Squamate Reptiles. *Mol. Ecol. Resour.* **2017**, *17*, e12–e24. [CrossRef] [PubMed]
204. Wiens, J.J.; Hutter, C.R.; Mulcahy, D.G.; Noonan, B.P.; Townsend, T.M.; Sites, J.W.; Reeder, T.W. Resolving the Phylogeny of Lizards and Snakes (Squamata) with Extensive Sampling of Genes and Species. *Biol. Lett.* **2012**, *8*, 1043–1046. [CrossRef] [PubMed]
205. Pyron, R.A.; Burbrink, F.T.; Wiens, J.J. A Phylogeny and Revised Classification of Squamata, Including 4161 Species of Lizards and Snakes. *BMC Evol. Biol.* **2013**, *13*, 93. [CrossRef] [PubMed]
206. Kocher, T.D.; Thomas, W.K.; Meyer, A.; Edwards, S.V.; Pääbo, S.; Villablanca, F.X.; Wilson, A.C. Dynamics of Mitochondrial DNA Evolution in Animals: Amplification and Sequencing with Conserved Primers. *Proc. Natl. Acad. Sci. USA* **1989**, *86*, 6196–6200. [CrossRef] [PubMed]
207. Hillis, D.M.; Moritz, C.; Mable, B.K. *Molecular Systematics*, 2nd ed.; Sinauer Associates, Inc.: Sunderland, MA, USA, 1996; ISBN 978-0-87893-282-5.
208. Avise, J.C. *Phylogeography: The History and Formation of Species*; Harvard University Press: Cambridge, MA, USA, 2000; ISBN 978-0-674-66638-2.
209. Ashman, L.G.; Bragg, J.G.; Doughty, P.; Hutchinson, M.N.; Bank, S.; Matzke, N.J.; Oliver, P.; Moritz, C. Diversification across Biomes in a Continental Lizard Radiation. *Evolution* **2018**, *72*, 1553–1569. [CrossRef] [PubMed]
210. Blair, C.; Bryson, R.W., Jr.; García-Vázquez, U.O.; Nieto-Montes De Oca, A.; Lazcano, D.; Mccormack, J.E.; Klicka, J. Phylogenomics of Alligator Lizards Elucidate Diversification Patterns across the Mexican Transition Zone and Support the Recognition of a New Genus. *Biol. J. Linn. Soc.* **2022**, *135*, 25–39. [CrossRef]
211. Blom, M.P.K.; Bragg, J.G.; Potter, S.; Moritz, C. Accounting for Uncertainty in Gene Tree Estimation: Summary-Coalescent Species Tree Inference in a Challenging Radiation of Australian Lizards. *Syst. Biol.* **2017**, *66*, 352–366. [CrossRef]

212. Blom, M.P.K.; Matzke, N.J.; Bragg, J.G.; Arida, E.; Austin, C.C.; Backlin, A.R.; Carretero, M.A.; Fisher, R.N.; Glaw, F.; Hathaway, S.A.; et al. Habitat Preference Modulates Trans-Oceanic Dispersal in a Terrestrial Vertebrate. *Proc. Royal Soc. B* **2019**, *286*, 20182575. [CrossRef]
213. Bragg, J.G.; Potter, S.; Afonso Silva, A.C.; Hoskin, C.J.; Bai, B.Y.H.; Moritz, C. Phylogenomics of a Rapid Radiation: The Australian Rainbow Skinks. *BMC Evol. Biol.* **2018**, *18*, 15. [CrossRef]
214. Brennan, I.G.; Lemmon, A.R.; Lemmon, E.M.; Portik, D.M.; Weijola, V.; Welton, L.; Donnellan, S.C.; Keogh, J.S. Phylogenomics of Monitor Lizards and the Role of Competition in Dictating Body Size Disparity. *Syst. Biol.* **2021**, *70*, 120–132. [CrossRef]
215. Bryson, R.W., Jr.; Linkem, C.W.; Pavón-Vázquez, C.J.; Nieto-Montes de Oca, A.; Klicka, J.; McCormack, J.E. A Phylogenomic Perspective on the Biogeography of Skinks in the *Plestiodon brevirostris* Group Inferred from Target Enrichment of Ultraconserved Elements. *J. Biogeogr.* **2017**, *44*, 2033–2044. [CrossRef]
216. Domingos, F.M.C.B.; Colli, G.R.; Lemmon, A.; Lemmon, E.M.; Beheregaray, L.B. In the Shadows: Phylogenomics and Coalescent Species Delimitation Unveil Cryptic Diversity in a Cerrado Endemic Lizard (Squamata: *Tropidurus*). *Mol. Phylogenet. Evol.* **2017**, *107*, 455–465. [CrossRef] [PubMed]
217. Freitas, S.; Westram, A.M.; Schwander, T.; Arakelyan, M.; Ilgaz, Ç.; Kumlutas, Y.; Harris, D.J.; Carretero, M.A.; Butlin, R.K. Parthenogenesis in *Darevskia* Lizards: A Rare Outcome of Common Hybridization, not a Common Outcome of Rare Hybridization. *Evolution* **2022**, *76*, 899–914. [CrossRef] [PubMed]
218. Garcia-Porta, J.; Irisarri, I.; Kirchner, M.; Rodríguez, A.; Kirchhof, S.; Brown, J.L.; MacLeod, A.; Turner, A.P.; Ahmadzadeh, F.; Albaladejo, G.; et al. Environmental Temperatures Shape Thermal Physiology as Well as Diversification and Genome-Wide Substitution Rates in Lizards. *Nat. Commun.* **2019**, *10*, 4077. [CrossRef]
219. Grummer, J.A.; Morando, M.M.; Avila, L.J.; Sites, J.W.; Leaché, A.D. Phylogenomic Evidence for a Recent and Rapid Radiation of Lizards in the Patagonian *Liolaemus fitzingerii* Species Group. *Mol. Phylogenet. Evol.* **2018**, *125*, 243–254. [CrossRef]
220. Morando, M.; Olave, M.; Avila, L.J.; Sites, J.W.; Leaché, A.D. Phylogenomic Data Resolve Higher-Level Relationships within South American *Liolaemus* Lizards. *Mol. Phylogenet. Evol.* **2020**, *147*, 106781. [CrossRef]
221. Panzera, A.; Leaché, A.D.; D'Elía, G.; Victoriano, P.F. Phylogenomic Analysis of the Chilean Clade of *Liolaemus* Lizards (Squamata: Liolaemidae) Based on Sequence Capture Data. *PeerJ* **2017**, *5*, e3941. [CrossRef]
222. Ramírez-Reyes, T.; Blair, C.; Flores-Villela, O.; Piñero, D.; Lathrop, A.; Murphy, R. Phylogenomics and Molecular Species Delimitation Reveals Great Cryptic Diversity of Leaf-Toed Geckos (Phyllodactylidae: *Phyllodactylus*), Ancient Origins, and Diversification in Mexico. *Mol. Phylogenet. Evol.* **2020**, *150*, 106880. [CrossRef]
223. Reilly, S.B.; Stubbs, A.L.; Arida, E.; Karin, B.R.; Arifin, U.; Kaiser, H.; Bi, K.; Iskandar, D.T.; McGuire, J.A. Phylogenomic Analysis Reveals Dispersal-Driven Speciation and Divergence with Gene Flow in Lesser Sunda Flying Lizards (Genus *Draco*). *Syst. Biol.* **2022**, *71*, 221–241. [CrossRef]
224. Reilly, S.B.; Karin, B.R.; Stubbs, A.L.; Arida, E.; Arifin, U.; Kaiser, H.; Bi, K.; Hamidy, A.; Iskandar, D.T.; McGuire, J.A. Diverge and Conquer: Phylogenomics of Southern Wallacean Forest Skinks (Genus: *Sphenomorphus*) and Their Colonization of the Lesser Sunda Archipelago. *Evolution* **2022**, *76*, 2281–2301. [CrossRef]
225. Reynolds, R.G.; Miller, A.H.; Pasachnik, S.A.; Knapp, C.R.; Welch, M.E.; Colosimo, G.; Gerber, G.P.; Drawert, B.; Iverson, J.B. Phylogenomics and Historical Biogeography of West Indian Rock Iguanas (Genus *Cyclura*). *Mol. Phylogenet. Evol.* **2022**, *174*, 107548. [CrossRef] [PubMed]
226. Rodriguez, Z.B.; Perkins, S.L.; Austin, C.C. Multiple Origins of Green Blood in New Guinea Lizards. *Sci. Adv.* **2018**, *4*, eaao5017. [CrossRef] [PubMed]
227. Schools, M.; Kasprowicz, A.; Blair Hedges, S. Phylogenomic Data Resolve the Historical Biogeography and Ecomorphs of Neotropical Forest Lizards (Squamata, Diploglossidae). *Mol. Phylogenet. Evol.* **2022**, *175*, 107577. [CrossRef] [PubMed]
228. Singhal, S.; Hoskin, C.J.; Couper, P.; Potter, S.; Moritz, C. A Framework for Resolving Cryptic Species: A Case Study from the Lizards of the Australian Wet Tropics. *Syst. Biol.* **2018**, *67*, 1061–1075. [CrossRef] [PubMed]
229. Skipwith, P.L.; Bi, K.; Oliver, P.M. Relicts and Radiations: Phylogenomics of an Australasian Lizard Clade with East Gondwanan Origins (Gekkota: Diplodactyloidea). *Mol. Phylogenet. Evol.* **2019**, *140*, 106589. [CrossRef] [PubMed]
230. Tucker, D.B.; Hedges, S.B.; Colli, G.R.; Pyron, R.A.; Sites, J.W., Jr. Genomic Timetree and Historical Biogeography of Caribbean Island Ameiva Lizards (*Pholidoscelis*: Teiidae). *Ecol. Evol.* **2017**, *7*, 7080–7090. [CrossRef] [PubMed]
231. Wood, P.L.; Guo, X.; Travers, S.L.; Su, Y.-C.; Olson, K.V.; Bauer, A.M.; Grismer, L.L.; Siler, C.D.; Moyle, R.G.; Andersen, M.J.; et al. Parachute Geckos Free Fall into Synonymy: Gekko Phylogeny, and a New Subgeneric Classification, Inferred from Thousands of Ultraconserved Elements. *Mol. Phylogenet. Evol.* **2020**, *146*, 106731. [CrossRef]
232. Zozaya, S.M.; Teasdale, L.C.; Tedeschi, L.G.; Higgie, M.; Hoskin, C.J.; Moritz, C. Initiation of Speciation across Multiple Dimensions in a Rock-Restricted, Tropical Lizard. *Mol. Ecol.* **2022**, *32*, 680–695. [CrossRef]
233. Bernstein, J.M.; Ruane, S. Maximizing Molecular Data from Low-Quality Fluid-Preserved Specimens in Natural History Collections. *Front. Ecol. Evol.* **2022**, *10*, 581. [CrossRef]
234. Blair, C.; Bryson, R.W., Jr.; Linkem, C.W.; Lazcano, D.; Klicka, J.; McCormack, J.E. Cryptic Diversity in the Mexican Highlands: Thousands of UCE Loci Help Illuminate Phylogenetic Relationships, Species Limits and Divergence Times of Montane Rattlesnakes (Viperidae: *Crotalus*). *Mol. Ecol. Resour.* **2019**, *19*, 349–365. [CrossRef]
235. Chen, X.; Lemmon, A.R.; Lemmon, E.M.; Pyron, R.A.; Burbrink, F.T. Using Phylogenomics to Understand the Link between Biogeographic Origins and Regional Diversification in Ratsnakes. *Mol. Phylogenet. Evol.* **2017**, *111*, 206–218. [CrossRef] [PubMed]

236. Esquerré, D.; Donnellan, S.; Brennan, I.G.; Lemmon, A.R.; Moriarty Lemmon, E.; Zaher, H.; Grazziotin, F.G.; Keogh, J.S. Phylogenomics, Biogeography, and Morphometrics Reveal Rapid Phenotypic Evolution in Pythons after Crossing Wallace's Line. *Syst. Biol.* **2020**, *69*, 1039–1051. [CrossRef] [PubMed]
237. Hallas, J.M.; Parchman, T.L.; Feldman, C.R. Phylogenomic Analyses Resolve Relationships among Garter Snakes (*Thamnophis*: Natricinae: Colubridae) and Elucidate Biogeographic History and Morphological Evolution. *Mol. Phylogenet. Evol.* **2022**, *167*, 107374. [CrossRef] [PubMed]
238. Li, J.; Liang, D.; Zhang, P. Simultaneously Collecting Coding and Non-Coding Phylogenomic Data Using Homemade Full-Length CDNA Probes, Tested by Resolving the High-Level Relationships of Colubridae. *Front. Ecol. Evol.* **2022**, *10*, 969581. [CrossRef]
239. Myers, E.A.; Mulcahy, D.G.; Falk, B.; Johnson, K.; Carbi, M.; de Queiroz, K. Interspecific Gene Flow and Mitochondrial Genome Capture during the Radiation of Jamaican *Anolis* Lizards (Squamata; Iguanidae). *Syst. Biol.* **2022**, *71*, 501–511. [CrossRef] [PubMed]
240. Natusch, D.J.D.; Esquerré, D.; Lyons, J.A.; Hamidy, A.; Lemmon, A.R.; Lemmon, E.M.; Riyanto, A.; Keogh, J.S.; Donnellan, S. Phylogenomics, Biogeography and Taxonomic Revision of New Guinean Pythons (Pythonidae, *Leiopython*) Harvested for International Trade. *Mol. Phylogenet. Evol.* **2021**, *158*, 106960. [CrossRef] [PubMed]
241. Nikolakis, Z.L.; Orton, R.W.; Crother, B.I. Fine-Scale Population Structure within an Eastern Nearctic Snake Complex (*Pituophis melanoleucus*). *Zool. Scr.* **2022**, *51*, 133–146. [CrossRef]
242. Ruane, S.; Austin, C.C. Phylogenomics Using Formalin-Fixed and 100+ Year-Old Intractable Natural History Specimens. *Mol. Ecol. Resour.* **2017**, *17*, 1003–1008. [CrossRef]
243. Burbrink, F.T.; Grazziotin, F.G.; Pyron, R.A.; Cundall, D.; Donnellan, S.; Irish, F.; Keogh, J.S.; Kraus, F.; Murphy, R.W.; Noonan, B.; et al. Interrogating Genomic-Scale Data for Squamata (Lizards, Snakes, and Amphisbaenians) Shows no Support for Key Traditional Morphological Relationships. *Syst. Biol.* **2020**, *69*, 502–520. [CrossRef]
244. Singhal, S.; Colston, T.J.; Grundler, M.R.; Smith, S.A.; Costa, G.C.; Colli, G.R.; Moritz, C.; Pyron, R.A.; Rabosky, D.L. Congruence and Conflict in the Higher-Level Phylogenetics of Squamate Reptiles: An Expanded Phylogenomic Perspective. *Syst. Biol.* **2021**, *70*, 542–557. [CrossRef]
245. Streicher, J.W.; Wiens, J.J. Phylogenomic Analyses of More than 4000 Nuclear Loci Resolve the Origin of Snakes among Lizard Families. *Biol. Lett.* **2017**, *13*, 20170393. [CrossRef]
246. Shaffer, H.B.; McCartney-Melstad, E.; Near, T.J.; Mount, G.G.; Spinks, P.Q. Phylogenomic Analyses of 539 Highly Informative Loci Dates a Fully Resolved Time Tree for the Major Clades of Living Turtles (Testudines). *Mol. Phylogenet. Evol.* **2017**, *115*, 7–15. [CrossRef]
247. Baird, N.A.; Etter, P.D.; Atwood, T.S.; Currey, M.C.; Shiver, A.L.; Lewis, Z.A.; Selker, E.U.; Cresko, W.A.; Johnson, E.A. Rapid SNP Discovery and Genetic Mapping Using Sequenced RAD Markers. *PLoS ONE* **2008**, *3*, e3376. [CrossRef]
248. Davey, J.W.; Blaxter, M.L. RADSeq: Next-Generation Population Genetics. *Brief. Funct. Genomics* **2010**, *9*, 416–423. [CrossRef] [PubMed]
249. Etter, P.D.; Bassham, S.; Hohenlohe, P.A.; Johnson, E.A.; Cresko, W.A. SNP Discovery and Genotyping for Evolutionary Genetics Using RAD Sequencing. In *Molecular Methods for Evolutionary Genetics*; Orgogozo, V., Rockman, M.V., Eds.; Methods in Molecular Biology; Humana Press: Totowa, NJ, USA, 2011; pp. 157–178. ISBN 978-1-61779-228-1.
250. Peterson, B.K.; Weber, J.N.; Kay, E.H.; Fisher, H.S.; Hoekstra, H.E. Double Digest RADseq: An Inexpensive Method for De Novo SNP Discovery and Genotyping in Model and Non-Model Species. *PLoS ONE* **2012**, *7*, e37135. [CrossRef]
251. Arnold, B.; Corbett-Detig, R.B.; Hartl, D.; Bomblies, K. RADseq Underestimates Diversity and Introduces Genealogical Biases Due to Nonrandom Haplotype Sampling. *Mol. Ecol.* **2013**, *22*, 3179–3190. [CrossRef]
252. Cariou, M.; Duret, L.; Charlat, S. Is RAD-seq Suitable for Phylogenetic Inference? An In Silico Assessment and Optimization. *Ecol. Evol.* **2013**, *3*, 846–852. [CrossRef] [PubMed]
253. Dunn, C.W.; Hejnol, A.; Matus, D.Q.; Pang, K.; Browne, W.E.; Smith, S.A.; Seaver, E.; Rouse, G.W.; Obst, M.; Edgecombe, G.D.; et al. Broad Phylogenomic Sampling Improves Resolution of the Animal Tree of Life. *Nature* **2008**, *452*, 745–749. [CrossRef] [PubMed]
254. Bi, K.; Vanderpool, D.; Singhal, S.; Linderoth, T.; Moritz, C.; Good, J.M. Transcriptome-Based Exon Capture Enables Highly Cost-Effective Comparative Genomic Data Collection at Moderate Evolutionary Scales. *BMC Genom.* **2012**, *13*, 403. [CrossRef] [PubMed]
255. Bi, K.; Linderoth, T.; Vanderpool, D.; Good, J.M.; Nielsen, R.; Moritz, C. Unlocking the Vault: Next-Generation Museum Population Genomics. *Mol. Ecol.* **2013**, *22*, 6018–6032. [CrossRef]
256. Jarvis, E.D.; Mirarab, S.; Aberer, A.J.; Li, B.; Houde, P.; Li, C.; Ho, S.Y.W.; Faircloth, B.C.; Nabholz, B.; Howard, J.T.; et al. Whole-Genome Analyses Resolve Early Branches in the Tree of Life of Modern Birds. *Science* **2014**, *346*, 1320–1331. [CrossRef] [PubMed]
257. Shaffer, H.B.; Toffelmier, E.; Corbett-Detig, R.B.; Escalona, M.; Erickson, B.; Fiedler, P.; Gold, M.; Harrigan, R.J.; Hodges, S.; Luckau, T.K.; et al. Landscape Genomics to Enable Conservation Actions: The California Conservation Genomics Project. *J. Hered.* **2022**, *113*, 577–588. [CrossRef] [PubMed]
258. Grismer, J.L.; Escalona, M.; Miller, C.; Beraut, E.; Fairbairn, C.W.; Marimuthu, M.P.A.; Nguyen, O.; Toffelmier, E.; Wang, I.J.; Shaffer, H.B. Reference Genome of the Rubber Boa, *Charina bottae* (Serpentes: Boidae). *J. Hered.* **2022**, *113*, 641–648. [CrossRef]

259. Todd, B.D.; Jenkinson, T.S.; Escalona, M.; Beraut, E.; Nguyen, O.; Sahasrabudhe, R.; Scott, P.A.; Toffelmier, E.; Wang, I.J.; Shaffer, H.B. Reference Genome of the Northwestern Pond Turtle, *Actinemys marmorata*. *J. Hered.* **2022**, *113*, 624–631. [CrossRef]
260. Wood, D.A.; Richmond, J.Q.; Escalona, M.; Marimuthu, M.P.A.; Nguyen, O.; Sacco, S.; Beraut, E.; Westphal, M.; Fisher, R.N.; Vandergast, A.G.; et al. Reference Genome of the California Glossy Snake, *Arizona elegans occidentalis*: A Declining California Species of Special Concern. *J. Hered.* **2022**, *113*, 632–640. [CrossRef] [PubMed]
261. Nurk, S.; Koren, S.; Rhie, A.; Rautiainen, M.; Bzikadze, A.V.; Mikheenko, A.; Vollger, M.R.; Altemose, N.; Uralsky, L.; Gershman, A.; et al. The Complete Sequence of a Human Genome. *Science* **2022**, *376*, 44–53. [CrossRef]
262. Bravo, G.A.; Antonelli, A.; Bacon, C.D.; Bartoszek, K.; Blom, M.P.K.; Huynh, S.; Jones, G.; Knowles, L.L.; Lamichhaney, S.; Marcussen, T.; et al. Embracing Heterogeneity: Coalescing the Tree of Life and the Future of Phylogenomics. *PeerJ* **2019**, *7*, e6399. [CrossRef]
263. McCormack, J.E.; Faircloth, B.C.; Crawford, N.G.; Gowaty, P.A.; Brumfield, R.T.; Glenn, T.C. Ultraconserved Elements Are Novel Phylogenomic Markers That Resolve Placental Mammal Phylogeny When Combined with Species-Tree Analysis. *Genome Res.* **2012**, *22*, 746–754. [CrossRef]
264. Karl, S.A.; Avise, J.C. PCR-Based Assays of Mendelian Polymorphisms from Anonymous Single-Copy Nuclear DNA: Techniques and Applications for Population Genetics. *Mol. Biol. Evol.* **1993**, *10*, 342–361. [CrossRef]
265. Thomson, R.C.; Wang, I.J.; Johnson, J.R. Genome-Enabled Development of DNA Markers for Ecology, Evolution and Conservation. *Mol. Ecol.* **2010**, *19*, 2184–2195. [CrossRef]
266. Reilly, S.B.; Marks, S.B.; Jennings, W.B. Defining Evolutionary Boundaries across Parapatric Ecomorphs of Black Salamanders (*Aneides flavipunctatus*) with Conservation Implications. *Mol. Ecol.* **2012**, *21*, 5745–5761. [CrossRef] [PubMed]
267. Stephens, M.; Smith, N.J.; Donnelly, P. A New Statistical Method for Haplotype Reconstruction from Population Data. *Am. J. Hum. Genet.* **2001**, *68*, 978–989. [CrossRef]
268. Scheet, P.; Stephens, M. A Fast and Flexible Statistical Model for Large-Scale Population Genotype Data: Applications to Inferring Missing Genotypes and Haplotypic Phase. *Am. J. Hum. Genet.* **2006**, *78*, 629–644. [CrossRef] [PubMed]
269. Andermann, T.; Fernandes, A.M.; Olsson, U.; Töpel, M.; Pfeil, B.; Oxelman, B.; Aleixo, A.; Faircloth, B.C.; Antonelli, A. Allele Phasing Greatly Improves the Phylogenetic Utility of Ultraconserved Elements. *Syst. Biol.* **2019**, *68*, 32–46. [CrossRef] [PubMed]
270. Huang, J.; Bennett, J.; Flouri, T.; Leaché, A.D.; Yang, Z. Phase Resolution of Heterozygous Sites in Diploid Genomes Is Important to Phylogenomic Analysis under the Multispecies Coalescent Model. *Syst. Biol.* **2022**, *71*, 334–352. [CrossRef] [PubMed]
271. Card, D.C.; Shapiro, B.; Giribet, G.; Moritz, C.; Edwards, S.V. Museum Genomics. *Annu. Rev. Genet.* **2021**, *55*, 633–659. [CrossRef] [PubMed]
272. Edwards, S.V.; Cloutier, A.; Baker, A.J. Conserved Nonexonic Elements: A Novel Class of Marker for Phylogenomics. *Syst. Biol.* **2017**, *66*, 1028–1044. [CrossRef] [PubMed]
273. Jiang, X.; Edwards, S.V.; Liu, L. The Multispecies Coalescent Model Outperforms Concatenation Across Diverse Phylogenomic Data Sets. *Syst. Biol.* **2020**, *69*, 795–812. [CrossRef]
274. Edwards, S.V. Is a New and General Theory of Molecular Systematics Emerging? *Evol. Int. J. Org. Evol.* **2009**, *63*, 1–19. [CrossRef]
275. Brito, P.H.; Edwards, S.V. Multilocus Phylogeography and Phylogenetics Using Sequence-Based Markers. *Genetica* **2009**, *135*, 439–455. [CrossRef]
276. Beerli, P.; Mashayekhi, S.; Sadeghi, M.; Khodaei, M.; Shaw, K. Population Genetic Inference with MIGRATE. *Curr. Protoc. Bioinforma.* **2019**, *68*, e87. [CrossRef]
277. Flouri, T.; Jiao, X.; Rannala, B.; Yang, Z. A Bayesian Implementation of the Multispecies Coalescent Model with Introgression for Phylogenomic Analysis. *Mol. Biol. Evol.* **2020**, *37*, 1211–1223. [CrossRef]
278. Chojnowski, J.L.; Kimball, R.T.; Braun, E.L. Introns Outperform Exons in Analyses of Basal Avian Phylogeny Using Clathrin Heavy Chain Genes. *Gene* **2008**, *410*, 89–96. [CrossRef] [PubMed]
279. Reddy, S.; Kimball, R.T.; Pandey, A.; Hosner, P.A.; Braun, M.J.; Hackett, S.J.; Han, K.-L.; Harshman, J.; Huddleston, C.J.; Kingston, S.; et al. Why Do Phylogenomic Data Sets Yield Conflicting Trees? Data Type Influences the Avian Tree of Life More than Taxon Sampling. *Syst. Biol.* **2017**, *66*, 857–879. [CrossRef] [PubMed]
280. Armstrong, J.; Hickey, G.; Diekhans, M.; Fiddes, I.T.; Novak, A.M.; Deran, A.; Fang, Q.; Xie, D.; Feng, S.; Stiller, J.; et al. Progressive Cactus Is a Multiple-Genome Aligner for the Thousand-Genome Era. *Nature* **2020**, *587*, 246–251. [CrossRef] [PubMed]
281. Guarracino, A.; Heumos, S.; Nahnsen, S.; Prins, P.; Garrison, E. ODGI: Understanding Pangenome Graphs. *Bioinformatics* **2022**, *38*, 3319–3326. [CrossRef]
282. Garrison, E.; Guarracino, A. Unbiased Pangenome Graphs. *Bioinformatics* **2022**, *39*, btac743. [CrossRef]
283. Smith, S.D.; Pennell, M.W.; Dunn, C.W.; Edwards, S.V. Phylogenetics Is the New Genetics (for Most of Biodiversity). *Trends Ecol. Evol.* **2020**, *35*, 415–425. [CrossRef]
284. Pease, J.B.; Haak, D.C.; Hahn, M.W.; Moyle, L.C. Phylogenomics Reveals Three Sources of Adaptive Variation during a Rapid Radiation. *PLoS Biol.* **2016**, *14*, e1002379. [CrossRef]
285. McLean, C.Y.; Reno, P.L.; Pollen, A.A.; Bassan, A.I.; Capellini, T.D.; Guenther, C.; Indjeian, V.B.; Lim, X.H.; Menke, D.B.; Schaar, B.T.; et al. Human-Specific Loss of Regulatory DNA and the Evolution of Human-Specific Traits. *Nature* **2011**, *471*, 216–219. [CrossRef]
286. Hiller, M.; Schaar, B.T.; Indjeian, V.B.; Kingsley, D.M.; Hagey, L.R.; Bejerano, G. A "Forward Genomics" Approach Links Genotype to Phenotype Using Independent Phenotypic Losses among Related Species. *Cell. Rep.* **2012**, *2*, 817–823. [CrossRef] [PubMed]

287. Marcovitz, A.; Jia, R.; Bejerano, G. "Reverse Genomics" Predicts Function of Human Conserved Noncoding Elements. *Mol. Biol. Evol.* **2016**, *33*, 1358–1369. [CrossRef] [PubMed]
288. Marcovitz, A.; Turakhia, Y.; Gloudemans, M.; Braun, B.A.; Chen, H.I.; Bejerano, G. A Novel Unbiased Test for Molecular Convergent Evolution and Discoveries in Echolocating, Aquatic and High-Altitude Mammals. *bioRxiv* **2017**, *1*, 170985. [CrossRef]
289. Sackton, T.B.; Grayson, P.; Cloutier, A.; Hu, Z.; Liu, J.S.; Wheeler, N.E.; Gardner, P.P.; Clarke, J.A.; Baker, A.J.; Clamp, M.; et al. Convergent Regulatory Evolution and Loss of Flight in Paleognathous Birds. *Science* **2019**, *364*, 74–78. [CrossRef] [PubMed]
290. Muntané, G.; Farré, X.; Rodríguez, J.A.; Pegueroles, C.; Hughes, D.A.; de Magalhães, J.P.; Gabaldón, T.; Navarro, A. Biological Processes Modulating Longevity across Primates: A Phylogenetic Genome-Phenome Analysis. *Mol. Biol. Evol.* **2018**, *35*, 1990–2004. [CrossRef] [PubMed]
291. Kowalczyk, A.; Partha, R.; Clark, N.L.; Chikina, M. Pan-Mammalian Analysis of Molecular Constraints Underlying Extended Lifespan. *eLife* **2020**, *9*, e51089. [CrossRef]
292. Kolora, S.R.R.; Owens, G.L.; Vazquez, J.M.; Stubbs, A.; Chatla, K.; Jainese, C.; Seeto, K.; McCrea, M.; Sandel, M.W.; Vianna, J.A.; et al. Origins and Evolution of Extreme Life Span in Pacific Ocean Rockfishes. *Science* **2021**, *374*, 842–847. [CrossRef]
293. Treaster, S.; Deelen, J.; Daane, J.M.; Murabito, J.; Karasik, D.; Harris, M.P. Convergent Genomics of Longevity in Rockfishes Highlights the Genetics of Human Life Span Variation. *Sci. Adv.* **2023**, *9*, eadd2743. [CrossRef]
294. Roscito, J.G.; Sameith, K.; Kirilenko, B.M.; Hecker, N.; Winkler, S.; Dahl, A.; Rodrigues, M.T.; Hiller, M. Convergent and Lineage-Specific Genomic Differences in Limb Regulatory Elements in Limbless Reptile Lineages. *Cell Rep.* **2022**, *38*, 110280. [CrossRef] [PubMed]
295. Lartillot, N.; Poujol, R. A Phylogenetic Model for Investigating Correlated Evolution of Substitution Rates and Continuous Phenotypic Characters. *Mol. Biol. Evol.* **2011**, *28*, 729–744. [CrossRef]
296. Hu, Z.; Sackton, T.B.; Edwards, S.V.; Liu, J.S. Bayesian Detection of Convergent Rate Changes of Conserved Noncoding Elements on Phylogenetic Trees. *Mol. Biol. Evol.* **2019**, *36*, 1086–1100. [CrossRef] [PubMed]
297. Kowalczyk, A.; Meyer, W.K.; Partha, R.; Mao, W.; Clark, N.L.; Chikina, M. RERconverge: An R Package for Associating Evolutionary Rates with Convergent Traits. *Bioinformatics* **2019**, *35*, 4815–4817. [CrossRef] [PubMed]
298. Yan, H.; Hu, Z.; Thomas, G.; Edwards, S.V.; Sackton, T.B.; Liu, J.S. PhyloAcc-GT: A Bayesian Method for Inferring Patterns of Substitution Rate Shifts and Associations with Binary Traits under Gene Tree Discordance. *bioRxiv* **2022**, *1*, 2022.12.23.521765. [CrossRef]
299. Shine, R. Life-History Evolution in Reptiles. *Annu. Rev. Ecol. Evol. Syst.* **2005**, *36*, 23–46. [CrossRef]
300. Brandley, M.C.; Huelsenbeck, J.P.; Wiens, J.J. Rates and Patterns in the Evolution of Snake-Like Body Form in Squamate Reptiles: Evidence for Repeated Re-Evolution of Lost Digits and Long-Term Persistence of Intermediate Body Forms. *Evolution* **2008**, *62*, 2042–2064. [CrossRef] [PubMed]

 animals

Article

First Insights on the Karyotype Diversification of the Endemic Malagasy Leaf-Toed Geckos (Squamata: Gekkonidae: *Uroplatus*)

Marcello Mezzasalma [1,*], Elvira Brunelli [1], Gaetano Odierna [2,*] and Fabio Maria Guarino [2]

[1] Department of Biology, Ecology and Earth Science, University of Calabria, Via P. Bucci 4/B, 87036 Rende, Italy
[2] Department of Biology, University of Naples Federico II, Via Cinthia 26, 80126 Naples, Italy
* Correspondence: m.mezzasalma@gmail.com (M.M.); gaetanodierna@gmail.com (G.O.)

Simple Summary: The geckos of the genus *Uroplatus* include peculiar endemic species to Madagascar. Even though they have been the subject of several morphological and molecular studies, karyological analyses have been performed only on *U. phantasticus*, leaving the chromosomal diversity of the genus completely unexplored. In this study, we performed a preliminary molecular analysis and a comparative cytogenetic study providing the first karyotype description of eight species of *Uroplatus* and an assessment of their karyological variability. We found chromosome diversity in the species studied in terms of total chromosome number (2n = 34–38), localization of loci of Nucleolar Organizer Regions (NORs) (alternatively on the 2nd, 6th, 10th or 16th pair), heterochromatin composition and occurrence of heteromorphic sex chromosome pairs. Adding our newly generated data to those available from the literature, we show that in the genus *Uroplatus*, as well as in a larger group of phylogenetically related gecko genera, chromosome diversification mainly occurred toward a reduction in the chromosome number by means of chromosome fusions and translocation of NOR-bearing chromosomes. We also hypothesize that the diversification of sex chromosome systems occurred independently in different genera.

Abstract: We provide here the first karyotype description of eight *Uroplatus* species and a characterization of their chromosomal diversity. We performed a molecular taxonomic assessment of several *Uroplatus* samples using the mitochondrial 12S marker and a comparative cytogenetic analysis with standard karyotyping, silver staining (Ag-NOR) and sequential C-banding + Giemsa, +Chromomycin A3 (CMA_3), +4′,6-diamidino-2-phenylindole (DAPI). We found chromosomal variability in terms of chromosome number (2n = 34–38), heterochromatin composition and number and localization of loci or Nucleolar Organizer Regions (NORs) (alternatively on the 2nd, 6th, 10th or 16th pair). Chromosome morphology is almost constant, with karyotypes composed of acrocentric chromosomes, gradually decreasing in length. C-banding evidenced a general low content of heterochromatin, mostly localized on pericentromeric and telomeric regions. Centromeric bands varied among the species studied, resulting in CMA_3 positive and DAPI negative or positive to both fluorochromes. We also provide evidence of a first putative heteromorphic sex chromosome system in the genus. In fact, in *U. alluaudi* the 10th pair was highly heteromorphic, with a metacentric, largely heterochromatic W chromosome, which was much bigger than the Z. We propose an evolutionary scenario of chromosome reduction from 2n = 38 to 2n = 34, by means of translocations of microchromosomes on larger chromosomes (often involving the NOR-bearing microchromosomes). Adding our data to those available from the literature, we show that similar processes characterized the evolutionary radiation of a larger gecko clade. Finally, we hypothesize that sex chromosome diversification occurred independently in different genera.

Keywords: evolution; karyotype; NORs; Madagascar; reptiles; sex chromosomes

Citation: Mezzasalma, M.; Brunelli, E.; Odierna, G.; Guarino, F.M. First Insights on the Karyotype Diversification of the Endemic Malagasy Leaf-Toed Geckos (Squamata: Gekkonidae: *Uroplatus*). *Animals* **2022**, *12*, 2054. https://doi.org/10.3390/ani12162054

Academic Editor: Ettore Olmo

Received: 12 July 2022
Accepted: 9 August 2022
Published: 12 August 2022

1. Introduction

Madagascar is one of the world's "hottest" biodiversity hotspots and an ideal region to better understand complex evolutionary dynamics [1–3]. The Malagasy reptile fauna comprises more than 430 terrestrial endemic squamate species and nine different families (Boidae, Lamprophiidae, Typhlopidae, Agamidae, Chamaeleonidae, Gekkonidae, Gerrhosauridae, Opluridae and Scincidae) [4,5]. Among them, the family Gekkonidae includes 11 genera (*Blaesodactylus* Boettger, 1893, *Ebenavia* (Boettger, 1878), *Geckolepis* Grandidier, 1867, *Gehyra* (Wiegmann, 1834), *Hemidactylus* Oken, 1817, *Lygodactylus* Gray, 1864, *Matoatoa* Nussbaum, Raxworthy & Pronk, 1998, *Paragehyra* Angel, 1929, *Paroedura* Günther, 1879, *Phelsuma* Gray, 1825 and *Uroplatus* Duméril, 1806), with a total of more than 100 species currently described [5].

However, even if recent research started to better define the phylogeny and the taxonomy of many different groups, only a small fraction of species has been studied with cytogenetic methods, despite an increasing evidence that their species diversity is reflected at the karyotype level [6–12].

This applies also to the geckos of the genus *Uroplatus*, which have been the subject of several morphological and molecular studies (see e.g., [13–23]), but only *U. phantasticus* (Boulenger, 1888) has a known karyotype, leaving the chromosome diversity of the genus completely unexplored. Overall, the karyotypes of geckos exhibit a wide variability in terms of the total number of chromosomes, number of uni-armed and bi-armed chromosomes, localization of different chromosome markers and presence or absence of differentiated sex chromosomes [6,8,9,24]. In *U. phantasticus*, the karyotype is composed of 2n = 36, all acrocentric chromosomes, Nucleolar Organizer Regions (NORs) on the second pair and absence of differentiated sex chromosomes [24].

The genus *Uroplatus* currently includes 21, mostly nocturnal, forest-dwelling species, which are overall widespread in Madagascar and surrounding islands (such as Nosy Be), with the exception of the arid southern spiny forest and regions 2400 m asl [15]. The genus also includes several regional endemic and candidate species which are awaiting formal description, highlighting that the species diversity is currently underestimated (e.g., [15,23]).

In this paper we performed a preliminary molecular taxonomic analysis and a comparative cytogenetic study with standard karyotyping, Ag-NOR staining and sequential C-banding on different *Uroplatus* samples from distinct Malagasy areas. We provide the first karyotype description of eight species of the genus and a characterization of their chromosomal diversity. Then, superimposing our newly generated karyological data on available phylogenies [23,25] and comparing our results with available literature data on evolutionary related gecko species [6,8,24,26], we hypothesize that a progressive reduction in the chromosome number (with the formation of metacentric chromosomes and the translocation of NORs) is a common evolutionary trend in different genera.

We also provide a first record of a putative heteromorphic sex chromosome system in the genus and hypothesize that sex chromosome diversification occurred multiple times, independently in the phylogenetically related genera *Paroedura*, *Lygodactylus* and *Christinus* Wells & Wellington, 1983.

2. Material and Methods

2.1. Sampling

We examined 13 samples of 8 different species of the genus *Uroplatus*. The samples were collected during fieldwork in 1999–2004 by various collaborators and no animal was sampled during the realization of this study. Taxonomic attribution, field number, sex, and origin of all the samples analysed in this study are provided in Table 1.

After capture, animals were injected with a 0.5 mg/mL colchicine solution (0.1 mL/10 g body weight). Tissue samples (intestine, spleen and gonads) were incubated for 30 min in hypotonic solution (KCl 0.075 M + sodium citrate 0.5%, 1:1), fixed and conserved in Carnoy's solution (methanol and acetic acid, 3:1). The fixed material was preserved at 4 °C and transferred to the laboratory of University of Naples Federico II where it was processed as described below.

Table 1. Specimens analysed in this study. FN = field number. Max identity = Maximum identity scores with deposited homologous sequences.

Species	FN	Sex	Locality	Max Identity
U. alluaudi Mocquard, 1894	GA 476	female	Montagne d'Ambre	100% vs. KF160464
U. henkeli Böhme & Ibisch, 1990	GA 477	male	Montagne d'Ambre	99.3% vs. JX205281
U. henkeli	GA 1099	male	Montagne d'Ambre	99.3% vs. JX205281
U. ebenaui (Boettger, 1879)	FGMV 2205	female	Manongarivo	99.4% vs. JX205278
U. ebenaui	GA 1100	female	NA	99.4% vs. JX205278
U. fiera Ratsoavina, Ranjanaharisoa, Glaw, Raselimanana, Miralles & Vences, 2015	FGMV 3097	male	Fiherenana region	100% vs. JX205263
U. fiera	GA 140	juvenile	Fiherenana region	100% vs. JX205263
U. finiavana Ratsoavina, Louis Jr., Crottini, Randrianiaina, Glaw & Vences, 2011	FGMV 3084	male	Montagne d'Ambre	100% vs. MW035835
U. finiavana	GA 1100	juvenile	Montagne d'Ambre	100% vs. MW035835
U. fimbriatus (Schneider, 1797)	FGMV 2234	male	NA	99.5% vs. AB612276
U. prope *guentheri* Mocquard, 1908	GA 328	male	Marofandilia	96.8% vs. EU596688
U. prope *guentheri*	GA 329	male	Marofandilia	96.8% vs. EU596688
U. pietschmanni Böhle & Schönecker, 2004	FAZC 11627	male	NosyBe	99.7% vs. EU596687

2.2. Molecular Analysis

DNA was extracted from tissue samples following Sambrook et al. [27]. A fragment of about 450 bp of the mitochondrial 12S rRNA gene was amplified using the primer pair 12Sa 5′-AAACTGGGATTAGATACCCCACTAT−3′ and 12Sb 5′-GAGGGTGAGGGCGGTG-TGT−3′ [28]. This marker was chosen considering its wide use on *Uroplatus* geckos and the number of available sequences in public repositories [13,15–23].

PCR was conducted in 25 µL using the following parameters: initial denaturation at 94 °C for 5 min, followed by 36 cycles at 94 °C for 30 s, 55 °C for 30 s, 72 °C for 45 s and a final extension for 7 min at 72 °C. Amplicons were sequenced on an automated sequencer ABI 377 (Applied Biosystems, Foster City, CA, USA) using BigDye Terminator 3.1 (Applied Biosystems, Foster City, CA, USA).

Chromatograms were manually checked and edited using Chromas Lite 2.6.6 (Technelysium Pty Ltd., Brisbane, Australia) and BioEdit 7.2.6.1 [29]. All newly determined sequences were deposited in GenBank (accession numbers: OP094031-OP094043).

For taxonomic attribution, the newly determined sequences were compared with available homologous traits deposited in GenBank which were used in previous phylogenetic and taxonomic studies on the genus *Uroplatus* (see e.g., [13–23].

This preliminary analysis allowed us to perform a taxonomic assessment of the collected samples as reported in Table 1. Given the maximum identity scores between the specimens analysed in this work and deposited sequences of *Uroplatus* used in previous taxonomic studies (99.3–100%), we are confident in the taxonomic attribution provided in Table 1. A notable exception is represented by the specimens GA 328 and GA 329, which are here reported as *U.* prope *guentheri* (Table 1) based on their maximum identity score (96.8%) with a previously deposited homologous sequence of *U. guentheri*, (AN EU596688). Considering the pairwise distance threshold usually used for species identification in squamates for the 12S (3–4%) see e.g., [30,31], it is therefore possible that the samples GA 328 and GA 329 represent an undescribed lineage of *Uroplatus*, but more focused morphological and molecular analyses employing a combination of mitochondrial and nuclear markers should be performed to better assess the taxonomic placement of these samples. This result is not surprising considering the significant number of newly described *Uroplatus* species in the last years and the molecular identification of different undescribed lineages (e.g., [20,23]).

2.3. Cytogenetic Analysis

Metaphase plates were obtained from tissues sampled during previous fieldwork (see above) using the air-drying method as described in Mezzasalma et al. [32].

Chromosomes were stained with conventional colorations (5% Giemsa solution at pH 7), silver staining (Ag-NOR) [33], C-banding according to Sumner [34] and sequential C-banding + Chromomycin A3 (CMA_3), +4′,6-diamidino-2-phenylindole (DAPI). following Mezzasalma et al. [35].

Karyotype reconstruction was performed after scoring at least five plates per sample and chromosomes were classified following Levan et al. [36].

3. Results

Cytogenetic Analysis

Our chromosome analysis showed the occurrence of karyological variability among the studied samples in terms of chromosome number, number and chromosome location of loci of NORs, pattern of heterochromatin and the occurrence of a putative heteromorphic sex chromosome pair.

Chromosome number varied from 2n = 34 (in *U.* prope *guentheri*) to 2n = 38 of (in *U. ebenaui* and *U. fiera*). A karyotype of 2n = 36 was the most common condition in the samples studied and shown by five different species (*U. alluaudi*, *U. finiavana*, *U. fimbriatus*, *U. henkeli* and *U. pietschmanni*). The karyotypes of all the analysed specimens were composed of all acrocentric chromosomes, gradually decreasing in length. The only exception was represented by the studied female of *U. alluaudi*, whose karyotype showed a heteromorphic pair (10th pair) including an acrocentric chromosome which was distinctively shorter than a metacentric chromosome. This pair, also in consideration of C-banding results (see below), can be considered as a putative heteromorphic sex chromosome pair with female heterogamety (ZZ/ZW) (Figure 1).

In three species (*U. alluaudi*, *U. guentheri*, and *U. pietschmanni*), loci of NORs were localised in a telomeric position on the chromosomes of the 2nd pair. In two species (*U. fimbriatus* and *U. henkeli*), loci of NORs were in a peritelomeric position on the 6th chromosome pair. In U. *finiavana* NORs were on the chromosomes of the 10th and 16th pair, while in *U. ebenaui* NORs were localised on the chromosomes of the 16th pair. Loci NORs were peculiar in *U. fiera*, residing on pericentromeric regions of the chromosomes of the 2nd pair and on one of the chromosomes of the 16th pair (Figure 1).

Given the quantity and quality of metaphase plates, sequential C-banding + CMA_3 + DAPI + Giemsa was successfully performed only in *U. ebenaui*, *U. finiavana*, *U. pietschmanni* and *U. alluaudi*.

C-banding evidenced a low content of heterochromatin in the species studied, with the occurrence of heterochromatic regions on pericentromeric and telomeric regions of almost all chromosomes of all the studied taxa. Nevertheless, although generally barely visible with fluorochromes, centromeric bands varied among different species by being CMA_3 positive and DAPI negative (in U. *ebenaui* and *U. finiavana*) or positive to either CMA_3 and DAPI (in *U. pietschmanni*) (Figure 2). In *U. alluaudi*, C-banding evidenced thin centromeric heterochromatic bands in several chromosome pairs, which were positive to both CMA_3 and DAPI (Figure 3). Interestingly, the larger (metacentric) chromosome of the heteromorphic pair were completely heterochromatic, positive to both fluorochromes and was therefore identified as a putative W sex chromosome (Figure 3). Because the Z chromosome did not show any distinctive heterochromatic pattern after C-banding, allowing its unambiguously identification among different autosome pairs, the ZW pair was tentatively assigned to the 10th chromosome pair (see Discussion).

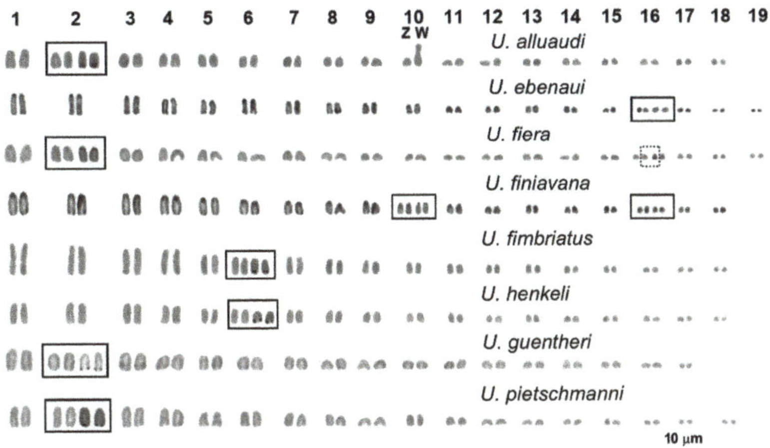

Figure 1. Giemsa stained karyotypes of the studied taxa. Insets include the NOR-bearing pair.

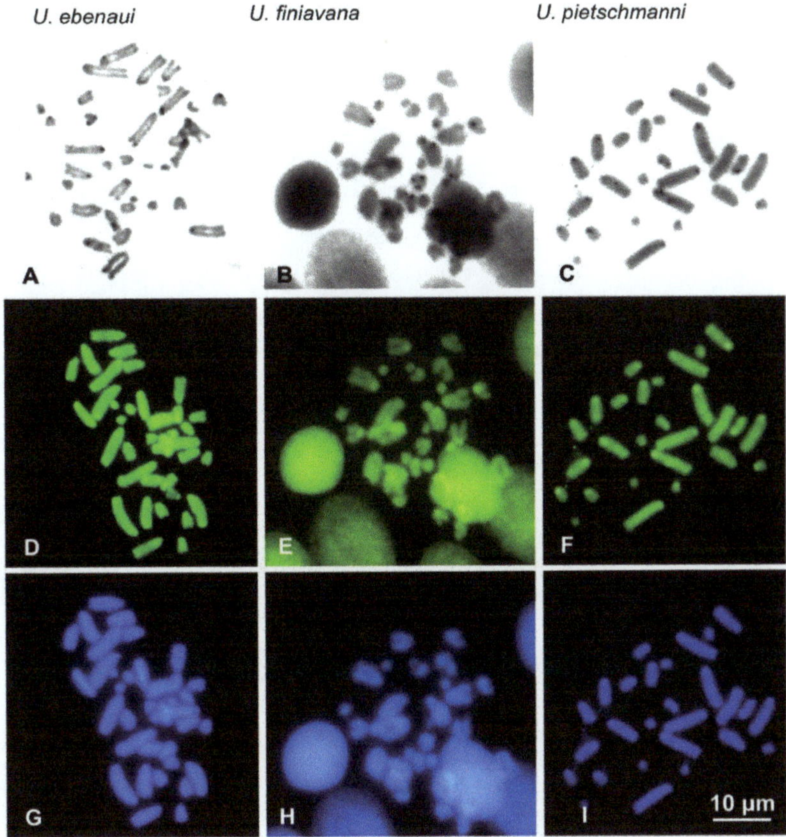

Figure 2. Metaphase plates of *U. ebenaui* (**A**,**D**,**G**), *U. finiavana* (**B**,**E**,**H**) and *U. pietschmanni* (**C**,**F**,**I**) sequentially stained with C-banding + Giemsa (**A**–**D**) + CMA_3 (**D**–**F**) + DAPI (**G**–**I**).

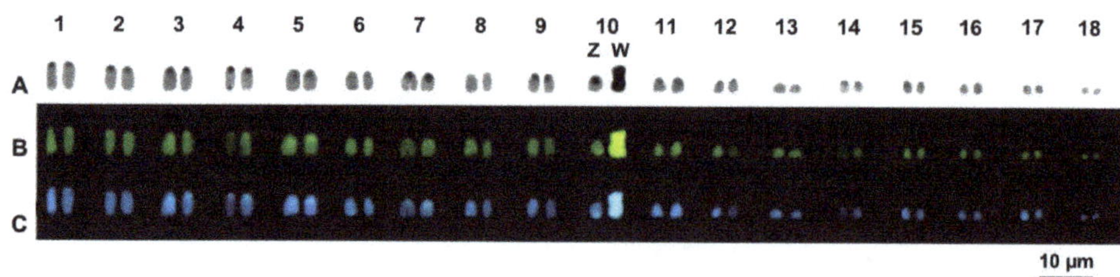

Figure 3. Karyotype of *U. alluaudi* sequentially stained with C-banding + Giemsa (**A**), +CMA_3 (**B**) and +DAPI (**C**).

4. Discussion

Our cytogenetic analysis provided the first karyotype description of eight Malagasy gecko species of *Uroplatus* and represents the first step in describing the karyological variability of the genus, as well as a new contribution to reconstruct chromosomal evolutionary dynamics in a larger clade of leaf-toed geckos.

Overall, we found that the chromosomal diversity in *Uroplatus* mostly encompasses the total chromosome number (from 2n = 34 to 38), a different localization of loci of NORs and the raising of putative heteromorphic sex chromosomes. Chromosome morphology resulted almost invariably acrocentric in the genus with the exception of a large metacentric chromosome found in *U. alluaudi*, here considered as the W sex chromosome (see below).

Taking into account different karyological features which are considered plesiomorphic in squamates (high total number of chromosomes, number of dot-shaped microchromosomes and loci on NORs on the smallest pairs (see e.g., [37–41]), the karyotype of *U. ebenaui* (2n = 38, with NORs on one of the smallest pair) should be considered as a primitive state in *Uroplatus*. From karyotypes with a similar structure, the chromosomal diversification in the genus probably proceeded toward a progressive reduction in the total chromosome number (2n = 36 in *U. phantasticus*, *U. alluaudi*, *U. finiavana*, *U. fimbriatus*, *U. henkeli* and *U. pietschmanni* and 2n = 34 in *U.* prope *guentheri*) ([24] this study) by means of chromosome fusions and translocations of chromosomes of the smallest pairs (Figure 4).

The variability of loci of NORs also plays an important role in the karyotype diversification of the genus *Uroplatus*. In fact, rDNA gene clusters are considered recombination "hotspots" and can induce significant evolutionary changes by means of their translocation among different genomic regions and/or the differential inactivation of different loci [9,40,41]. In *Uroplatus*, the traslocation of NORs probably occurred among different chromosomes, from those of the smallest pairs (16th and 10th pair in *U. finiavana* and *U. ebenaui*) to middle-sized (6th pair in *U. fimbriatus* and *U. henkeli*) and large chromosomes (2nd pair in *U. phantasticus*, *U. alluaudi*, *U. guentheri*, and *U. pietschmanni*) ([24] this study) (Figure 4). The condition displayed by *U. fiera* (NORs on the 2nd pair and an extra, unpaired locus, on one of the chromosomes of the 16th pair), is quite rare in reptiles, but similar configurations have been documented in Lacertidae, Opluridae, Leiocephalidae and Helodermatidae (see e.g., [37–40,42–45]).

More in general, the karyotypes of the *Uroplatus* species studied here resemble those of the phylogenetically related Malagasy leaf-toed geckos of the genera *Paroedura*, *Ebenavia*, *Phelsuma*, *Matoatoa* and the Australian genus *Christinus*. To highlight karyological affinities and differences between these phylogenetically related genera we superimposed the haploid karyograms of the studied samples of *Uroplatus*, as well as those available from the literature, to the phylogentic tree by Pyron et al. [25], adding the intrageneric relationships of the *U. ebenaui* species group by Ratsoavina et al. [23] (Figure 5).

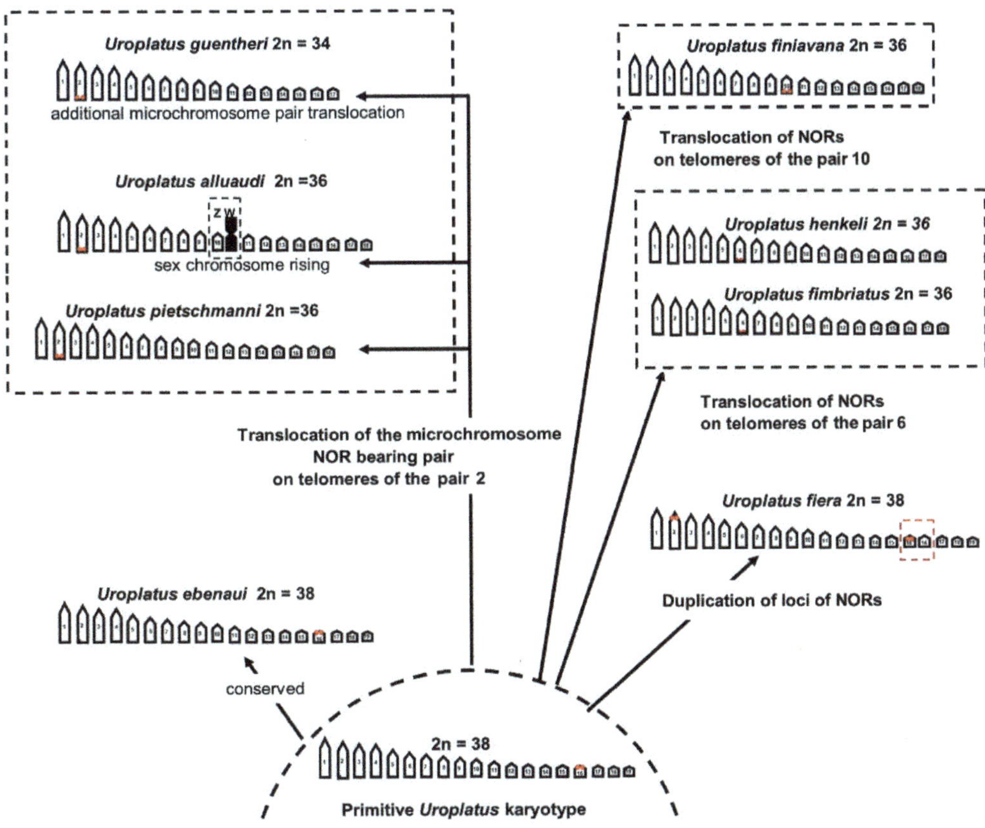

Figure 4. Hypothesized scenario of chromosome diversification in *Uroplatus*.

Similarly, to what has been previously described within *Uroplatus* (see above), *Lygodactylus* [8], *Matoatoa* [44], *Paroedura* and *Christinus* [6,24,26], the whole group seems to be characterized by an overall reduction in the chromosome number and the independent acquisition of derivate chromosome features. In fact, all these genera display a karyotype composed of 2n = 34–42 mostly acrocentric chromosomes, the progressive formation of metacentric chromosomes by means of chromosome fusions in karyotypes with a reduced chromosome number (in e.g., *Lygodactylus*, *Matoatoa*, *Paroedura* and *Christinus*) and/or the translocation of small NOR-bearing chromosomes on larger chromosomes (in e.g., *Uroplatus*, *Matoatoa* and *Ebenavia*) (see Figure 5).

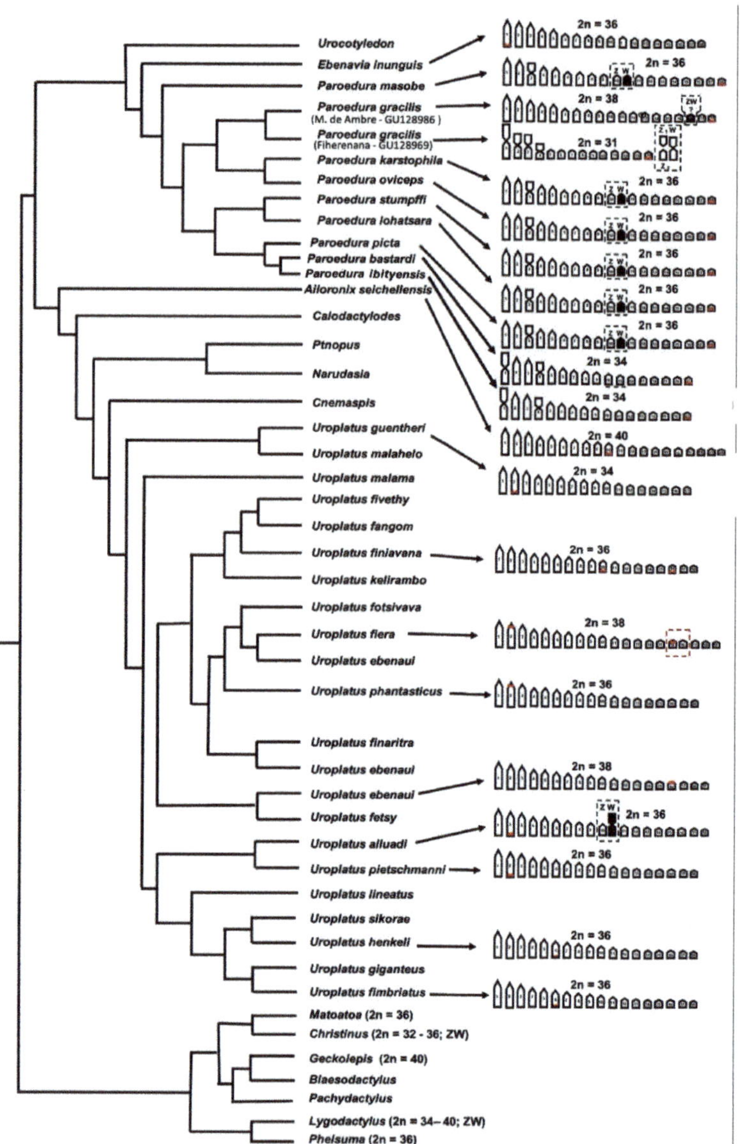

Figure 5. Hypothesized scenario of chromosome diversification in phylogenetically related leaf-toed geckos. Available karyological data from the literature [6,8,9,24,26,44,46–48] are superimposed on the phylogenetic tree by Pyron et al. [25], with relationships of the *U. ebenaui* group by Ratsoavina et al. [23].

We highlight that this group of geckos provides an example of an early stage of the transition between "symmetrical" (mostly composed by acrocentric chromosomes and without a clear distinction between macro- and microchromosomes) and "asymmetrical" karyotypes (with a high number of biarmed chromosomes and a clear distinction between macro- and microchromosomes) [45], which is hypothesized to represent a major evolutionary trend of the karyological diversification of squamates [39,49].

Another interesting outcome of our cytogenetic analysis is the first record in *Uroplatus* of a putative heteromorphic sex chromosome system (ZW in *U. alluaudi*). However, only a single female was studied in this work and more karyological data on males and females of *U. alluaudi* should be gathered in order to confirm this observation. Nevertheless, we highlight that the occurrence of a sex chromosome system is the most robust explanation of the heteromorphic pair found in the female specimen here studied. Notably, the largely heterochromatic W chromosome found in *U. alluaudi* is much bigger than the Z, a condition rarely observed in squamates, e.g., in *Clelia clelia* (Daudin, 1803) and *Phisalixella variabilis* (Boulenger, 1896) [7,50].

Bigger dimensions of the heteromorphic chromosome (Y/W) usually indicate its relatively recent diversification by means of heterochromatin addition and amplification, which is usually followed by the degeneration of the Y/W chromosome, down to the size of a microchromosome [39,51–53]. The lack of other heteromorphic sex chromosomes in the other *Uroplatus* species studied so far, also seems to support the relatively recent origin of the heteromorphic pair in *U. ebenaui.*

In phylogenetically related geckos, heteromorphic sex chromosome systems are not a common feature and are known mainly in *Paroedura* (different species with ZW and Z_1Z_2W chromosomes), *C. marmoratus* (ZW as the 4th pair) and *L. tuberosus* (ZW as the 1st pair) [6,8,24,26].

Reptiles are a well-known model organism in the study of sex chromosome diversification and include species with temperature dependent sex determination (TSD) and genetic sex determination (GSD) with either male or female heterogamety (see e.g., [39,51–58]).

According to the most supported hypotheses, the process of sex chromosome differentiation begins when a sex determining locus rises in one of the two homomorphic proto-sex chromosomes which are at this step cytogenetically undetectable with standard and banding methods [41,52–56]. The next step of the diversification of the proto-Y/W is the suppression of recombination in the region containing the sex-determining locus by means of an inversion or a progressive heterochromatin addition. This eventually leads to the evolutionary isolation of the Y/W chromosome and to its progressive degeneration. At intermediate and final stages of its diversification, the Y/W chromosome appears dimensionally distinguishable from the X/Z and/or largely heterochromatic [32,41,51,56].

In the gecko clade considered here, sex chromosome diversification seems to have followed different pathways in different genera. Diversification by progressive addition of heterochromatin probably occurred in *L. tuberorus*, six *Paroedura* species and *U. alluaudi.* In fact, the W chromosomes of these species show different levels of heterochomatinization; with pseudoautosomal regions (*L. tuberosus*) [8], largely heterochromatic but homomorphic (in *Paroedura*) [6,24,57] or heteromorphic and heterochromatic (*U. alluaudi*) (this study).

The alternative model has been proposed for *C. marmoratus*, whose euchromatic, submetacentric W started its diversification from the Z by means of an inversion [26], while the multiple sex chromosome system of *P. gracilis* from Fiherenena (2n = 31, with Z_1Z_1W) probably originated from an autosome-sex chromosome fusion [24].

It should also be noted that, excluding *Paroedura*, most species and genera (Figure 5) of the gecko clade considered here do not show any heteromorphic or heterochromatic sex chromosomes, suggesting their early diversification stage ([6,8] this study).

In *Paroedura* species with known heteromorphic W chromosomes, the sex chromosome pair is always the 10th, and chromosome painting with Z-specific markers showed pair homology among different species [6,57]. However, Z-specific markers are absent in other species of the genus without differentiated sex chromosomes, as well as in *E. inunguis*, which represents the sister clade to *Paroedura*.

The other species of the clade with known sex chromosome systems show their localization on different pairs. In *U. alluaudi* the Z chromosome is not easily distinguishable from different autosome pairs, and we tentatively described it as the 10th pair only based on its dimension. In two other genera, *L. tuberosus* shows sex chromosomes on the first pair, while they are on the 4th pair in *C. marmoratus* [8,26]. These evidences seem to suggest the independent origin (non-homology) of sex chromosome pairs in these different gecko genera (Figure 5), but more focused analysis with molecular cytogenetics are needed to confirm this hypothesis.

5. Conclusions

We provide here the first karyotype description of eight gecko species of the genus *Uroplatus*, which varied in terms of chromosome number (2n = 34–38), localization of loci of NORs (alternatively on the 2nd, 6th, 10th or 16th pair), heterochromatin composition and the occurrence of a putative heteromorphic sex chromosome pair.

Considering the occurrence of chromosome characters which are considered plesiomorphic in squamates, we hypothesise a karyotype of 2n = 38 with NORs on one of the smallest pairs as the primitive condition in *Uroplatus*. Progressive chromosome rearrangements eventually led to karyotypes with a lower chromosome number (2n = 34–36) and NORs on medium or large chromosomes.

Overall, the karyotypes of the *Uroplatus* species studied here resemble those of phylogenetically related leaf-toed geckos, including *Paroedura*, *Ebenavia*, *Phelsuma* and *Matoatoa* and the Australian genus *Christinus*. We show that the whole group is characterized by a tendency toward a reduction in the chromosome number (from 2n = 42 to 2n = 34), the formation of metacentric chromosomes and/or the translocation of NORs on middle-sized or large chromosomes.

We also found a first case of a putative heteromorphic sex chromosome pair in *Uroplatus* (ZW in *U. alluaudi*), with a largely heterochromatic W chromosome which is much bigger than the Z. We discuss similarities and differences of sex chromosome diversification in phylogenetically related taxa (different *Paroedura* species, *L. tuberosus* and *C. marmoratus*), hypothesizing that the rise of non-homologous sex chromosomes occurred independently in different genera.

Author Contributions: Conceptualization, M.M.; methodology, M.M. and G.O.; software, M.M.; validation, M.M. and G.O.; formal analysis, M.M. and G.O.; investigation, M.M.; resources, M.M., G.O. and F.M.G.; data curation, M.M., G.O., F.M.G. and E.B.; writing—original draft preparation, M.M.; writing—review and editing, M.M., G.O., F.M.G. and E.B.; visualization, M.M., G.O., F.M.G. and E.B. All authors have read and agreed to the published version of the manuscript.

Funding: This research received no external funding.

Institutional Review Board Statement: For this study we used samples already collected for other previously published studies with the approval of institutional committees and no further sampling was performed.

Informed Consent Statement: Not applicable, as this research did not involve humans.

Data Availability Statement: Newly generated cytogenetic data are available within this manuscript. DNA sequences are available on GenBank (accession numbers: OP094031-OP094043).

Acknowledgments: We are grateful to Malagasy authorities for granting research and export permits. We thank Gennaro Aprea, Frank Glaw and Franco Andreone for providing us the tissue samples. This contribution is dedicated to the memory of our friend and colleague Teresa Capriglione.

Conflicts of Interest: The authors declare no conflict of interest.

References

1. Myers, N.; Mittermeier, R.A.; Mittermeier, C.G.; da Fonseca, G.A.B.; Kent, J. Biodiversity hot-spots for conservation priorities. *Nature* **2000**, *403*, 853–858. [CrossRef] [PubMed]
2. Ganzhorn, J.U.; Lowry, P.P.; Schatz, G.E.; Sommer, S. The biodiversity of Madagascar: One of the world's hottest hotspots on its way out. *Oryx* **2001**, *35*, 346–348. [CrossRef]
3. Vences, M.; Wollenberg, K.C.; Vieites, D.R.; Lees, D.C. Madagascar as a model region of species diversification. *Trends Ecol. Evol.* **2009**, *24*, 456–465. [CrossRef] [PubMed]
4. Glaw, F.; Vences, M. *A Field Guide to the Amphibians and Reptiles of Madagascar*, 3rd ed.; Vences, M., Glaw, F., Eds.; Verlags GbR: Köln, Germany, 2007; p. 496.

5. Uetz, P.; Freed, P.; Hošek, J. (Eds.) The Reptile Database. Available online: http://www.reptile-database.org (accessed on 9 June 2022).
6. Koubová, M.; Pokorná, M.J.; Rovatsos, M.; Farkačová, K.; Altmanová, M.; Kratochvíl, L. Sex determination in Madagascar geckos of the genus *Paroedura* (Squamata: *Gekkonidae*): Are differentiated sex chromosomes indeed so evolutionary stable? *Chromosom. Res.* **2014**, *22*, 441–452. [CrossRef]
7. Mezzasalma, M.; Andreone, F.; Branch, W.R.; Glaw, F.; Guarino, F.M.; Nagy, Z.T.; Odierna, G.; Aprea, G. Chromosome evolution in pseudoxyrhophiine snakes from Madagascar: A wide range of karyotypic variability. *Biol. J. Linn. Soc.* **2014**, *112*, 450–460. [CrossRef]
8. Mezzasalma, M.; Andreone, F.; Aprea, G.; Glaw, F.; Odierna, G.; Guarino, F.M. Molecular phylogeny, biogeography and chromosome evolution of Malagasy dwarf geckos of the genus *Lygodactylus* (Squamata, *Gekkonidae*). *Zool. Scr.* **2017**, *46*, 42–54. [CrossRef]
9. Mezzasalma, M.; Andreone, F.; Glaw, F.; Odierna, G.; Petraccioli, A.; Guarino, F.M. Chromosome aneupolyploidy in an endemic Malagasy gecko (Gekkonidae: *Geckolepis*). *Salamandra* **2018**, *54*, 56–62.
10. Mezzasalma, M.; Andreone, F.; Glaw, F.; Guarino, F.M.; Odierna, G.; Petraccioli, A.; Picariello, O. Changes in heterochromatin content and ancient chromosome fusion in the endemic Malagasy boid snakes *Sanzinia* and *Acrantophis* (Squamata: Serpentes). *Salamandra* **2019**, *55*, 140–144.
11. Rovatsos, M.; Pokorná, M.J.; Altmanová, M.; Kratochvíl, L. Female heterogamety in Madagascar chameleons (Squamata: Chamaeleonidae: *Furcifer*): Differentiation of sex and neo-sex chromosomes. *Sci. Rep.* **2014**, *5*, 13196. [CrossRef]
12. Rovatsos, M.; Altmanová, M.; Johnson Pokorná, M.; Velenský, P.; Sánchez Baca, A.; Kratochvíl, L. Evolution of karyotypes in chameleons. *Genes* **2017**, *8*, 382. [CrossRef]
13. Glaw, F.; Kosuch, J.; Henkel, F.W.; Sound, P.; Böhme, W. Genetic and morphological variation of the leaf-tailed gecko *Uroplatus fimbriatus* from Madagascar, with description of a new giant species. *Salamandra* **2006**, *42*, 129144.
14. Greenbaum, E.; Bauer, A.M.; Jackman, T.R.; Vences, M.; Glaw, F. A phylogeny of the enigmatic Madagascan geckos of the genus *Uroplatus* (Squamata: Gekkonidae). *Zootaxa* **2007**, *1493*, 41–51. [CrossRef]
15. Raxworthy, C.J.; Pearson, R.G.; Zimkus, B.M.; Reddy, S.; Deo, A.J.; Nussbaum, R.A.; Ingram, C.M. Continental speciation in the tropics: Contrasting biogeographic patterns of divergence in the *Uroplatus* leaf-tailed gecko radiation of Madagascar. *J. Zool.* **2008**, *275*, 423–440. [CrossRef]
16. Ratsoavina, F.M.; Louis, E.E., Jr.; Crottini, A.; Randrianiaina, R.-D.; Glaw, F.; Vences, M. A new leaf tailed gecko species from northern Madagascar with a preliminary assessment of molecular and morphological variability in the *Uroplatus ebenaui* group. *Zootaxa* **2011**, *3022*, 39–57. [CrossRef]
17. Ratsoavina, F.M.; Vences, M.; Louis, E.E., Jr. Phylogeny and phylogeography of the Malagasy leaf-tailed geckos in the *Uroplatus ebenaui* group. *Afr. J. Herpetol.* **2012**, *61*, 143–158. [CrossRef]
18. Ratsoavina, F.M.; Raminosoa, N.R.; Louis, E.E., Jr.; Raselimanana, A.P.; Glaw, F.; Vences, M. An overview of Madagascar's leaf tailed geckos (genus *Uroplatus*): Species boundaries, candidate species and review of geographical distribution based on molecular data. *Salamandra* **2013**, *49*, 115–148.
19. Ratsoavina, F.M.; Ranjanaharisoa, F.A.; Glaw, F.; Raselimanana, A.P.; Miralles, A.; Vences, M. A new leaf-tailed gecko of the *Uroplatus ebenaui* group (Squamata: Gekkonidae) from Madagascar's central eastern rainforests. *Zootaxa* **2015**, *4006*, 143–160. [CrossRef]
20. Ratsoavina, F.M.; Gehring, P.-S.; Scherz, M.D.; Vieites, D.R.; Glaw, F.; Vences, M. Two new species of leaf-tailed geckos (*Uroplatus*) from the Tsaratanana mountain massif in northern Madagascar. *Zootaxa* **2017**, *4347*, 446–464. [CrossRef]
21. Ratsoavina, F.M.; Raselimanana, A.P.; Scherz, M.D.; Rakotoarison, A.; Razafindraibe, J.H.; Glaw, F.; Vences, M. Finaritra! A new leaf-tailed gecko (*Uroplatus*) species from Marojejy National Park in north-eastern Madagascar. *Zootaxa* **2019**, *4545*, 563–577. [CrossRef]
22. Ratsoavina, F.M.; Scherz, M.D.; Tolley, K.A.; Raselimanana, A.P.; Glaw, F.; Vences, M. A new species of *Uroplatus* (Gekkonidae) from Ankarana National Park, Madagascar, of remarkably high genetic divergence. *Zootaxa* **2019**, *4683*, 84–96. [CrossRef]
23. Ratsoavina, F.M.; Glaw, F.; Raselimanana, A.P.; Rakotoarison, A.; Vieites, D.R.; Hawlitschek, O.; Vences, M.; Scherz, M.D. Towards completion of the species inventory of small-sized leaf-tailed geckos: Two new species of *Uroplatus* from northern Madagascar. *Zootaxa* **2020**, *4895*, 251–271. [CrossRef] [PubMed]
24. Aprea, G.; Andreone, F.; Fulgione, D.; Petraccioli, A.; Odierna, G. Chromosomal Rearrangements Occurred Repeatedly and Independently during Species Diversification in Malagasy Geckos, genus *Paroedura*. *Afr. Zool.* **2013**, *48*, 96–108. [CrossRef]
25. Pyron, R.A.; Burbrink, F.T.; Wiens, J.J. A phylogeny and revised classification of Squamata, including 4161 species of lizards and snakes. *BMC Evol. Biol.* **2013**, *13*, 93. [CrossRef] [PubMed]
26. King, M.; Rofe, R. Karyotypic variation in the Australian Gekko *Phyllodactylus marmoratus* (Gray) (Gekkonidae: Reptilia). *Chromosoma* **1976**, *54*, 75–87. [CrossRef]
27. Sambrook, J.; Fritsch, E.F.; Maniatis, T. *Molecular Cloning: A Laboratory Manual*, 2nd ed.; Cold Spring Harbor Lab Press: New York, NY, USA, 1989.
28. Kocher, T.D.; Thomas, W.K.; Meyer, A.; Edwards, S.V.; Paabo, S.; Villablanca, F.X.; Wilson, A.C. Dynamics of mitochondrial DNA evolution in animals: Amplification and sequencing with conserved primers. *Proc. Natl. Acad. Sci. USA* **1989**, *86*, 6196–6200. [CrossRef] [PubMed]

29. Hall, T.A. BioEdit: A user-friendly biological sequence alignment editor and analysis program for Windows 95/98/NT. *Nucleic Acids Symp. Ser.* **1999**, *41*, 95–98.
30. Jeong, T.J.; Jun, J.; Han, S.; Kim, H.T.; Oh, K.; Kwak, M. DNA barcode reference data for the Korean herpetofauna and their applications. *Mol. Ecol. Resour.* **2013**, *13*, 1019–1032. [CrossRef]
31. Nugraha, F.A.D.; Fatchiyah, F.; Smith, E.N.; Kurniawan, N. Phylogenetic analysis of colubrid snakes based on 12S rDNA reveals distinct lineages of Dendrelaphis pictus (Gmelin, 1789) populations in Sumatra and Java. *Biodiversitas* **2018**, *19*, 303–310. [CrossRef]
32. Mezzasalma, M.; Visone, V.; Petraccioli, A.; Odierna, G.; Capriglione, T.; Guarino, F.M. Non-random accumulation of LINE1-like sequences on differentiated snake W chromosomes. *J. Zool.* **2016**, *300*, 67–75. [CrossRef]
33. Howell, W.M.; Black, D.A. Controlled silver-staining of nucleolus organizer regions with a protective colloidal developer: A 1-step method. *Experientia* **1980**, *36*, 1014–1015. [CrossRef]
34. Sumner, A.T. A simple technique for demonstrating centromeric heterochromatin. *Exp. Cell Res.* **1972**, *75*, 304–306. [CrossRef]
35. Mezzasalma, M.; Andreone, F.; Odierna, G.; Guarino, F.M.; Crottini, A. Comparative cytogenetics on eight Malagasy Mantellinae (Anura, Mantellidae) and a synthesis of the karyological data on the subfamily. *Comp. Cytogenet.* **2022**, *16*, 1–17. [CrossRef]
36. Levan, A.; Fredga, K.; Sandberg, A.A. Nomenclature for Centromeric Position on Chromosomes. *Hereditas* **1964**, *52*, 201–220. [CrossRef]
37. Porter, C.A.; Hamilton, M.J.; Sites, J.W., Jr.; Baker, R.J. Location of ribosomal DNA in chromosomes of squamate reptiles: Systematic and evolutionary implications. *Herpetologica* **1991**, *47*, 271–280.
38. Altmanová, M.; Rovatsos, M.; Kratochvíl, L.; Johnson Pokorná, M. Minute Y chromosomes and karyotype evolution in Madagascan iguanas (Squamata: Iguania: Opluridae). *Biol. J. Linn. Soc.* **2016**, *118*, 618–633. [CrossRef]
39. Mezzasalma, M.; Guarino, F.M.; Odierna, G. Lizards as Model Organisms of Sex Chromosome Evolution: What We Really Know from a Systematic Distribution of Available Data? *Genes* **2021**, *12*, 1341. [CrossRef]
40. Stults, D.M.; Killen, M.W.; Williamson, E.P.; Hourigan, J.S.; Vargas, H.D.; Arnold, S.M.; Moscow, J.A.; Pierce, A.J. Human rRNA Gene Clusters Are Recombinational Hotspots in Cancer. *Cancer Res.* **2009**, *69*, 9096–9104. [CrossRef]
41. McStay, B. Nucleolar organizer regions: Genomic 'dark matter' requiring illumination. *Genes Dev.* **2016**, *30*, 1598–1610. [CrossRef]
42. Odierna, G.; Capriglione, T.; Olmo, E.; Cardone, A.; Rosati, C. The karyology of some South African lacertids belonging to three genera *Heliobolus*, *Meroles*, and *Pedioplanis*. *J. Afr. Zool.* **1990**, *104*, 541–547.
43. Odierna, G.; Olmo, E.; Capriglione, T.; Caputo, V. Karyological Differences between *Lacerta lepida* and *Lacerta pater*. *S. Am. J. Herpetol.* **1990**, *24*, 97–99. [CrossRef]
44. Mezzasalma, M.; Guarino, F.; Loader, S.; Odierna, G.; Streicher, J.; Cooper, N. First karyological analysis of the endemic Malagasy phantom gecko *Matoatoa brevipes* (Squamata: Gekkonidae). *Acta Herpetol.* **2020**, *15*, 137–141. [CrossRef]
45. White, M.J.D. *Animal Cytology and Evolution*, 3rd ed.; Cambridge University Press: Cambridge, UK, 1973.
46. Aprea, G.; Odierna, G.; Capriglione, T.; Caputo, V.; Morescalchi, A.; Olmo, E. Heterochromatin and NOR distribution in the chromosomes of six gekkonid species of the genus *Phelsuma* (Squamata: *Gekkonidae*). *J. Afr. Zool.* **1996**, *119*, 341–349.
47. Volobouev, V.; Ineich, I. A chromosome banding study of *Ailuronyx seychellensis* (Reptilia, Gekkonidae). *J. Herpetol.* **1994**, *28*, 267–270. [CrossRef]
48. King, M.; King, D. An additional chromosome race of *Phyllodactyllus marmoratus* (Gray) (Reptilia: Gekkonidae) and its phylogenetic implications. *Aust. J. Zool.* **1977**, *25*, 667–672. [CrossRef]
49. Olmo, E. Trends in the evolution of reptilian chromosomes. *Integr. Comp. Biol.* **2008**, *48*, 486–493. [CrossRef]
50. Beçak, W. Constititução cromossomica e mecanismo de determinação do sexo em ofidios sudamericanos. I. Aspectos cariotipicos. *Mem. Do Inst. Butantan* **1965**, *32*, 37–78.
51. Wright, A.E.; Dean, R.; Zimmer, F.; Mank, J.E. How to make a sex chromosome. *Nat. Commun.* **2016**, *7*, 12087. [CrossRef]
52. Ohno, S. *Sex Chromosomes and Sex-Linked Genes*; Springer: Berlin, Germany, 1967.
53. Charlesworth, B. The evolution of sex chromosomes. *Science* **1991**, *251*, 1030–1033. [CrossRef]
54. Pallotta, M.M.; Turano, M.; Ronca, R.; Mezzasalma, M.; Petraccioli, A.; Odierna, G.; Capriglione, T. Brain Gene Expression is Influenced by Incubation Temperature During Leopard Gecko (*Eublepharis macularius*) Development. *J. Exp. Zool. Mol. Dev. Evol.* **2017**, *328*, 360–370. [CrossRef]
55. Sidhom, M.; Said, K.; Chatti, N.; Guarino, F.M.; Odierna, G.; Petraccioli, A.; Picariello, O.; Mezzasalma, M. Karyological characterization of the common chameleon (*Chamaeleo chamaeleon*) provides insights on the evolution and diversification of sex chromosomes in Chamaeleonidae. *Zoology* **2020**, *141*, 125738. [CrossRef]
56. Charlesworth, D. Sex differences in fitness and selection for centric fusions between sex-chromosomes and autosomes. *Genet. Res.* **1980**, *35*, 205–214. [CrossRef] [PubMed]
57. Rovatsos, M.; Farkačová, K.; Altmanová, M.; Johnson Pokorná, M.J.; Kratochvíl, L. The rise and fall of differentiated sex chromosomes in geckos. *Mol. Ecol.* **2019**, *28*, 3042–3052. [CrossRef] [PubMed]
58. Thépot, D. Sex Chromosomes and Master Sex-Determining Genes in Turtles and Other Reptiles. *Genes* **2021**, *12*, 1822. [CrossRef] [PubMed]

Article

Chromosome Evolution of the *Liolaemus monticola* (Liolaemidae) Complex: Chromosomal and Molecular Aspects

Madeleine Lamborot [1,*], Carmen Gloria Ossa [2,3], Nicolás Aravena-Muñoz [4], David Véliz [1,5] and Raúl Araya-Donoso [6,*]

1 Departamento de Ciencias Ecológicas, Facultad de Ciencias, Universidad de Chile, Santiago 7800003, Chile
2 Instituto de Biología, Facultad de Ciencias, Universidad de Valparaíso, Valparaíso 2361845, Chile
3 Centro de Investigación y Gestión de Recursos Naturales, Universidad de Valparaíso, Valparaíso 2361845, Chile
4 Liceo Alberto Blest Gana, Conchalí 8551270, Chile
5 Centro de Ecología y Manejo Sustentable de Islas Oceánicas, Coquimbo 1780000, Chile
6 School of Life Sciences, Arizona State University, Tempe, AZ 85287, USA
* Correspondence: mlamboro@uchile.cl (M.L.); raul.araya.d@asu.edu (R.A.-D.)

Simple Summary: Chromosome variation is highly relevant for evolution because chromosomal mutations can influence speciation. Here, we assessed population cytogenetics in *Liolaemus monticola*, a lizard endemic to Chile that consists of several chromosomal races with highly polymorphic chromosome rearrangements. We sampled individuals from the northernmost distribution of the species to obtain chromosomes and mitochondrial gene sequences and compared the samples to previously published data from other populations across the distribution. Our results show the existence of seven differentiated races of *L. monticola*, each with unique chromosome characteristics and high levels of polymorphism. Interestingly, the geographical delimitation of the races is associated with the presence of rivers that could represent barriers to gene flow. Thus, our study highlights the importance of chromosomal mutations for population differentiation, and in turn, speciation.

Abstract: Chromosomal rearrangements can directly influence population differentiation and speciation. The *Liolaemus monticola* complex in Chile is a unique model consisting of several chromosome races arranged in a latitudinal sequence of increasing karyotype complexity from south to north. Here, we compared chromosomal and mitochondrial cytochrome b data from 15 localities across the northern geographic distribution of *L. monticola*. We expanded the distribution of the previously described Multiple Fissions race (re-described as MF2), in the Coastal range between the Aconcagua River and the Petorca River, and described a new Multiple Fissions 1 (MF1) race in the Andean range. Both races present centric fissions in pairs 1 and 2, as well as a pericentric inversion in one fission product of pair 2 that changes the NOR position. Additionally, we detected a new chromosomal race north of the Petorca River, the Northern Modified 2 (NM2) race, which is polymorphic for novel centric fissions in pairs 3 and 4. Our results increase the number of chromosomal races in *L. monticola* to seven, suggesting a complex evolutionary history of chromosomal rearrangements, population isolation by barriers, and hybridization. These results show the relevant role of chromosome mutations in evolution, especially for highly speciose groups such as *Liolaemus* lizards.

Keywords: centric fissions; chromosome rearrangements; cytochrome b; population cytogenetics; speciation

Citation: Lamborot, M.; Ossa, C.G.; Aravena-Muñoz, N.; Véliz, D.; Araya-Donoso, R. Chromosome Evolution of the *Liolaemus monticola* (Liolaemidae) Complex: Chromosomal and Molecular Aspects. *Animals* **2022**, *12*, 3372. https://doi.org/10.3390/ani12233372

Academic Editor: Ettore Olmo

Received: 31 October 2022
Accepted: 27 November 2022
Published: 30 November 2022

1. Introduction

Many closely related plant and animal taxa differ in their chromosomal characteristics [1–4]. Thus, mutations or chromosomal rearrangements (CR) may play an important role in speciation [5–10]. Chromosomal variation (CV) can be mediated by different structural and/or numerical CR, such as Robertsonian translocations (centric fissions and centric fusions), inversions, and translocations. Several models have proposed that CR are causal to genic diversification between populations and therefore facilitate speciation [2,4,9–11]. Indeed, CR can usually produce strong evolutionary effects by preventing or reducing the fertility of hybrids, creating a barrier to genetic exchange, and influencing the differentiation of individuals within a population [4,8,9,12,13]. Chromosome variation is of special interest in highly radiated groups, where CR have been proposed as one of the genetic mechanisms associated with high speciation rates, as reported for some reptiles [5,14]. For instance, wide CV has been reported for the pleurodont iguanians *Anolis* [15–17], *Sceloporus* [5,14,18,19], and *Liolaemus* [20–23], and these complex chromosomal rearrangements could be exceptionally relevant for the high speciation described in these groups.

Among iguanians, Liolaemidae is a large and diverse monophyletic family of lizards, endemic to South America, that has been classified into three genera: *Ctenoblepharis*, *Phymaturus*, and *Liolaemus* [24–26]. The genus *Liolaemus* is by far the most diverse and highly speciose group in Liolaemidae, with over 292 species widely distributed in southern South America that are characterized by high ecological, chromosomal, genetic, and morphological diversity [23,25]. Thus, the genus *Liolaemus* is a suitable model to study evolutionary biology [23,27,28]. *Liolaemus* shows extensive karyotypic variation, and most taxa karyotyped to date have six pairs of metacentric macrochromosomes and 20–22 microchromosomes (2n = 32–34). The submetacentric chromosome pair 2 has visible secondary constrictions in the long arms, which correspond to the nucleolar organizing region (NOR). These features are considered the ancestral state in *Liolaemus* and primitive among iguanians [6,20,22,23,29–32]. However, some species present increased chromosome numbers derived from the plesiomorphic karyotypes, which could have originated via Robertsonian rearrangements or polyploidy [23,32].

One chromosomally derived group in *Liolaemus* is the *L. monticola* complex. This complex is endemic to Chile and is widely distributed throughout the Andes, Coastal, and Transversal Mountain ranges, from 320 to 2000 masl, and between 31° and 35° south latitudes [33]. *L. monticola* is a suitable model to study chromosomal evolution because it displays different CR with elevated polymorphism and diploid chromosome numbers ranging from 2n = 32 to 44 [23]. At least five chromosomal races, whose complexity increases from south to north, have been recognized (Figure 1). The primitive race (P), 2n = 32 (considered ancestral in the *Liolaemus*), is located in the southernmost distribution [34]. The Southern race (S), 2n = 34, is located on the Andes and Coastal Mountain ranges between the Lontué River and the Maipo River [35,36]. The Southern race is characterized by stable fixed translocations between pairs 5 and 7 and one microchromosome pair addition. The Northern race (N), 2n = 38–40, is located from the northern Maipo River and Yeso River to the southern bank of the Aconcagua River [35,36]. The Northern race added centric fissions in pairs 3 and 4, a pericentric inversion in one fission product of chromosome 3, and one pair of microchromosomes. The Northern Modified 1 race (NM1), 2n = 38–40, is located north of the Aconcagua River and east of the Rocín River [37]. The Northern Modified 1 race includes two polymorphisms: an enlarged chromosome in pair 6 and a pericentric inversion in pair 7. Finally, the Multiple Fission race, 2n = 42–44, is found in the "Hierro Viejo" locality between the Aconcagua River and the Petorca River [38]. The Multiple Fission race presents novel polymorphic centric fissions for pairs 1 and 2 and a polymorphic pericentric inversion in one fission product of pair 2.

The patterns of CV between races can be used as genetic markers to improve our understanding of evolution in this species complex. Preliminary studies have shown consistency between the variation detected with mitochondrial *cytb* gene sequences [39], allozymes [40], and morphological analyses [33,41,42] and the south-to-north gradient of chromosomal differentiation. Rivers have been proposed as the main biogeographic barriers that separated and restricted gene flow between the races, especially in association with Pleistocene climatic changes [23,33–42]. Moreover, chromosomal hybrid zones and introgression between the races have been reported for the *L. monticola* complex in these heterogenous and arid environments [35,36,43,44].

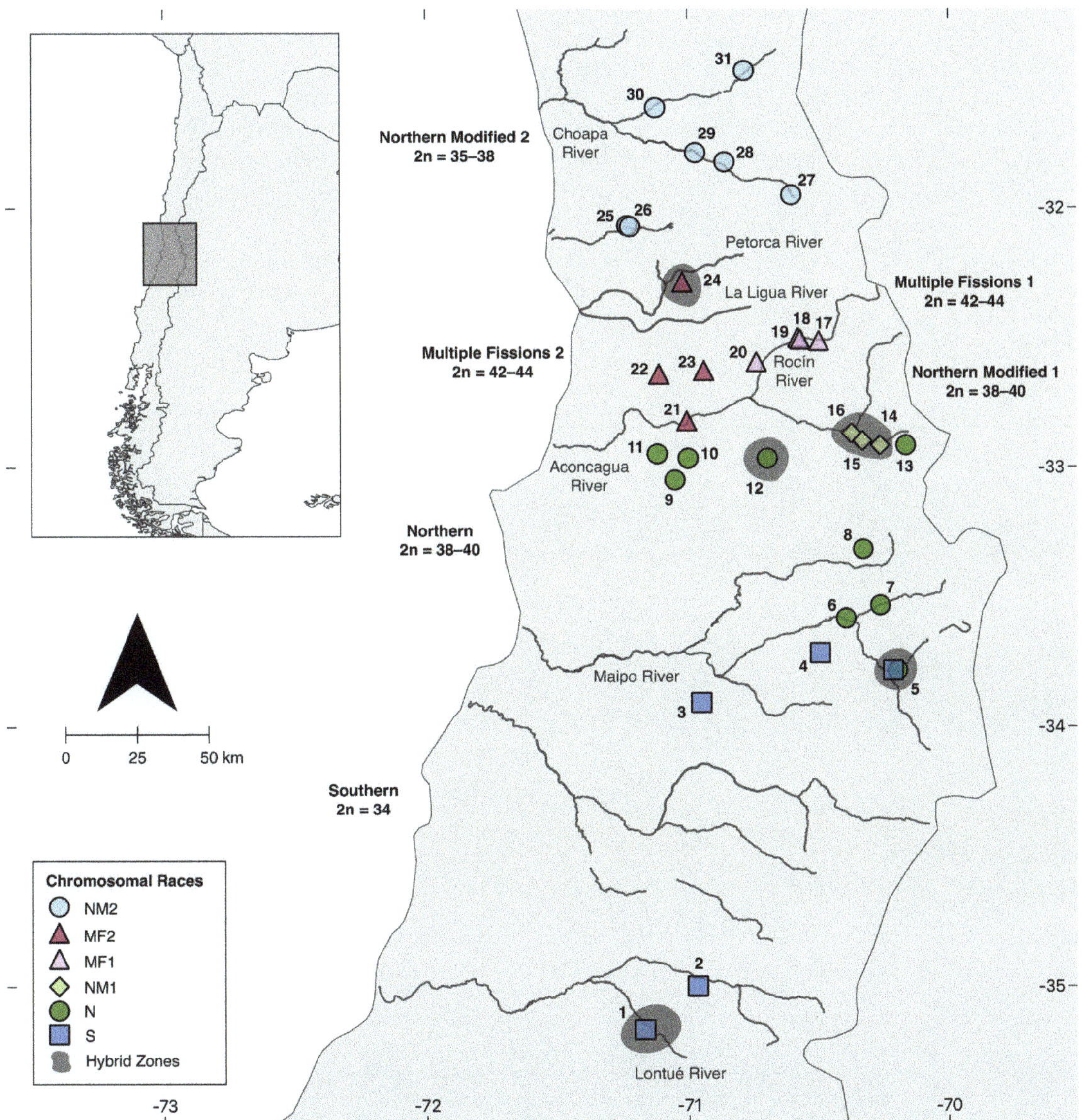

Figure 1. Distribution map of locality samples for six *L. monticola* chromosomal races. Detected potential hybridization zones are shaded. Data include new sampling as well as previously described chromosome races.

Here, we assessed the patterns of chromosomal and mitochondrial genetic variation in *L. monticola* from fifteen localities northward from the Aconcagua River, which corresponds to the northern geographic distribution of the complex (Figure 1). We complemented our analysis with previously published chromosome data and mitochondrial *cytb* sequences from several populations across the species' range. Our aim was to infer the underlying processes that explain the patterns of chromosome geographical distribution observed across the *L. monticola* chromosomal races.

2. Materials and Methods

2.1. Sampling and Data Collection

All *L. monticola* individuals (n = 251) were collected between the spring and fall from 1990 to 2003 (Collecting permit SAG Res: N° 688–3095). We collected lizards from fifteen sites north of the Aconcagua River (Figure 1, Table 1), including the previously published "Hierro Viejo" locality (24) [38], expanding the geographic sampling ~15,000 km^2. We also added representative localities from the "Southern", "Northern" [35,36], and "Northern Modified 1" [37] races. Lizards were euthanized using a 0.001 g/g urethane 1% injection in the pineal eye, and tissues were obtained for chromosome analysis and DNA extraction. Voucher specimens were deposited in the collection of the Evolutive Cytogenetic Laboratory, Facultad de Ciencias, Universidad de Chile (CUCH). Catalogue identification and sampling locality coordinates for each karyotyped lizard are listed in Table S1.

Table 1. Sampling locations for the different races of *L. monticola*. See Table S1 for a full list of sampled individuals.

Race	Location	Race	Location
Southern, 2n = 34	1. Lontué	Northern Modified 1, 2n = 38–40	16. Colorado River North
	2. Los Queñes	Multiple Fissions 1, 2n = 42–44	17. Rocín River
	3. Cantillana		18. Los Patos
	4. Clarillo River		19. Chalaco
	5. Yeso River South		20. Lo Vicuña
Northern, 2n = 38–40	5. Yeso River North	Multiple Fissions 2, 2n = 42–44	21. Cabrería
	6. El Manzano		22. El Soldado
	7. El Alfalfal		23. Mina Cerrillos
	8. Yerba Loca		24. Hierro Viejo
	9. El Roble	Northern Modified 2, 2n = 35–38	25. Culimó
	10. La Dormida		26. Tilama
	11. La Campana		27. Tranquilla
	12. Chacabuco		28. Chillepín
	13. Saladillo		29. Salamanca
Northern Modified 1, 2n = 38–40	14. Blanco River		30. Illapel
	15. Colorado River South		31. Santa Virginia

2.2. Cytogenetic Analyses

Chromosomes were obtained from bone marrow, liver, spleen, and testes using the colchicine-hypotonic pretreated air-drying technique and stained with Giemsa following Lamborot [35]. Selected metaphase plates from each specimen were photographed with a Leitz-Ortholux microscope. Several karyotypes were constructed from enlarged photographs, which were used to score the chromosomal morphology as "genotypes" for the first six macrochromosome complements and the microchromosome pair 7. Chromosome alleles were coded following Lamborot [43]: The ancestral non-fissioned chromosomes were coded "A", and the metacentric fission rearrangements were coded "B". Inversions of the ancestral bi-armed chromosomes were coded as "C", whereas inversions of the fission product in chromosome pair 2 were coded as "D". The enlarged chromosome, present in some individuals for pair 6, was coded as "E". Additionally, the novel fission products detected in pairs 3 and 4 from the northernmost locations were coded as "F" and "G", respectively. Additional observations of spermatocytes at diakinesis, chiasmata, and metaphases II were also made.

The coded genotypes were analyzed in R 4.2.1 [45]. Rogers' genetic distances [46] were calculated between representative sampling locations using the 'adegenet' package [47]. A UPGMA dendrogram was generated based on genetic distances, using the 'phangorn' package [48], to assess the relationship between the new sampled sites and previously described chromosomal races. Additionally, we used the genotyped chromosome alleles to determine the population cytogenetic structure with three independent runs of the STRUCTURE v2.3.4 software [49], with a burn-in of 10,000 plus 100,000 MCMC iterations, an admixture ancestry model, and correlated allele frequencies. The most likely number of genetic clusters was estimated with Evanno's method using the STRUCTURE harvester tool [50]. The UPGMA dendrogram and STRUCTURE suggested the existence of six chromosomal groups within the studied samples. Therefore, further analyses were conducted on these six *L. monticola* chromosomal races. Allele frequencies for each chromosomal pair were calculated for all races using the 'hierfstat' package [51]. The allelic richness, expected and observed heterozygosity, and inbreeding coefficient (F_{IS}) were also calculated for each race using 'hierfstat'. A Chi-square test was used to analyze the heterozygote frequencies of each chromosomal pair expected under Hardy–Weinberg equilibrium (HWE) using the 'pegas' package [52].

2.3. *Cytochrome b Sequencing and Analysis*

The genomic DNA was obtained from the tissues of 59 individuals across the *L. monticola* distribution (Table S1) following the salt extraction method by Aljanabi & Martínez [53]. Amplification of an 800 bp fragment of the mitochondrial *cytb* gene was performed using GLUDG and CB3 primers [54] under the PCR conditions described by Torres-Pérez et al. [39]. We further added 35 previously published *cytb* sequences from the S, N, NM1, and MF races (Table S1) [39], making a total of 95 sequences representing six chromosomal races. Sequences were visually curated and aligned using the CodonCode Aligner 10.0 (CodonCode Corporation, Dedham, MA, USA).

To assess the phylogenetic relationships between chromosomal races, we performed a Maximum Likelihood phylogenetic inference using RaxML v8 [55] in the CIPRES Science Gateway [56]. A sequence of *L. fuscus* was added as an outgroup. The node support values were obtained from 1000 bootstrap replicates. The trees were visualized using Fig Tree v.1.4.2 [57]. To further explore the mitochondrial relationships between the races, a neighbor-joining haplotype network was constructed using PopART v3 [58]. The F_{ST} index between chromosomal races (based on *cytb* data) was calculated in R using 'hierfstat'. Additionally, we calculated the number of different haplotypes, the number of segregating sites, haplotype diversity, and nucleotide diversity using the 'pegas' package in R. Finally, deviation from neutrality was assessed with a Tajima's D test for each race using 'pegas'.

3. Results

3.1. *Patterns of Chromosomal Variation*

3.1.1. Andean Range between the Aconcagua River and the Petorca River: The Multiple Fission 1 (MF1) Race, 2n = 42–44

The lizards from four localities (17–20) revealed 39 unique karyotypes (from 54 samples), with a 2n = 42 to 44, 18–20 macrochromosomes, and 24 microchromosomes (Figure 2). Lizards from this area presented polymorphisms for centric fissions on pairs 1 and 2 (Figure 3a,b). Pair 2 fission was highly polymorphic: some individuals presented the NOR at the tip of the long arm of the fission product (genotypes BB and AB), whereas other individuals presented a pericentric inversion in the fission product with the NOR on the tip of the short arm (genotypes AD, BD, and DD). Pair 3 retained the fission polymorphism present in the other races, whereas pair 4 was homozygous for the fission products (Figure 3c,d). The submetacentric pair 5 was homozygous, whereas the enlarged pair 6 and the pericentric inversion of microchromosome pair 7 were polymorphic (Figure 3f,g). In the meiotic diakinesis plates, all the bivalent pairs presented one or two terminal chiasmata, and the trivalent(s) exhibited end-to-end pairing chromosomes (Figure S1).

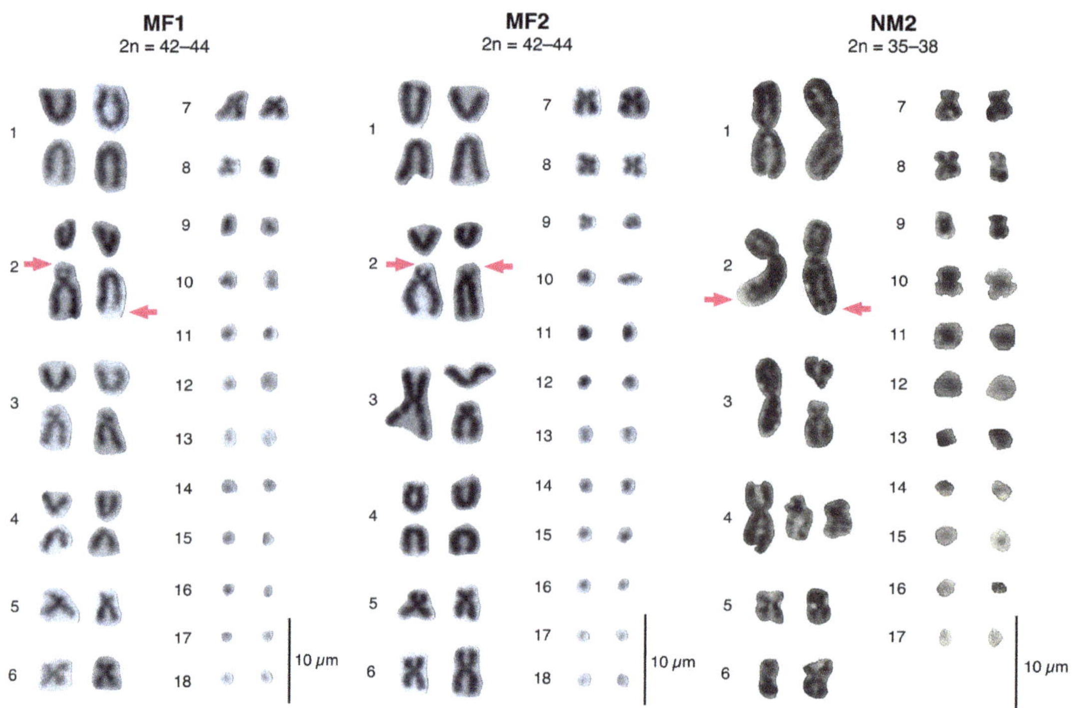

Figure 2. Representative karyotypes for the newly described races of *L. monticola*. Arrow indicates the NOR position on chromosome pair 2. Individuals whose chromosomes are shown are: L2180 (MF1 race), L1362 (MF2 race), and L730 (NM2 race).

3.1.2. Coastal Range between the Aconcagua River and the Petorca River: The Multiple Fissions 2 (MF2) Race, 2n = 42–44

The lizards from localities 21–23 and the previously published "Hierro Viejo" (24) exhibited 25 unique karyotypes from 57 individuals, with a 2n = 42 to 44, 18–20 macrochromosomes, and 24 microchromosomes (Figure 2). Similar to the MF1 race, pair 1 showed a centric fission (Figure 3a), which was only polymorphic in the "Hierro Viejo" locality. The pair 2 fission presented the inversion product with the NOR at the tip of the short arm (DD, Figure 3b), and the AD heterozygotes were only present in the "Hierro Viejo". Pair 3 retained the polymorphism present in the other races (Figure 3c). All lizards were homozygous for the fission of pair 4 and the submetacentric pair 5 (Figure 3d). Pair 6 was polymorphic for an enlarged chromosome, and pair 7 was polymorphic for a pericentric inversion (Figure 3f,g). All bivalent pairs showed one or two terminal chiasmata on the meiotic diakinesis plates, and the trivalent(s) showed end-to-end pairing chromosomes (Figure S1).

3.1.3. Northward from the Petorca River: The Northern Modified 2 Race, 2n = 35–38

In the northernmost locations (25–31), we detected two unique karyotypes from six individuals with a 2n = 35 to 38, 13–16 macrochromosomes, and 22 microchromosomes (Figure 2). The metacentric pairs 1 and 2 were monomorphic (Figure 3a,b). A novel polymorphism was detected for pair 3, in which one of the fission products was acrocentric and the other was submetacentric (Figure 3c). The populations also presented a new polymorphism for pair 4, consisting of two new submetacentric fission products (Figure 3d). All populations were monomorphic for the submetacentric pair 5. The enlarged pair 6 retained the polymorphism, and the metacentric pair 7 was monomorphic (Figure 3f,g). In the diakinesis arrays, pairs 1 and 2 had similar sizes and presented two to three chiasmata per bivalent. The fission pairs 3 and/or 4 of polymorphic lizards showed one or two linear trivalent(s) with two terminal chiasmata each. The bivalents for pairs 5, 6, and 7 presented two terminal chiasmata each (Figure S1).

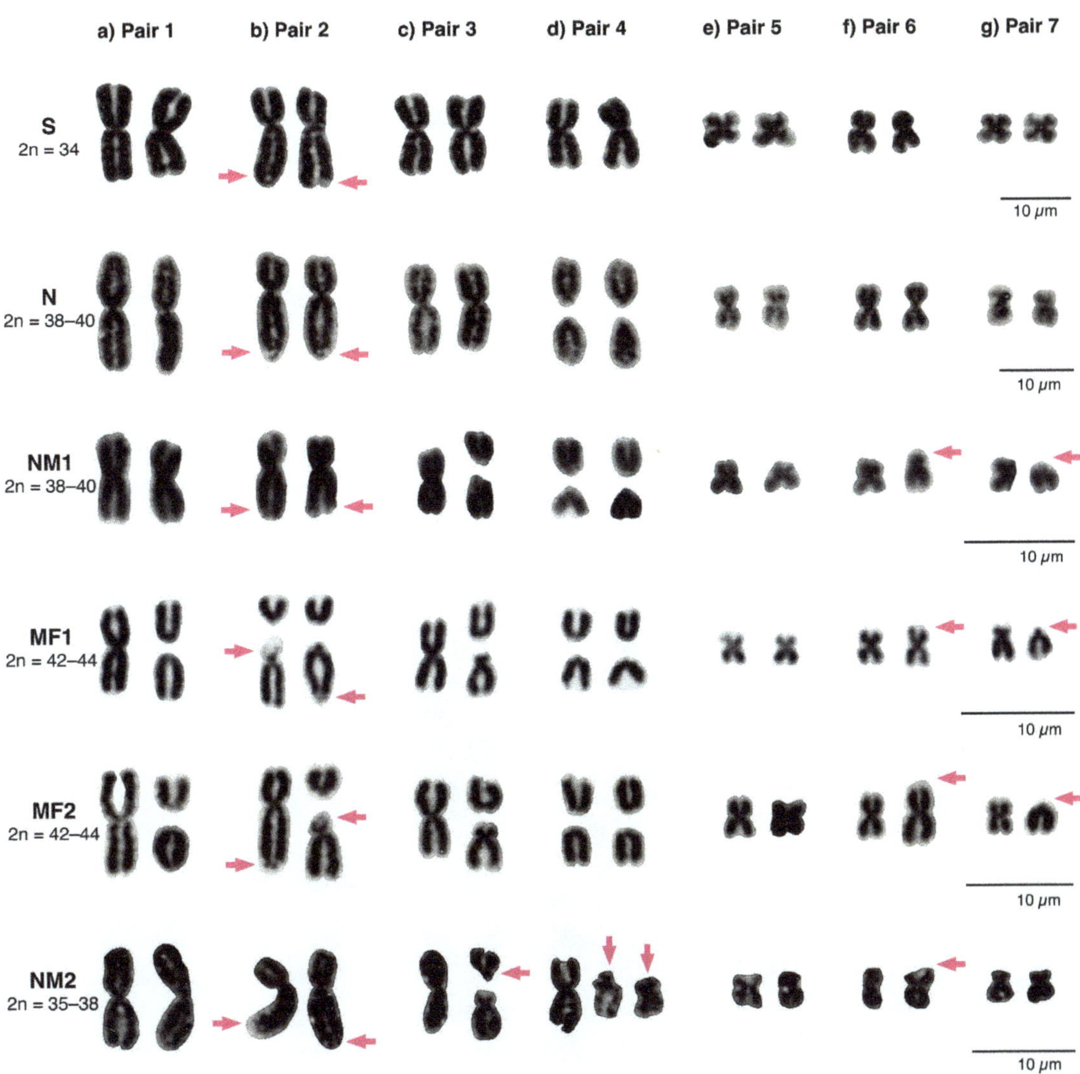

Figure 3. Representation of partial karyotypes for the six macrochromosome pairs (pair 1 to 6) and microchromosome pair 7 of the six chromosome races of *L. monticola*. (**a**) Metacentric pair 1. For the following cases, the arrow depicts; (**b**) the NOR position in pair 2; (**c**) the subacrocentric fission product in pair 3; (**d**) the submetacentric chromosomes in pair 4; (**e**) monomorphic for the submetacentric pair 5; (**f**) the enlarged pair 6; and (**g**) the pericentric inversion in pair 7. Individuals whose chromosomes are shown are: L1085 (S race), L547 (N race), L2651 (NM1 race), L2180 (MF1 race), L1362 (MF2 race), and L730 (NM2 race).

3.2. Population Cytogenetics

The UPGMA based on Rogers' genetic distance differentiated two main groups within *L. monticola* (Figure 4a). In the first group, the northernmost localities of the distribution (NM2) clustered with the Southern race, and they were associated with a second cluster containing the N and NM1 races. A second group included locations from the Multiple Fissions 1 and 2 races and was divided into two clusters corresponding to the Andean (MF1) and Coastal (MF2) ranges, respectively. The structure analysis was consistent with this result, which identified the most likely number of resulting genetic groups as K = 4 (Figure 4b). The first group consisted of locations from the MF1 and MF2 races, the second group included the Southern race plus individuals from the NM2 race, and the other two groups clustered individuals from the N and NM1 races, respectively. Therefore, our results identified six differentiated chromosomal races within the analyzed locations that were distinguished by their geographic distribution, chromosome rearrangements, and genetic-chromosomal population variability. Further analyses were performed based on these six chromosomal races.

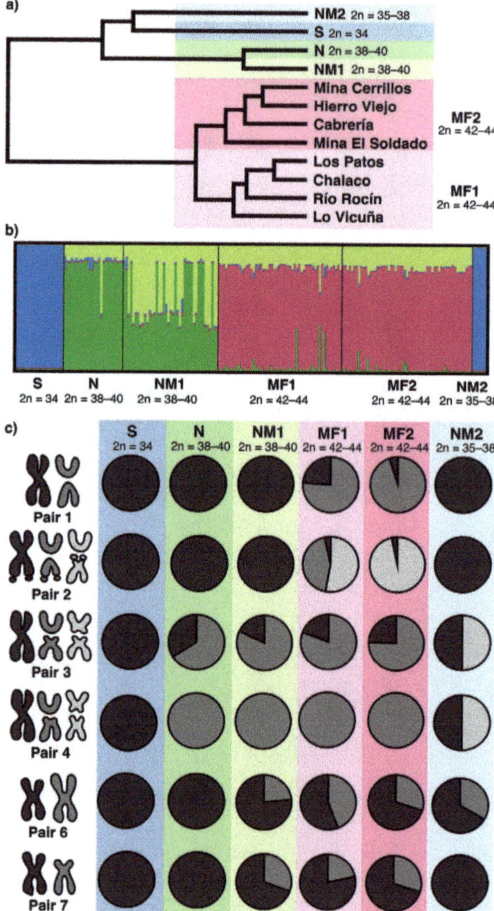

Figure 4. Population cytogenetic analysis of *L. monticola*. (**a**) UPGMA dendrogram based on Rogers' genetic distances between localities, showing the six chromosomal races organized geographically from left to right. (**b**) Structure plot based on chromosome alleles for K = 4. (**c**) Allele frequencies for each CR of the first seven chromosome pairs in all races.

Allele frequencies were variable between the populations (Figure 4c). All individuals from the Southern race, 2n = 34, were fixed for the ancestral non-fissioned chromosomes in all pairs. The Northern race, 2n = 38–40, was characterized by a high frequency of the fission polymorphism on pair 3, and the fixation of the pair 4 fission. In the NM1 race, 2n = 38–40, an enlarged pair 6 and a pericentric inversion in pair 7 were characteristic of new rearrangements. The Multiple Fission races were characterized by high-frequency polymorphisms on the fission products of pairs 1 and 2. The main difference between both MF races was that those in the Andean locations (MF1, 2n = 42–44) contained the fission product of pair 2 with and without the inversion (52.77% and 43.52% of frequency, respectively), whereas the inversion of the fission product was almost fixed (96.49% of allele frequency) in the coastal locations (MF2, 2n = 42–44). Finally, the NM2 race, 2n = 35–38, presented a high frequency of the novel rearrangements for pairs 3 and 4, while also sharing the polymorphism for an enlarged pair 6.

The highest allelic richness values were found in both the MF races, whereas the lowest allelic richness values were found in the S race (Table 2). Heterozygosity values were higher in the MF1 and NM2 populations and the lowest in the S race (Table 2). The inbreeding coefficient was significant and positive for the N and NM1 races, whereas it was significant and negative for the NM2 race. We detected significant deviations from the HWE on pairs 3, 4, 6, and 7 (Table 3). Pair 6 had heterozygote deficiency for the NM1 and MF2 races, and pair 7 exhibited heterozygote deficiency in the MF1 and MF2 races. The NM2 race showed heterozygote excess for pairs 3 and 4. The F_{ST} values between chromosomal races were consistent with the UPGMA. The highest F_{ST} values were detected between the S and the rest of the races, whereas the lowest values were detected between the N and NM1 races and between the MF1 and MF2 races (Table 4).

Table 2. Chromosome variability parameters in six populations of *L. monticola* (S: Southern; N: Northern; NM1: Northern Modified 1; MF1: Multiple Fissions 1; MF2: Multiple Fissions 2; NM2: Northern Modified 2). Allele richness (Ar), observed (Ho) and expected (He) heterozygosity, and inbreeding coefficient (F_{IS}), * $p < 0.05$.

	S	N	NM1	MF1	MF2	NM2
Ar	1	1.142	1.412	1.749	1.544	1.429
Ho	0	0.060	0.115	0.273	0.155	0.381
He	0	0.064	0.157	0.294	0.199	0.210
F_{IS}	-	0.058 *	0.273 *	0.078	0.126	−0.810 *

Table 3. Chi squared test *p*-values for Hardy–Weinberg equilibrium for the first seven chromosome pairs from each *L. monticola* chromosomal race (S: Southern; N: Northern; NM1: Northern Modified 1; MF1: Multiple Fissions 1; MF2: Multiple Fissions 2; NM2: Northern Modified 2). * $p < 0.05$.

	S	N	NM1	MF1	MF2	NM2
Pair 1	1	1	1	0.113	0.675	1
Pair 2	1	1	1	0.838	0.784	1
Pair 3	1	0.844	0.089	0.971	0.081	0.014 *
Pair 4	1	1	1	1	1	0.014 *
Pair 5	1	1	1	1	1	1
Pair 6	1	1	0.014 *	0.542	0.002 *	0.221
Pair 7	1	1	0.381	0.001 *	<0.001 *	1

3.3. *Mitochondrial cytb Sequence Analyses*

Thirty-seven unique haplotypes were detected among the 94 samples of representative *L. monticola* lizards. In both the neighbor-joining network and the maximum likelihood phylogeny, individuals clustered according to their chromosomal race and geographic origin (Figure 5). Nonetheless, the phylogenetic inference had low support values in general. Populations from the S race presented the highest divergence from all the other races in the phylogeny and were separated from the other races by 33 mutational steps in the haplotype network. The NM2 race and the "Hierro Viejo" locality (MF2) formed a supported monophyletic clade in the phylogenetic tree and clustered together in the haplotype network. However, the MF1, MF2, and NM1 races were monophyletically reciprocal to the NM2 and "Hierro Viejo" (Figure 5b). Some lizards presented a discordant position with respect to their chromosomal race in the phylogeny and the haplotype network. For example, the lizards from "Hierro Viejo" (MF2, site 24) clustered with the "Culimó" (NM2, site 25), the samples from "Saladillo" (N, site 13) were grouped with the races MF1 and MF2, and individuals from the NM1 race clustered within different races (Figure 5). Genetic distances (pairwise F_{ST} calculated from *cytb* data) between the races were consistent with geographic distance. F_{ST} values were lower between the NM1, MF1, and MF2 races and higher between the S race and the rest of the races (Table 4). The number of segregating sites was higher in the S and N races, the number of haplotypes and haplotype diversity was higher in the N race, and the nucleotide diversity was higher in the NM1 race, whereas the lowest values were found in the MF1 race. All chromosomal races showed negative values for the Tajima's D statistic, but the MF1 race was the only race that showed significant deviations from neutrality (Table 5).

Table 4. Differentiation between the six *Liolaemus monticola* chromosomal races (S: Southern; N: Northern; NM1: Northern Modified 1; MF1: Multiple Fissions 1; MF2: Multiple Fissions 2; NM2: Northern Modified 2). Above the diagonal, F_{ST} values obtained from chromosome "alleles"; below the diagonal *cytb* sequences data *: $p < 0.05$.

	S	N	NM1	MF1	MF2	NM2
S	-	0.853 *	0.712 *	0.680 *	0.777 *	0.653 *
N	0.332 *	-	0.151 *	0.497 *	0.645 *	0.652 *
NM1	0.355 *	0.036 *	-	0.443 *	0.590	0.530
MF1	0.422 *	0.176 *	0.073 *	-	0.123 *	0.552 *
MF2	0.351 *	0.078 *	0.028 *	0.159 *	-	0.684 *
NM2	0.349 *	0.228 *	0.194 *	0.380 *	0.167 *	-

Table 5. Genetic variability of *cytb* in the *Liolaemus monticola* chromosome race samples (S: Southern; N: Northern; NM1: Northern Modified 1; MF1: Multiple Fissions 1; MF2: Multiple Fissions 2; NM2: Northern Modified 2). Sample size (n), number of haplotypes (K), number of segregating sites (S), haplotype diversity (H), nucleotide diversity (π), and Tajima's D statistic for each *L. monticola* chromosomal race. *: $p < 0.05$.

	S	**N**	**NM1**	**MF1**	**MF2**	**NM2**
n	14	20	7	16	19	19
K	8	16	6	7	10	13
S	91	68	56	12	52	38
H	0.824	0.979	0.952	0.867	0.895	0.947
π	0.025	0.022	0.028	0.003	0.026	0.012
D	−1.949	−1.071	−1.086	−2.403 *	−0.731	−0.837

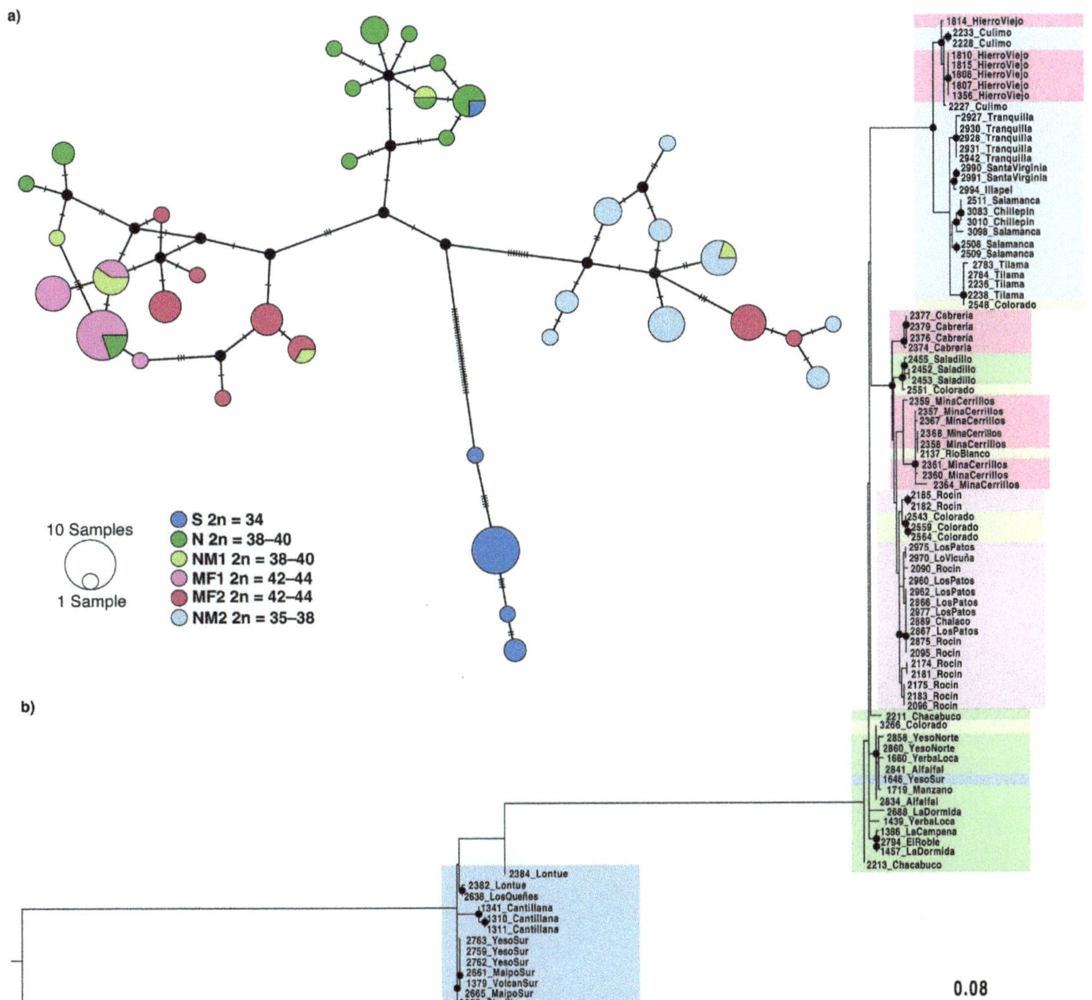

Figure 5. (**a**) Neighbor-joining haplotype network generated from the mitochondrial *cytb* gene with new and previously obtained sequences of *L. monticola*. Each line indicates a mutational step. (**b**) Maximum likelihood phylogenetic reconstruction for the *cytb* gene. Nodes with black filled circles indicate >70% bootstrap support. Note the presence of individuals with discordant positions in both analysis.

4. Discussion

This study is the first to analyze and compare individuals across nearly the whole distribution of *L. monticola*. We detected a high intraspecific CV that can be attributed to the concatenation of CR. In addition, we found long-lasting chromosome polymorphisms that can represent the "ghost of hybridization" events or that may be generated after different waves of colonization and recolonization in populations where de novo mutations have occurred. The addition of several samples from the northern geographic distribution increased the total number of chromosomal races to seven (the six analyzed here plus the not included primitive race), which enabled the recognition of two new chromosome races (MF1 and NM2) and the expansion of the previously described "MF" race (now MF2, [38]) from one to various localities. The existence of these races is supported by chromosome and *cytb* gene data and is consistent with previously published cytogenetic, morphological, allozyme, and mitochondrial evidence [35–44].

4.1. Patterns of Chromosomal Variation

4.1.1. The Multiple Fissions Races; MF1 and MF2

The chromosomal data supported the presence of two chromosome races north of the Aconcagua River up to the previously described "Hierro Viejo" locality (17–24) including the MF1, from the Andean Range (17–20), and the MF2, from the Coastal Range (21–24). The MF1 race is geographically and chromosomally intermediate between the NM1 and MF2 races. It retains all chromosome characteristics of the NM1, such as the fixation of the pair 4 fission, polymorphisms for a pair 3 fission, an enlarged pair 6 and a pericentric inversion in chromosome 7, and the same microchromosome number. However, the MF1 race adds three novel chromosomal rearrangements including polymorphic fissions in pairs 1 and 2 and a pericentric inversion in one of the fission products of pair 2 that changes the NOR position (Figures 2 and 3b). On the other hand, the MF2 race at the Coastal Range is considered the most derived of the *L. monticola* complex. It presents the same CR as the MF1 race, but it fixates the fissioned pair 1 (BB) and the fissioned pair 2 with the pericentric inversion (DD; Figure 2 and 3b) in all localities except for the unique "Hierro Viejo" population (see below).

The rearrangements of pair 2 associated with the NOR position constitute a very interesting feature of the Multiple Fissions races. All other races present chromosome 2 as submetacentric (AA) with the NOR at the tip of the long arm. The existence of different chromosome combinations (particularly AB, AD, and BD heterozygotes in the MF1 race), suggests that both chromosomal mutations were sequentially independent events. In this scenario, the pericentric inversion could have followed the centric fission (Figure S2), as described by Kolnicki [59], according to Todd's karyotypic fission theory, and it could be fixated by meiotic selection stabilizing the karyotype (e.g., mammals [60]). In *L. monticola*, we detected intermediate stages that show the transition from the ancestral pair 2 in the MF1 race to the derived form in the MF2 race. The detection of sequential chromosome mutations is usually not possible, as intermediate CR may remain in low frequency or be lost in natural populations. Porter and Sites [61] and Goyenechea et al. [62] reported similar variability patterns in pair 2 associated with the NOR position in *Sceloporus grammicus*. The *L. monticola* chromosome 2 evolution described here highlights the evolutionary importance of pericentric inversions to stabilize other chromosome rearrangements.

4.1.2. The Northern Modified 2 Race

Lizards from the northernmost distribution of the *L. monticola* complex (25–31) were assigned to a new NM2 race, 2n = 35–38, which included two novel fissions in pairs 3 and 4. Pair 3 was polymorphic in most lizards with a metacentric chromosome and two fission products, one acrocentric and the other submetacentric. This rearrangement is different from the acrocentric and subacrocentric fission products described for pair 3 in the other races (N, NM1, MF1, and MF2) [35–38]. Pair 4 also presented a unique polymorphism, including a metacentric chromosome and two submetacentric fission products, instead of the common acrocentric fission products detected in pair 4 for the other races (except for the polymorphism found in hybrids between the Southern, 2n = 34, and Northern, 2n = 38–40, races [36,44]). Interestingly, this race shared a similar karyotype with the Southern race (Figures 2 and 3) including a metacentric pair 1, a submetacentric pair 2, a metacentric pair 7, and 22 microcromosomes. This observation contrasts with the expected pattern of increased karyotype complexity from south to north. Furthermore, bivalents 1 and 2 in the NM2 race presented two to three chiasmata (Figure S1). The other races (N, NM1, MF1, and MF2) only present terminal chiasmata, except for the Southern race which shows several chiasmata in bivalents 1 and 2, as observed in the NM2 race [35,36].

4.2. Chromosome Polymorphism

We reported strikingly high chromosomal polymorphism levels across the *L. monticola* races (Figure 4, Table 2), which are comparable to the high CV and multiple chromosome races described in *Sceloporus* [61]. The fact that several polymorphisms have persisted in *L. monticola* in high frequency and in conformance with the HWE (Figure 4c, Table 3) suggests that these rearrangements may not be selectively disadvantageous for centric fission heterozygotes compared to homozygotes. For instance, Lamborot and Alvarez Sarret [36] demonstrated that the polymorphism in pair 3 from the Northern race does not appear to undergo abnormal meiotic segregation. In previous reports, the degree of polymorphism of pair 3 varies depending on the genetic background and its geographical origin [34,35,37,42]. In addition, the number of aneuploidies observed in metaphase II may be normal at the Andes Range (less than 5%), 10–23% at the Coast Range, and 26–32% for chromosome hybrids from the hybrid zone [36].

4.3. Riverine Barriers and Gene Flow

The species formation process requires the disruption of gene flow by geographic isolation of a panmictic population into two or more populations, thus allowing for the accrual of mutations. Therefore, limited gene flow can differentiate populations and originate races within a species. Both chromosome and mitochondrial *cytb* data supported the existence of differentiated races within *L. monticola*. These populations and races are spatially fragmented and could be considered a metapopulation system [63]. Interestingly, the geographic limits of chromosome races match the presence of rivers throughout Chile. The relevance of riverine barriers for gene flow restriction has been proposed for various taxa (e.g., [64]). Rivers in this area seem to be major barriers to gene flow. Indeed, the Maipo River and the Aconcagua River are proposed barriers to gene flow for *L. monticola* and other reptile species [33,41,42,65,66]. Our analysis showed deep mitochondrial *cytb* divergence between the Southern race and the other chromosome races (Figure 5, Table 4), as reported by Torres-Pérez et al. [39,67]. This result suggests that populations of *L. monticola* may have been geographically isolated for a long period of time, which could even correspond to incipient speciation.

Geological data for the middle Chilean Andes demonstrate that Pleistocene glaciations were extensive and that rivers could have been larger during past glaciation/deglaciation cycles [68–70]. In the last glaciation episodes, the glaciers on the transversal valleys (such as the Maipo Valley, Aconcagua Valley, and Petorca Valley) were particularly well developed. These glacial tongues in central Chile could have interrupted gene flow between chromosomal races prior to the development of rivers. Heusser [71] indicated that the Coastal Range was in general not influenced by glaciers, therefore populations may have found refuge in coastal locations and then recolonized the Andean locations. Our *cytb* results support this possibility by showing signatures of population expansion for all races (Table 5), consistent with Torres-Perez et al. [39]. Therefore, we propose that *L. monticola* had a complex evolutionary history, with rivers (such as the Petorca River, La Ligua River, Aconcagua River, and Maipo River) acting as barriers to gene flow and the Pleistocene glaciation cycles affecting the populations' evolutionary dynamics.

4.4. Hybridization between Chromosomal Races

Discordances were detected in our analyses, as some individuals showed *cytb* sequences related to other chromosome races (Figure 5). This pattern could be associated with introgression, retention of ancestral polymorphisms, or incomplete lineage sorting [65,72]. Parapatry in narrow secondary contact zones and hybridization have previously been proposed for this complex between the chromosomal races: P × S, S × N, N × NM1 (Figure 1). For example, localities nearby the Yeso River (5) constitute a hybrid zone between the Southern and Northern races [35]. The NM1 race, located between two tributaries of the Aconcagua River (Colorado River and the Juncal River, 14–16), was proposed to have a primary hybrid origin [37,40], and the "Chacabuco" (12) was described as a potential hybrid zone between the Coastal and Andean populations within the N race [43]. Thus, it is possible that individuals from different races can produce viable offspring, especially given some potential mechanisms that could stabilize meiosis and maintain stable recombination rates in hybrids [35].

The previously described "Hierro Viejo" population (24) [38], the northernmost population of the MF2 race, shared similar mitochondrial *cytb* sequences with individuals from the "Culimó" (25, NM2). This location can be considered unique among the other MF2 populations because it was the only population that presented AB heterozygotes (5 out of 29 lizards) for the pair 1 fission and AD heterozygotes (4 out of 29 lizards) for the fission and inversion products of pair 2 (both CR are fixed in the other MF2 populations; Figure 4c, Table S1). These polymorphisms could have been retained when diverging from the MF1 race. Alternatively, this karyotype and the discordant mitochondrial relationships for this population could be indicative of introgression between the MF2 and NM2 races, which at this location are only ~30 km apart but are separated by the Petorca River. One less plausible hypothesis is that these CR correspond to de novo mutations.

4.5. Model of Chromosome Evolution

We propose a sequence of events describing the chromosomal rearrangements occurring through the evolution of *L. monticola* races based on our results. Initially, individuals with a karyotype similar to the Southern race, 2n = 34, could have distributed throughout the entire geographic range of the complex. A first colonization event could have originated the Northern race, 2n = 38–40, with its characteristic rearrangements (homozygotic centric fission of pair 4 and centric fission polymorphism in pair 3), expanding northward from the Maipo River [35,36]. The Northern Modified 1 race, 2n = 38–40, probably originated from the Andean range of the N race, with a polymorphism for an enlarged chromosome 6 and a polymorphic pericentric inversion in chromosome 7 [37]. Then, the NM1 race gave rise to the Multiple Fissions 1 race, 2n = 42–44, with its new polymorphisms for fissions in pairs 1 and 2. Subsequently, the inversion product of the pair 2 was fixed (DD homozygotes) in the MF2 race stabilizing the inversion–noninversion heterozygote, as described above. To explain the unique rearrangements in the NM2 race, 2n = 35–38, and its similarity to the S race, we propose an independent process of colonization in which the ancient "Southern like" populations underwent centric fissions at the northernmost range of the complex distribution. Chromosome evidence supports the hypothesis of a colonization initiated by the Southern race that originated the other chromosomally derived races, which continued with processes of colonization, hybridization, and replacement [23]. This is consistent with morphological [33,41,42], alloenzyme [40], and mitochondrial genetic data [39,67].

The sequential chromosome rearrangements that originated all races, and the latitudinally arranged karyotypic variation (corroborated by the increased complexity from south to north, except for the NM2 race), resemble Hall's "cascade model of speciation" [14], the "chain process" [2], and the "primary chromosomal allopatry" [4] hypotheses, among others. Similar processes have been reported for chromosomal speciation of *Sceloporus* lizards [18,19], *Sorex* shrews [73], and *Ctenomys* rodents [74]. Geological complexity such as rivers, transversal mountain ranges, and latitudinal climatic gradients, may have also played an important role in restricting gene flow and promoting race differentiation. Moreover, when considering other *Liolaemus* species, there is a correlation between the chromosome number and environmental gradients where the number of centric fissions increases towards more arid and heterogeneous environments in northern Chile [21,32]. This highlights the relevance of understanding the adaptive potential of the Robertsonian-centric fissions for colonizing more challenging environments.

5. Conclusions

Our results present a unique opportunity to investigate incipient in situ evolution, recognizing the chromosome rearrangements that account for the different races found in *L. monticola* throughout its distribution. Further studies would help to unravel the mechanisms associated with the origin and fixation of these chromosome mutations in this complex, as well as their implications on fitness and population differentiation. Here, we have highlighted the importance of chromosomal rearrangements for the evolution and potential speciation in highly radiated groups such as *Liolaemus* lizards.

Supplementary Materials: The following supporting information can be downloaded at: https://www.mdpi.com/article/10.3390/ani12233372/s1, Table S1. Full data for the sampled individuals. Figure S1. Representative meiotic karyotype for the first seven chromosome pairs for lizards from the newly described chromosomal races of *Liolaemus monticola*. Figure S2. Model of evolution for the chromosome pair 2 in *Liolaemus monticola* in the Multiple Fission races.

Author Contributions: Conceptualization, M.L.; formal analysis R.A.-D., D.V. and C.G.O.; data curation N.A.-M. and M.L.; writing-original draft preparation, M.L., C.G.O., N.A.-M. and R.A.-D.; writing-review and editing M.L., D.V., N.A.-M. and R.A.-D.; funding acquisition M.L. All authors have read and agreed to the published version of the manuscript.

Funding: M.L. acknowledges grants 1030776 (FONDECYT) and DID ENL 07/09 (Universidad de Chile) from 2009. R.A.-D. was supported by the Doctoral scholarship 72200094 (ANID, Chile). D.V. acknowledges grant 1200589 (FONDECYT). C.G.O. acknowledges grant 11190305 (FONDECYT).

Institutional Review Board Statement: The animal study protocol was approved by the Institutional Committee for the Care and Use of Animals (CICUA) of the Universidad de Chile (protocol N° 2232-FCS-UCH, 11/25/2022).

Informed Consent Statement: Not applicable.

Data Availability Statement: Data used for this study and Genebank accession numbers are available in the Supplementary Material.

Acknowledgments: The authors thank S. Brito for her valuable field and laboratory assistance. We thank P. Astete and E. Paez for their contributions to this study, and M. Liempi for assistance with the manuscript. We thank D. Benson and S. Baty for their comments on the manuscript. The authors thank "Servicio Agrícola y Ganadero, Gobierno de Chile" (SAG) for collecting permits (N° 688-3095).

Conflicts of Interest: The authors declare no conflict of interest.

References

1. Bush, G.L.; Case, S.M.; Wilson, A.C.; Patton, J.L. Rapid speciation and chromosomal evolution in mammals. *Proc. Natl. Acad. Sci. USA* **1977**, *74*, 3942–3946. [CrossRef]
2. White, M.J.D. Chain processes in chromosomal speciation. *Syst. Zool.* **1978**, *27*, 285–298. [CrossRef]
3. King, M. Karyotypic evolution in *Gehyra* (Gekkonidae: Reptilia) IV. Chromosome change and speciation. *Genetica* **1984**, *64*, 101–114. [CrossRef]
4. King, M. *Species Evolution: The Role of Chromosome Change*; Cambridge University Press: New York, NY, USA, 1995; pp. X–Y.
5. Hall, W.P. Comparative population cytogenetics, speciation and evolution of the crevice-using species of *Sceloporus* (Sauria: Iguanidae). Ph.D. Thesis, Harvard University, Cambridge, MA, USA, 1973.
6. Paull, D.; Williams, E.E.; Hall, W.P. Lizard karyotypes from Galápagos Islands: Chromosomes in phylogeny and evolution. *Breviora* **1976**, *441*, 1–31.
7. Capanna, E. Robertsonian numerical variation in animal speciation: *Mus musculus*, an emblematic model. *Prog. Clin. Biol. Res.* **1982**, *96*, 155–177. [PubMed]
8. Sites, J.W., Jr.; Moritz, C. Chromosomal evolution and speciation revisited. *Syst. Zool.* **1987**, *36*, 153–174. [CrossRef]
9. Britton-Davidian, J. How do chromosomal changes fit in? *J. Evol. Biol.* **2001**, *14*, 872–873. [CrossRef]
10. Olmo, E. Rate of chromosome changes and speciation in reptiles. *Genetica* **2005**, *125*, 185–203. [CrossRef] [PubMed]
11. Ayala, F.J.; Coluzzi, M. Chromosome speciation: Humans, *Drosophila*, and mosquitoes. *Proc. Natl. Acad. Sci USA* **2005**, *102*, 6535–6542. [CrossRef] [PubMed]
12. Coyne, J.A.; Orr, H.A. The evolutionary genetics of speciation. *Philos. Trans. R. Soc. Lond. B Biol. Sci.* **1998**, *35*, 287–305. [CrossRef] [PubMed]
13. Rieseberg, L. Chromosomal rearrangements and speciation. *Trends Ecol. Evol.* **2001**, *16*, 351–358. [CrossRef] [PubMed]
14. Hall, W.P. Chromosome variation, genomics, speciation and evolution in *Sceloporus* lizards. *Cytogenet. Genome Res.* **2009**, *127*, 143–165. [CrossRef] [PubMed]
15. Castiglia, R.; Flores-Villela, O.; Bezerra, A.M.R.; Muñoz, A.; Gornung, E. Pattern of chromosomal changes in 'beta' *Anolis* (*Norops* group) (Squamata: Polychrotidae) depicted by an ancestral state analysis. *Zool. Stud.* **2013**, *52*, 60. [CrossRef]

16. Gamble, T.; Geneva, A.J.; Glor, R.E.; Zarkower, D. *Anolis* sex chromosomes are derived from a single ancestral pair. *Evolution* **2014**, *68*, 1027–1041. [CrossRef] [PubMed]
17. Giovannotti, M.; Trifonov, V.A.; Paoletti, A.; Kichigin, I.G.; O'Brien, P.C.M.; Kasai, F.; Giovagnoli, G.; Ng, B.L.; Ruggeri, P.; Nisi Cerioni, P.; et al. New insights into sex chromosome evolution in anole lizards (Reptilia, Dactyloidae). *Chromosoma* **2017**, *126*, 245–260. [CrossRef]
18. Leaché, A.D.; Sites, J.W., Jr. Chromosome evolution and diversification in North American spiny lizards (genus *Sceloporus*). *Cytogenet. Genome Res.* **2009**, *127*, 166–181. [CrossRef] [PubMed]
19. Leaché, A.D.; Banbury, B.L.; Linkem, C.W.; de Oca, A.N.M. Phylogenomics of a rapid radiation: Is chromosomal evolution linked to increased diversification in north american spiny lizards (Genus *Sceloporus*)? *BMC Evol. Biol.* **2016**, *16*, 1–16. [CrossRef]
20. Lamborot, M.; Espinoza, A.; Alvarez, E. Karyotypic variation in Chilean lizards of the genus *Liolaemus* (Iguanidae). *Experientia* **1979**, *35*, 593–595. [CrossRef]
21. Lamborot, M.; Alvarez-Sarret, E. Karyotypic characterization of some *Liolaemus* lizards in Chile (Iguanidae). *Genome* **1989**, *32*, 393–403. [CrossRef]
22. Aiassa, D.; Martori, R.; Gorla, N. Citogenética de los lagartos del género *Liolaemus* (Iguania: Liolaemidae) de América del Sur. *Cuad. Herpetol.* **2005**, *18*, 23–36.
23. Lamborot, M. Evolución cromosómica en reptiles. In *Herpetología de Chile*; Vidal, M.A., Labra, A., Eds.; Science Verlag: Santiago, Chile, 2008; pp. 161–193.
24. Etheridge, R. Redescription of *Ctenoblepharis adspersa* Tschudi, 1845, and the taxonomy of Liolaeminae (Reptilia: Squamata: Tropiduridae). *Am. Mus. Novit.* **1995**, *3142*, 34.
25. Esquerré, D.; Brennan, I.G.; Catullo, R.A.; Torres-Pérez, F.; Keogh, J.S. How mountains shape biodiversity: The role of the Andes in biogeography, diversification, and reproductive biology in South America's most species rich lizard radiation (Squamata: Liolaemidae). *Evolution* **2019**, *73*, 214–230. [CrossRef] [PubMed]
26. Abdala, C.S.; Laspiur, A.; Langstroth, R.P. Las especies del género *Liolaemus* (Liolaemidae). *Cuad. Herpetol.* **2021**, *35*, 193–222. [CrossRef]
27. Esquerré, D.; Ramírez-Álvarez, D.; Pavón-Vázquez, C.J.; Troncoso-Palacios, J.; Garín, C.F.; Keogh, J.S.; Leaché, A.D. Speciation across mountains: Phylogenomics, species delimitation and taxonomy of the *Liolaemus leopardinus* clade (Squamata, Liolaemidae). *Mol. Phylogenet. Evol.* **2019**, *139*, 106524. [CrossRef]
28. Araya-Donoso, R.; San Juan, E.; Tamburrino, Í.; Lamborot, M.; Veloso, C.; Véliz, D. Integrating genetics, physiology and morphology to study desert adaptation in a lizard species. *J. Anim. Ecol.* **2022**, *91*, 1148–1162. [CrossRef]
29. Gorman, G.C. The chromosomes of the Reptilia, a cytotaxonomic interpretation. In *Cytotaxonomy and Vertebrate Evolution*; Chiarelli, A.B., Capanna, E., Eds.; Academic Press: New York, NY, USA, 1972; pp. 349–424.
30. Espinoza, N.D.; Formas, J.R. Karyological pattern of two Chilean lizard species of the genus *Liolaemus* (Sauria: Iguanidae). *Experientia* **1976**, *32*, 299–301. [CrossRef]
31. Bertolotto, C.E.V.; Rodrigues, M.T.; Skuk, G.; Yonenaga-Yassuda, Y. Comparative cytogenetic analysis with differential staining in three species of *Liolaemus* (Squamata, Tropiduridae). *Hereditas* **1996**, *125*, 257–264. [CrossRef]
32. Lamborot, M. Chromosomal evolution and speciation in some Chilean lizards. *Evol. Biológica* **1993**, *7*, 133–151.
33. Lamborot, M.; Eaton, L.; Carrasco, B.A. The Aconcagua River as another barrier to *Liolaemus monticola* (Sauria: Iguanidae) chromosomal races of central Chile. *Rev. Chil. De Hist. Nat.* **2003**, *76*, 23–34. [CrossRef]
34. Lamborot, M.; Alvarez, E.; Campos, I.; Espinoza, A. Karyotypic characterization of three Chilean subspecies of *Liolaemus monticola*. *J. Hered.* **1981**, *72*, 328–334. [CrossRef]
35. Lamborot, M. Karyotypic variation among populations of *Liolaemus monticola* (Tropiduridae) separated by riverine barriers in the Andean Range. *Copeia* **1991**, *4*, 1044–1059. [CrossRef]
36. Lamborot, M.; Alvarez-Sarret, E. Karyotypic variation within and between populations of *Liolaemus monticola* (Tropiduridae) separated by the Maipo River in the coastal range of central Chile. *Herpetologica* **1993**, *49*, 435–449.
37. Lamborot, M.; Ossa, C.G.; Vásquez, M. Population cytogenetics of the "Northern Mod 1" chromosomal race of *Liolaemus monticola* Müller & Helmich (Iguanidae) from Central Chile. *Gayana* **2012**, *76*, 10–21.
38. Lamborot, M. A new derived and highly polymorphic chromosomal race of *Liolaemus monticola* (Iguanidae) from the 'Norte Chico' of Chile. *Chromosome Res.* **1998**, *6*, 247–254. [CrossRef]
39. Torres-Pérez, F.; Lamborot, M.; Boric-Bargetto, D.; Hernández, C.E.; Ortiz, J.C.; Palma, R.E. Plylogeography of a mountain lizard species: An ancient fragmentation process mediated by rivervine barriers in the *Liolaemus monticola* complex (Sauria: Liolaemidae). *J. Zool. Syst. Evol. Res.* **2007**, *45*, 72–81. [CrossRef]
40. Vásquez, M.; Torres-Pérez, F.; Lamborot, M. Genetic variation within and between four chromosomal races of *Liolaemus monticola* in Chile. *Herpetol. J.* **2007**, *17*, 149–160.
41. Lamborot, M.; Eaton, L.C. Concordance of morphological variation and chromosomal races in *Liolaemus monticola* (Tropiduridae) separated by riverine barriers in the Andes. *J. Zool. Syst. Evol. Res.* **1992**, *30*, 189–200. [CrossRef]
42. Lamborot, M.; Eaton, L. The Maipo River as a biogeographical barrier to *Liolaemus monticola* (Tropiduridae) in the mountain ranges of central Chile. *J. Zool. Syst. Evol. Res.* **1997**, *35*, 105–111. [CrossRef]
43. Lamborot, M. Karyotypic polymorphism and evolution within and between the *Liolaemus monticola* (Iguanidae) "northern 2n = 38–40" chromosome race populations in Central Chile. *Rev. Chil. Hist. Nat.* **2001**, *71*, 121–138. [CrossRef]

44. Aravena-Muñoz, N. Niveles de Introgresión Interracial en *Liolaemus monticola* (Iguanidae). Master's Thesis, Universidad de Chile, Santiago, Chile, 2012.
45. R Core Team. *R: A Language and Environment for Statistical Computing*; R Foundation for Statistical Computing: Vienna, Austria, 2022; Available online: https://www.R-project.org/ (accessed on 1 September 2022).
46. Rogers, J.S. Measures of genetic similarity and genetics distances. *Stud. Genet.* **1972**, *721*, 145–153.
47. Jombart, T. adegenet: A R package for the multivariate analysis of genetic markers. *Bioinformatics* **2008**, *24*, 1403–1405. [CrossRef] [PubMed]
48. Schliep, K.P. phangorn: Phylogenetic analysis in R. *Bioinformatics* **2011**, *27*, 592–593. [CrossRef] [PubMed]
49. Pritchard, J.K.; Matthew-Stephens, M.M.; Donnelly, P. Inference of population structure using multi-locus genotype data. *Genetics* **2000**, *155*, 945–959. [CrossRef] [PubMed]
50. Earl, D.A.; VonHoldt, B.M. STRUCTURE HARVESTER: A website and program for visualizing STRUCTURE output and implementing the Evanno method. *Conserv. Genet. Resour.* **2012**, *4*, 359–361. [CrossRef]
51. Goudet, J. Hierfstat, a package for R to compute and test hierarchical F-statistics. *Mol. Ecol. Notes* **2005**, *5*, 184–186. [CrossRef]
52. Paradis, E. pegas: An R package for population genetics with an integrated–modular approach. *Bioinformatics* **2010**, *26*, 419–420. [CrossRef]
53. Aljanabi, S.M.; Martinez, L. Universal and rapid salt-extraction of high quality genomic DNA for PCR-based techniques. *Nucleic Acids Res.* **1997**, *25*, 4692–4693. [CrossRef]
54. Palumbi, S.R. Nucleic acids II: The polymerase chain reaction. In *Molecular Systematics*, 2nd ed.; Hillis, D.M., Moritz, C., Mable, B.K., Eds.; Sinauer Associates, Inc.: Sunderland, MA, USA, 1996; pp. 205–247.
55. Stamatakis, A. RAxML version 8: A tool for phylogenetic analysis and post-analysis of large phylogenies. *Bioinformatics* **2014**, *30*, 1312–1313. [CrossRef]
56. Miller, M.A.; Pfeiffer, W.; Schwartz, T. Creating the CIPRES Science Gatewayfor inference of large phylogenetic trees. In Proceedings of the 2010 Gateway Computing Environments Workshop (GCE), New Orleans, LA, USA, 14 November 2010. [CrossRef]
57. Rambaut, A. FigTree v1.4.2, A Graphical Viewer of Phylogenetic Trees. 2004. Available online: http://tree.bio.ed.ac.uk/software/figtree/ (accessed on 15 September 2022).
58. Leigh, J.W.; Bryant, D. popart: Full-feature software for haplotype network construction. *Methods Ecol. Evol.* **2015**, *6*, 1110–1116. [CrossRef]
59. Kolnicki, R.L. Kinetochore reproduction in animal evolution: Cell biological explanation of karyotypic fission theory. *Proc. Natl. Acad. Sci. USA* **2000**, *97*, 9493–9497. [CrossRef]
60. Dobigny, G.; Britton-Davidian, J.; Robinson, T.J. Chromosomal polymorphism in mammals: An evolutionary perspective. *Biol. Rev.* **2017**, *92*, 1–21. [CrossRef] [PubMed]
61. Porter, C.A.; Sites, J.W., Jr. Evolution of *Sceloporus grammicus* complex (Sauria: Iguanidae) in central Mexico: Population cytogenetics. *Syst. Biol.* **1986**, *35*, 334–358. [CrossRef]
62. Goyenechea, I.; Mendoza-Quijano, F.; Flores-Villela, O.; Reed, K.M. Extreme chromosomal polytypy in a population of *Sceloporus grammicus* (Sauria: Phrynosomatidae) at Santuario Mapethé, Hidalgo, México. *J. Herpetol.* **1996**, *30*, 39–45. [CrossRef]
63. Hanski, I.; Gaggiotti, O. (Eds.) Metapopulation biology: Past, present, and future. In *Ecology, Genetics and Evolution of Metapopulations*; Academic Press: New York, NY, USA, 2004; pp. 3–22. [CrossRef]
64. Eriksson, J.; Hohmann, G.; Boesch, C.; Vigilant, L. Rivers influence the population genetic structure of bonobos (*Pan paniscus*). *Mol. Ecol.* **2004**, *13*, 3425–3435. [CrossRef] [PubMed]
65. Araya-Donoso, R.; Torres-Perez, F.; Lamborot, M.; Veliz, D. Hybridization and polyploidy in the weeping lizard *Liolaemus chiliensis* (Squamata: Liolaemidae). *Biol. J. Linn. Soc.* **2019**, *128*, 963–974. [CrossRef]
66. Sallaberry-Pincheira, N.; Garin, C.F.; González-Acuña, D.; Sallaberry, M.A.; Vianna, J.A. Genetic divergence of Chilean long-tailed snake (*Philodryas chamissonis*) across latitudes: Conservation threats for different lineages. *Divers. Distrib.* **2011**, *17*, 152–162. [CrossRef]
67. Torres-Pérez, F.; Boric-Bargetto, D.; Rodríguez-Valenzuela, E.; Escobar, C.; Palma, R.E. Molecular phylogenetic analyses reveal the importance of taxon sampling in cryptic diversity: *Liolaemus nigroviridis* and *L. monticola* (Liolaeminae) as focal species. *Rev. Chil. Hist. Nat.* **2017**, *90*, 5. [CrossRef]
68. Brüggen, J. *Fundamentos de la geología de Chile*; Instituto Geográfico Militar: Santiago, Chile, 1950; 382p.
69. Vuilleumier, B.S. Pleistocene changes in the fauna and flora of South America. *Science* **1971**, *173*, 771–780. [CrossRef]
70. Caviedes, C. *Geomorfología del Cuaternario del valle del Aconcagua, Chile Central*; Geographischen Institute der Albert-Ludwigs-Universität: Freiburg, Germany, 1972; p. 153.
71. Heusser, C.J. Late-Pleistocene pollen diagrams from the Province of Llanquihue, southern Chile. *Proc. Am. Philos. Soc.* **1966**, *110*, 269–305.
72. Olave, M.; Avila, L.J.; Sites, J.W., Jr.; Morando, M. Hybridization could be a common phenomenon within the highly diverse lizard genus *Liolaemus*. *J. Evol. Biol.* **2018**, *31*, 893–903. [CrossRef]
73. Biltueva, L.; Vorobieva, N.; Perelman, P.; Trifonov, V.; Volobouev, V.; Panov, V.; Ilyashenko, V.; Onischenko, S.; O'Brien, P.; Yang, F.; et al. Karyotype evolution of Eulipotyphla (Insectivora): The genome homology of seven *Sorex* species revealed by comparative chromosome painting and banding data. *Cytogenet. Genome Res.* **2011**, *135*, 51–64. [CrossRef] [PubMed]
74. Torgasheva, A.A.; Basheva, E.A.; Gomez Fernandez, M.J.; Mirol, P.; Borodin, P.M. Chromosomes and speciation in tuco-tuco (*Ctenomys*, Hystricognathi, Rodentia). *Russ. J. Genet. Appl.* **2017**, *7*, 350–357. [CrossRef]

Article

Conservation of Major Satellite DNAs in Snake Heterochromatin

Artem Lisachov [1,2,*], Alexander Rumyantsev [3], Dmitry Prokopov [3], Malcolm Ferguson-Smith [4] and Vladimir Trifonov [3]

1 Animal Genomics and Bioresource Research Unit (AGB Research Unit), Faculty of Science, Kasetsart University, 50 Ngamwongwan, Chatuchak, Bangkok 10900, Thailand
2 Institute of Cytology and Genetics, Russian Academy of Sciences, Siberian Branch, Novosibirsk 630090, Russia
3 Institute of Molecular and Cellular Biology, Russian Academy of Sciences, Siberian Branch, Novosibirsk 630090, Russia
4 Cambridge Resource Centre for Comparative Genomics, Department of Veterinary Medicine, University of Cambridge, Cambridge CB3 0ES, UK
* Correspondence: aplisachev@gmail.com

Simple Summary: In the present work, we describe the satellite DNA families that occur in the genomes of two snakes from different families: *Daboia russelii* (Viperidae) and *Pantherophis guttatus* (Colubridae). We show high conservation of nucleotide sequences and chromosomal localizations of these satellites, despite the widespread view that such genomic elements evolve very rapidly.

Abstract: Repetitive DNA sequences constitute a sizeable portion of animal genomes, and tandemly organized satellite DNAs are a major part of them. They are usually located in constitutive heterochromatin clusters in or near the centromeres or telomeres, and less frequently in the interstitial parts of chromosome arms. They are also frequently accumulated in sex chromosomes. The function of these clusters is to sustain the architecture of the chromosomes and the nucleus, and to regulate chromosome behavior during mitosis and meiosis. The study of satellite DNA diversity is important for understanding sex chromosome evolution, interspecific hybridization, and speciation. In this work, we identified four satellite DNA families in the genomes of two snakes from different families: *Daboia russelii* (Viperidae) and *Pantherophis guttatus* (Colubridae) and determine their chromosomal localization. We found that one family is localized in the centromeres of both species, whereas the others form clusters in certain chromosomes or subsets of chromosomes. BLAST with snake genome assemblies showed the conservation of such clusters, as well as a subtle presence of the satellites in the interspersed manner outside the clusters. Overall, our results show high conservation of satellite DNA in snakes and confirm the "library" model of satellite DNA evolution.

Keywords: Serpentes; Colubridae; Viperidae; sex chromosomes; repetitive DNA; centromere

Citation: Lisachov, A.; Rumyantsev, A.; Prokopov, D.; Ferguson-Smith, M.; Trifonov, V. Conservation of Major Satellite DNAs in Snake Heterochromatin. *Animals* **2023**, *13*, 334. https://doi.org/10.3390/ani13030334

Academic Editor: Ettore Olmo

Received: 7 December 2022
Revised: 11 January 2023
Accepted: 13 January 2023
Published: 17 January 2023

1. Introduction

Repetitive DNA sequences are a key component of eukaryotic genomes. There are several types of repeats, classified by their structure and sub-chromosomal localization. Interspersed and tandem repeats are recognized by their genomic organization. Interspersed repeats can be located in various regions of the genome, whereas tandem repeats are mostly organized into clusters in specific segments of chromosomes [1]. Satellite DNA sequences (satDNA) are among the most abundant types of tandem repeats. They are usually located in the C-positive heterochromatic blocks at centromeres, as well as in the pericentromeric, subtelomeric, and, more rarely, interstitial chromosomal regions [2]. Every eukaryotic genome usually contains several families of satDNAs, with each family having its specific localization. For example, centromeric heterochromatin is typically composed of the special centromeric satellite, whereas the pericentromeric heterochromatin blocks harbor the satellites of other families [3]. Some satDNA families and subfamilies occur at similar positions in all chromosomes (e.g., pan-centromeric repeats), whereas others are accumulated on a subset of chromosomes or even one specific chromosome (for example, a sex chromosome) [4]. Specific satellite families spread inside chromosomes and between chromosomes by means of ectopic recombination, gene conversion, and transposition with mobile genetic elements (TEs) [5–7].

Since satDNAs do not encode proteins, they were once viewed as "selfish", "junk DNA", and "genomic parasites". However, there is a growing body of evidence that satDNA clusters are technical elements of chromosomes that participate in regulating their structure and behavior during the cell cycle, i.e., condensation, decondensation, kinetochore formation, and meiotic pairing [8–10]. Depending on their function, satDNAs differ in their degree of conservation. While certain families are species-specific, others can be characteristic for the whole genus or taxonomic family [11–13]. It has been hypothesized that satDNA divergence may contribute to the constrained meiotic chromosome pairing in hybrids, thus directly affecting speciation [14]. This makes satDNA an important marker to study phylogenetics, genome evolution, and genome function in diverse animal groups.

In reptiles, satDNAs are poorly studied. A notable exception is the lizard family Lacertidae, in which numerous satellites have been identified and extensively studied [15–19]. Two satellites have been identified in Scincidae [20,21], and two satellite families have been found in Varanidae [22,23]. Four types of satDNAs are known from the Chinese softshell turtle (*Pelodiscus sinensis*) [24]. Recently, a high conservation of tandem repetitive DNAs has been demonstrated in crocodilians [25]. Snakes comprise nearly half of the total squamate diversity; however, data on their satDNAs are scarce. Four families of satDNAs were found in different snake species. The PFL-MspI satellite was isolated from *Protobothrops flavoviridis* (Crotalinae, Viperidae), located in the centromeric regions of its chromosomes. This satDNA is shared at least by *Gloydius blomhoffi* from the same subfamily Crotalinae, as shown by FISH and slot blot hybridization. The slot blot analysis did not reveal this satellite even in *Bitis arietans* (Viperinae, Viperidae), a member of the same family. The PBI-MspI satellite was found in *Python bivittatus*, *P. molurus*, and *Boa constrictor* by FISH and slot blot, indicating the conservation of this satellite at least at the Henophidia level. Lastly, the PBI-DdeI satellite was initially identified as a major centromeric satellite in *P. bivittatus*, whereas FISH and slot blot failed to detect this satellite in any other genus [26]. However, later the PBI-DdeI was found in a wide set of diverse snake species using PCR. In *Naja kaouthia*, this repeat was accumulated in the W chromosome [27]. Apparently, sequence divergence and/or low copy number may impede the detection of a satDNA by hybridization methods. Another repetitive sequence, BamHI-B4, is specific to the terminal part of the homolog of the *Anolis* chromosome 6 (ZZ/ZW chromosome in Caenophidia and XX/XY chromosome in *Python*) and is conserved in pythons, colubrids, and pit vipers [28].

Classical "wet" methods of satDNA isolation include the analysis of genomic fragments in gradient centrifugation and the digestion of genomic DNA with restriction enzymes, while a range of bioinformatic approaches have recently been suggested to search for tandemly arranged DNAs in genomic data. In the present work, we used the Tandem Repeat Analyzer software (TAREAN) [29] to identify satellite repeats in two species of snakes, *Daboia russelii* (Viperinae, Viperidae) and *Pantherophis guttatus* (Colubridae), from short genomic reads. This software de novo identifies tandem organized satellite repeats from raw Illumina reads of a genomic sample. We studied their chromosomal localization using FISH and analyzed the cross-species conservation using BLAST on the available snake genome assemblies. The genome assemblies of *Vipera latastei* (Viperinae, Viperidae) (rVipLat1.pri) and *V. ursinii* (rVipUrs1.1) were used for quantitative and localization analysis, since they have the best assembled repeat clusters among the available assemblies of snakes.

2. Materials and Methods

2.1. Cell Line Establishment and Karyotype Analysis

The *P. guttatus* and *D. russelii* cells were grown from fibroblasts obtained from the Cambridge Resource Center for Comparative Genomics, Department of Veterinary Medicine, UK. The cell cultures were provided to the Institute of Molecular and Cellular Biology, SB RAS, Russia for joint research. The cell lines of *P. guttatus* and *D. russelii* were deposited in the IMCB SB RAS cell bank ("The general collection of cell cultures", 0310-2016-0002). Chromosome suspensions from the cell cultures were obtained in the Laboratory of Comparative Genomics, IMCB SB RAS, Novosibirsk, Russia, as described previously [30,31].

2.2. Repetitive DNA Identification

DNA sequencing data were downloaded from the NCBI SRA database (accession number SRR5506741 for *D. russelii* genomic reads and SRR9596755 for *P. guttatus*) and used for the identification of tandemly arranged repeats. Filtering by quality and adapter trimming was performed using fastp 0.23.2 [32] with the parameters "–detect_adapter_for_pe -5 -3 -r -l 75". Trimmed reads were used in the analysis with the TAREAN 2.3.7 tool [29], which identified clusters of the most abundant tandemly arranged repeats. NCBI BLAST [33] was used to compare consensus tandem repeat sequences with available genome assemblies. RepBase was used to compare consensus tandem repeat sequences with available described repeat sequences [34]

2.3. Fluorescence In Situ Hybridization (FISH)

DNA of *P. guttatus* and *D. russelii* was extracted from the cell cultures using the standard phenol–chloroform technique. Primers for PCR amplification and labeling of seven probes were designed with PrimerQuestTool [35] (Table 1). PCR amplification was performed as described earlier [36]. Labeling was performed using PCR by incorporation of biotin-dUTP and digoxigenin-dUTP (Sigma, Darmstadt, Germany). FISH was performed in accordance with previously published protocols [37]. Images were captured using the VideoTest-FISH software (Imicrotec, New York, NY, USA) with a JenOptic charge-coupled device (CCD) camera (Jena, Germany) mounted on an Olympus BX53 microscope (Shinjuku, Japan). All images were processed in Adobe PhotoShop 2021 (Adobe, San Jose, CA, USA).

Table 1. Primers used to amplify satDNA in the current study.

Satellite	Primer Sequences
PGU-Sat-1	F 5′–TTTCAAGTACGAGCTTTCCC–3′ R 5′–GCTGAATTGAGCCCTACTG–3′
PGU-Sat-2	F 5′–GACACCAGGATGAGTTTCAG–3′ R 5′–TCCTGACCGTGGAGTAAA–3′
PGU-Sat-3	F 5′–CTTCCTCGGGCAGCAAA–3′ R 5′–GTAACAACGGATGCTAGAATGT–3′
DRU-Sat-1	F 5′–CCCGCCTGACCGAAGACC–3′ R 5′–GAGCTCTATCTGCAACGGG–3′
DRU-Sat-2	F 5′–ACCCCGAATCTCATTCTGGC–3′ R 5′–TCCTGATGCCGGGGTCAG–3′
DRU-Sat-3	F 5′–TTGTGTTTCTGGATCAATAACC–3′ R 5′–GCCTTTCCTGTATAATCCAAA–3′
DRU-Sat-5	F 5′–CAGAGCTGCTGGGAAGTG–3′ R 5′–GAGATCAATGAGGACCCCA–3′

3. Results

3.1. Tandem Repeat Identification

The TAREAN analysis revealed four high-confidence satellite repeats in the genome of *D. russelii* and three high-confidence satellite repeats in the genome of *P. guttatus*, which were named DRU-Sat-1, DRU-Sat-2, DRU-Sat-3, DRU-Sat-5, PGU-Sat-1, PGU-Sat-2, and PGU-Sat-3, respectively (Table 2). The satellites DRU-Sat-1 and PGU-Sat-1 were found to belong to the same family, while the satellites DRU-Sat-2, PGU-Sat-2, and PGU-Sat-3 belonged to another family. Interestingly, DRU-Sat-2 and PGU-Sat-2 shared a high level of similarity and were more distantly related to PGU-Sat-3 (Table 3, File S1).

Table 2. Putative satellites revealed by TAREAN in the genomes of *Daboia russelii* and *Pantherophis guttatus*.

Sattelite Name	Monomer Size (bp)	Genome Proportion, %	GC Content, %	Species	Accession Number
DRU-Sat-1	168	0.3	42.3	*Daboia russelii* (Russell's viper)	OP820475
DRU-Sat-2	170	0.13	35.9	—//—	OP820476
DRU-Sat-3	64	0.012	37.5	—//—	OP820477
DRU-Sat-5	147	0.025	44.9	—//—	OP820478
PGU-Sat-1	167	0.31	42.5	*Pantherophis guttatus* (Corn snake)	OP820479
PGU-Sat-2	187	0.085	37.4	—//—	OP820480
PGU-Sat-3	169	0.043	39.1	—//—	OP820481

Table 3. The p-distances between the consensus sequences of the PGU-Sat-2/PGU-Sat-3/DRU-Sat-2 family.

	PGU-Sat-2	PGU-Sat-3	DRU-Sat-2
PGU-Sat-2	-	—//—	—//—
PGU-Sat-3	0.278	-	—//—
DRU-Sat-2	0.262	0.226	-

3.2. FISH Analysis

The karyotypes of the studied specimens comprised 36 chromosomes (eight pairs of macrochromosomes and 10 pairs of microchromosomes) with pairs of heteromorphic Z and W chromosomes. This is a typical snake karyotype that corresponds to the previously described karyotypes of these species [38,39]. In *P. guttatus*, the satellite PGU-Sat-1 was localized in the centromeric regions of macrochromosomes and in several microchromosomes. It was also localized in the DAPI-positive interstitial band of the W chromosome (Figure 1). The PGU-Sat-2 and PGU-Sat-3 satellite types, despite belonging to the same family, showed strikingly different chromosomal localizations. The PGU-Sat-2 satellite was mapped to the same DAPI-positive band in the W chromosome and in the pericentromeric region of one small acrocentric macrochromosome. It was also present in certain pairs of microchromosomes, being extensively amplified in one pair (Figures 1, 2 and S1). The PGU-Sat-3 satellite tended to be localized in microchromosomes, but not in all pairs. It was colocalized with PGU-Sat-2 in the pericentromeric region of one small acrocentric macrochromosome, and it was also present in the pericentromeric region of the q-arm of the chromosome 2 and in the terminal region of the p-arm of the W chromosome (Figure 2).

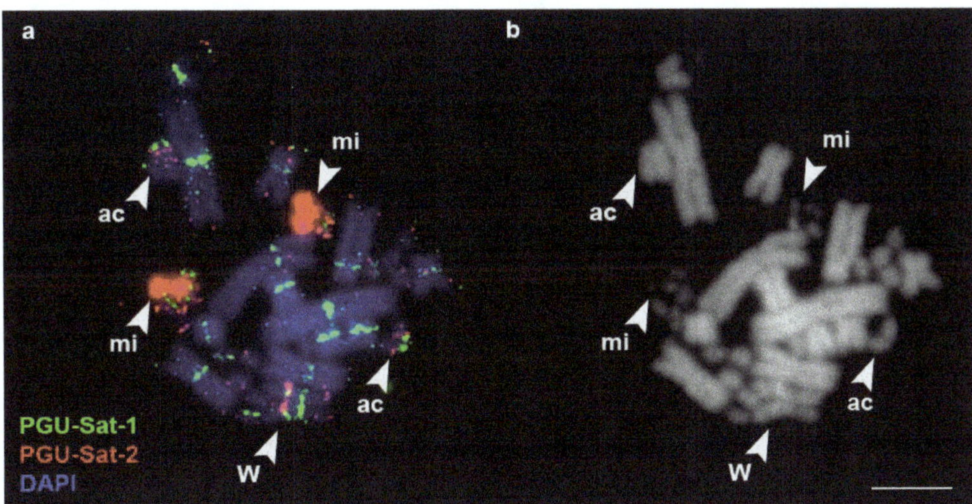

Figure 1. Localization of the satellites PGU-Sat-1 and PGU-Sat-2 in the chromosomes of *P. guttatus*. mi: the PGU-Sat-2 bearing microchromosome; ac: the acrocentric macrochromosome with both PGU-Sat-1 and PGU-Sat-2 signals; W: the W chromosome. (**a**) Merged image; (**b**) DAPI channel. Scale bar: 10 μm.

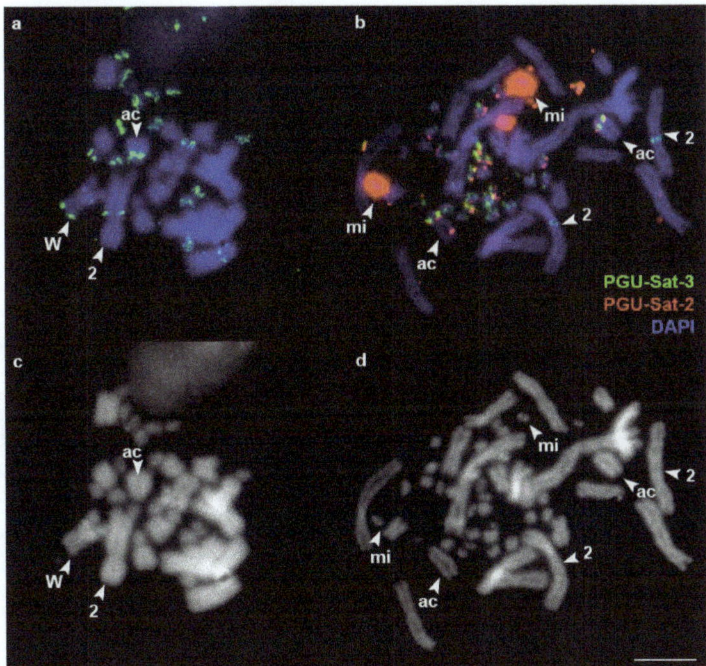

Figure 2. Localization of the satellites PGU-Sat-2 and PGU-Sat-3 in the chromosomes of *P. guttatus*. mi: the PGU-Sat-2 bearing microchromosome; ac: the acrocentric macrochromosome with both PGU-Sat-2 and PGU-Sat-3 signals; 2: chromosome 2; W: the W chromosome. (**a**,**b**) Merged images; (**c**,**d**) DAPI channel. Scale bar: 10 μm.

In *D. russelii*, the satellite DRU-Sat-1 was localized in the centromeric areas of all chromosomes. The satellite DRU-Sat-2 was localized in the p-arm of the chromosome 1 and the q-arm of chromosome 2 (Figure 3). The satellite DRU-Sat-3 was chromosome-specific and showed a band in one pair of microchromosomes (Figure 4). The satellite DRU-Sat-5 was amplified throughout the whole length of the W chromosome (Figure 5).

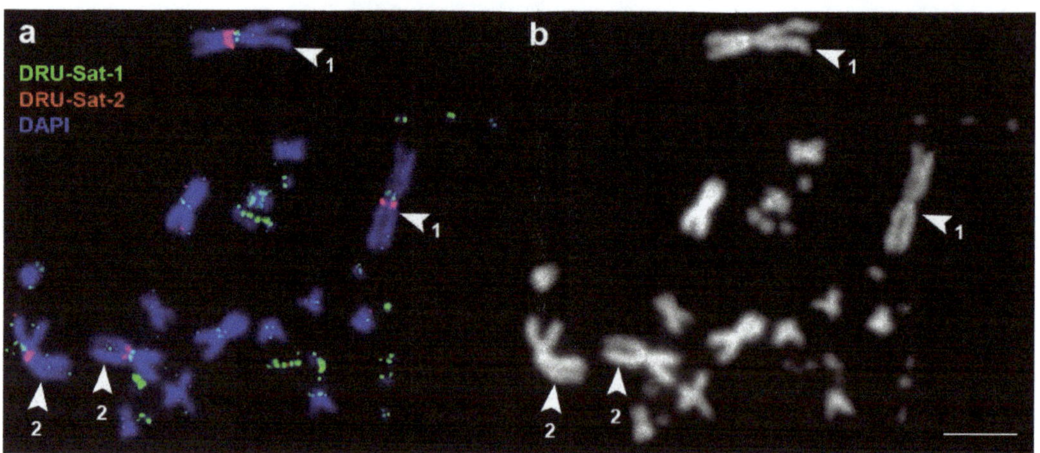

Figure 3. Localization of the satellites DRU-Sat-1 and DRU-Sat-2 in the chromosomes of *D. russelii.* 1: chromosome 1; 2: chromosome 2. (**a**) Merged image; (**b**) DAPI channel. Scale bar: 10 μm.

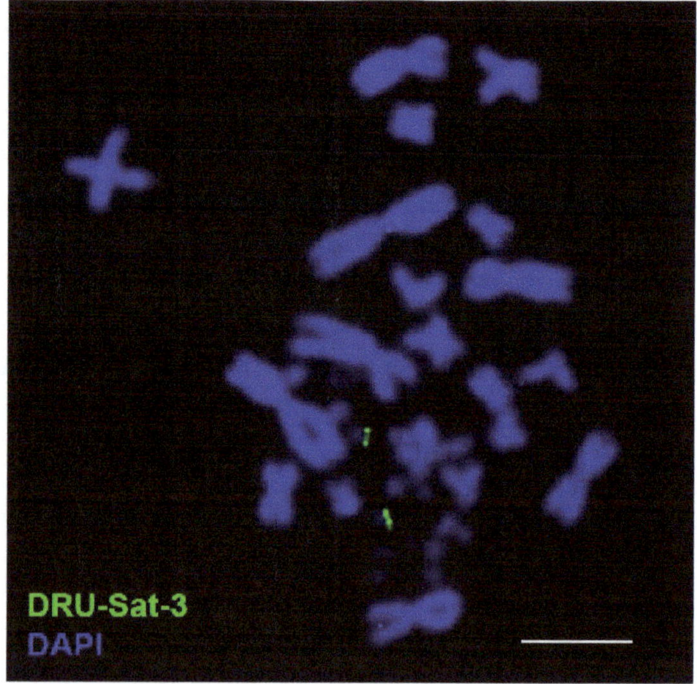

Figure 4. Localization of the satellite DRU-Sat-3 in the chromosomes of *D. russelii.* Scale: 10 μm.

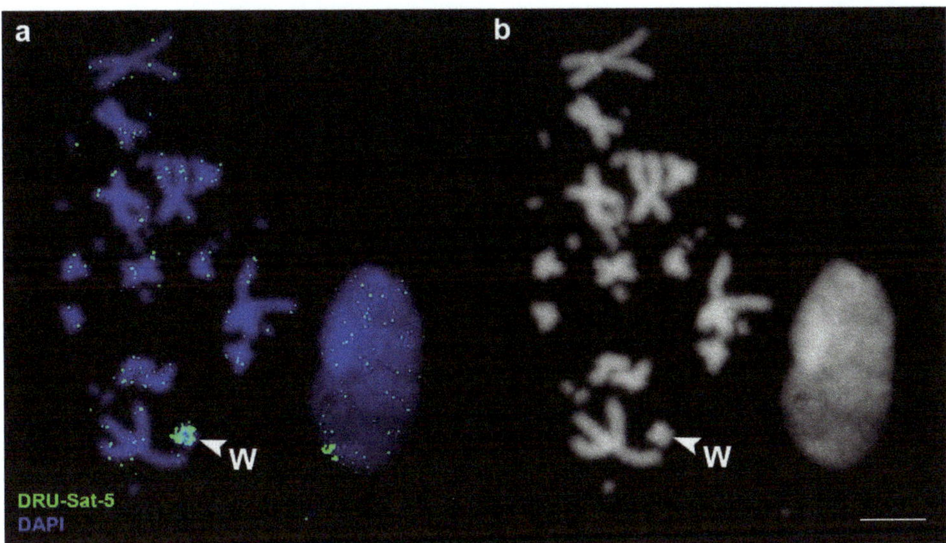

Figure 5. Localization of the satellite DRU-Sat-5 in the chromosomes of *D. russelii*. W: the W chromosome. (**a**) Merged image; (**b**) DAPI channel. Scale bar: 10 μm.

3.3. BLAST Analysis

Even though none of the detected satellites were found in the nr/nt NCBI database by BLAST, we found that the DRU-Sat-1/PGU-Sat-1 satellite belongs to the same family as PFL-MspI, which was described earlier [26,27]. We did not reveal homology between the other detected satDNAs and any of the previously described snake repetitive elements. However, we detected all the satellite families found in this work in the RefSeq genomes of other snakes by BLAST. The DRU-Sat-1/PGU-Sat-1, DRU-Sat-2/PGU-Sat-2/PGU-Sat-3 and DRU-Sat-3 satellites were found in various higher snakes, namely, *Protobothrops mucrosquamatus*, *Crotalus tigris* (Crotalinae, Viperidae), *Notechis scutatus*, *Pseudonaja textilis* (Elapidae), *Thamnophis sirtalis*, and *Thamnophis elegans* (Colubridae). Interestingly, BLAST revealed the DRU-Sat-3 satellite in *Pantherophis guttatus*, whereas TAREAN did not. The DRU-Sat-5 satellite was not found in any genome assemblies except those of Viperidae.

The alignment of DRU-Sat-1/PGU-Sat-1 to the genome assembly of *V. latastei* revealed its high copy number in all chromosome scaffolds except 15 and 17 (from 8877 in scaffold 3 to 358 in the scaffold Z), with predominantly medial localization, possibly corresponding to the centromere. The percentage identity between DRU-Sat-1/PGU-Sat-1 and the *V. latastei* sequences did not vary between the scaffolds and was between 95% and 97% for DRU-Sat-1.

The DRU-Sat-2/PGU-Sat-2/PGU-Sat-3 satellite was present in the scaffolds 1–3, Z, and 5–10, being the most abundant in scaffolds 2, 3, and 5. The copy numbers were 17,561, 2923, and 2522, respectively, in contrast to 101 in scaffold 1, where it was the second most abundant. In scaffolds 2, 3, and 5, this satellite was accumulated in clusters surrounding the centromere, possibly corresponding to the pericentromeric C bands. The copies in scaffolds 2, 3, and 5 had higher similarity to DRU-Sat-2 than the copies located in the scaffolds where this satellite was less abundant (percent of identity 91.86–94.08% versus 71.97–90.8%).

The satellite DRU-Sat-3 was present in 983 copies in scaffold 16 and was clustered in the subterminal position of the p-arm, if the DRU-Sat-1/PGU-Sat-1 cluster is considered as the centromere. Scaffold 2, where it was the second most abundant, harbored only 12 copies. The copies located in scaffold 16 had up to 96.88% identity with DRU-Sat-3, whereas the copies from other scaffolds had 75.86–93.1% identity. In the assembly of *V.*

ursinii, scaffold 15 with 459 copies and similar cluster localization was the only scaffold where DRU-Sat-3 was found.

Lastly, satellite DRU-Sat-5 showed a small copy number in the genome assembly of *V. latastei* (up to 117 in the scaffold 1) and did not show any clustering. The W chromosome was not present in the assembly. In the assembly of *V. ursinii*, DRU-Sat-5 was accumulated in the W chromosome (1998 copies), whereas, in the autosomes, it had no more than 15 copies per chromosome.

4. Discussion

The satDNAs DRU-Sat-1/PGU-Sat-1, DRU-Sat-2/PGU-Sat-2/PGU-Sat-3, and DRU-Sat-3 are found in a wide range of higher snake genomes, which means that they originated at least in the common ancestor of Viperidae and Colubridae at ~42 MYA [40]. In contrast, DRU-Sat-5 is apparently younger as it is restricted to the Viperidae. Since it is present in both Viperinae and Crotalinae, its estimated age is therefore around 31 MY [40]. Previously, a more ancient snake satDNA, PBI-DdeI, which is shared by Henophidia and Caenophidia, was described [27]. In most species, it has a low copy number and probably lacks the tandem organization pattern; therefore, it is detectable only by PCR and BLAST with good-quality genome assemblies, and not by FISH and slot blot [26,27]. This satellite was also not detected by TAREAN in our work, although it is probably present in the genomes of the species studied, since TAREAN detects only highly repeated and tandemly organized elements. Possibly, PBI-DdeI is dispersed and low-copy in the genomes of *D. russelii* and *P. guttatus*. These findings challenge the common conception that satDNAs evolve very rapidly and are usually restricted to one species or a narrow phylogenetic clade, since the "recent appearance" may in fact mean a "recent rise in copy number" of an ancient satellite [41]. According to the concept known as the "library" model of satDNA evolution, animal genomes usually contain many diverse families of satDNAs (the "library"), only a few of which are highly amplified. During phylogenesis and speciation, the "library" experiences dynamic evolution, with satDNA families rising and decreasing in copy number, which leads to contrasting satDNA profiles in related species despite the qualitative conservation of the satDNA repertoires [11].

PGU-Sat-2 strongly indicates a pair of microchromosomes and may represent a considerable part of this chromosomes content. This is in contrast to avian microchromosomes, which are usually gene-rich and heterochromatin-poor. The revealed accumulation makes the PGU-Sat-2 probe a convenient tool for microchromosome identification.

The distribution of BLAST hits of the detected satellites in the genome assemblies of *V. latastei* and *V. ursinii* was similar to that observed in the FISH results for *D. russelii*. Specifically, DRU-Sat-1/PGU-Sat-1, which belong to the same family as the previously described PFL-MspI satellite, represent a centromeric repeat, DRU-Sat-2/PGU-Sat-2/PGU-Sat-3 is located in the pericentromeric clusters in a subset of macrochromosomes, DRU-Sat-3 is accumulated in one pair of microchromosomes, and DRU-Sat-5 is accumulated in the W chromosome. This result indicates that this satellite landscape at least predates the divergence between *Vipera* and *Daboia*, which occurred around 15 MYA [40]. We suppose that PCR for the DRU-Sat-5 marker may serve as a molecular sexing method for at least *Vipera* and *Daboia*. It should be further tested in other species of Viperidae.

5. Conclusions

In this work, we described four satellite DNA families in snake genomes, and revealed their chromosomal localization using FISH and BLAST in chromosome-level genome assemblies. Three of these four families are completely novel. We show that three families are conserved in Colubridae and Viperidae, whereas one is characteristic for Viperidae. In two satellite families, the pattern of chromosomal localization is conserved in both Colubridae and Viperidae, and, in two families, it is conserved in *Daboia* and *Vipera*. Our results indicate that, despite the common opinion that satellite DNA evolves extremely quickly and is usually species- or genus-specific, ancient repeat families are not rare. This corroborates the "library" model of the satellite DNA evolution, which supposes that diverse types of satellites may coexist in the genome, and that the common view of their very rapid appearance and disappearance may be due to their changes in copy number.

Supplementary Materials: The following supporting information can be downloaded at https://www.mdpi.com/article/10.3390/ani13030334/s1: File S1. FASTA alignment file showing sequence similarity of DRU-Sat-2, PGU-Sat-2, and PGU-Sat-3. Localization of the satellite PGU-Sat-2 (red) in the chromosomes of *P. guttatus* with reduced exposure time to show the signal in the PGU-Sat-2-bearing microchromosome in more detail. (a) Merged image; (b) DAPI channel. Scale bar: 10 μm.

Author Contributions: Conceptualization, V.T.; methodology, V.T., D.P. and A.L.; software, D.P.; validation, A.R., D.P. and V.T.; formal analysis, D.P. and A.R.; investigation, A.R.; resources, V.T. and M.F.-S.; data curation, D.P.; writing—original draft preparation, A.L.; writing—review and editing, A.L., V.T., A.R., D.P. and M.F.-S.; visualization, A.R. and A.L.; supervision, V.T.; project administration, V.T.; funding acquisition, V.T. All authors read and agreed to the published version of the manuscript.

Funding: This work was supported by RFBR grant No 19-54-26017 and by the Ministry of Science and Higher Education of the Russian Federation grant number FWNR-2022-0015.

Institutional Review Board Statement: Not applicable.

Informed Consent Statement: Not applicable.

Data Availability Statement: The consensus sequences of the identified satDNAs are deposited in GenBank.

Acknowledgments: Computational resources were provided by the ELIXIR-CZ project (LM2015047), part of the international ELIXIR infrastructure.

Conflicts of Interest: The authors declare no conflict of interest.

References

1. Biscotti, M.A.; Olmo, E.; Heslop-Harrison, J.S. Repetitive DNA in Eukaryotic Genomes. *Chromosome Res.* **2015**, *23*, 415–420. [CrossRef] [PubMed]
2. Garrido-Ramos, M.A. Satellite DNA: An Evolving Topic. *Genes* **2017**, *8*, 230. [CrossRef] [PubMed]
3. Plohl, M.; Luchetti, A.; Meštrović, N.; Mantovani, B. Satellite DNAs between Selfishness and Functionality: Structure, Genomics and Evolution of Tandem Repeats in Centromeric (Hetero) Chromatin. *Gene* **2008**, *409*, 72–82. [CrossRef]
4. Navajas-Pérez, R.; de La Herrán, R.; Jamilena, M.; Lozano, R.; Rejón, C.R.; Rejón, M.R.; Garrido-Ramos, M.A. Reduced Rates of Sequence Evolution of Y-Linked Satellite DNA in Rumex (Polygonaceae). *J. Mol. Evol.* **2005**, *60*, 391–399. [CrossRef] [PubMed]
5. Meštrović, N.; Mravinac, B.; Pavlek, M.; Vojvoda-Zeljko, T.; Šatović, E.; Plohl, M. Structural and Functional Liaisons between Transposable Elements and Satellite DNAs. *Chromosome Res.* **2015**, *23*, 583–596. [CrossRef] [PubMed]
6. Šatović, E.; Plohl, M. Tandem Repeat-Containing MITEs in the Clam Donax Trunculus. *Genome Biol. Evol.* **2013**, *5*, 2549–2559. [CrossRef]
7. Satović, E.; Vojvoda Zeljko, T.; Luchetti, A.; Mantovani, B.; Plohl, M. Adjacent Sequences Disclose Potential for Intra-Genomic Dispersal of Satellite DNA Repeats and Suggest a Complex Network with Transposable Elements. *BMC Genom.* **2016**, *17*, 997. [CrossRef]
8. Jagannathan, M.; Cummings, R.; Yamashita, Y.M. A Conserved Function for Pericentromeric Satellite DNA. *Elife* **2018**, *7*, e34122. [CrossRef]
9. Spangenberg, V.; Losev, M.; Volkhin, I.; Smirnova, S.; Nikitin, P.; Kolomiets, O. DNA Environment of Centromeres and Non-Homologous Chromosomes Interactions in Mouse. *Cells* **2021**, *10*, 3375. [CrossRef] [PubMed]
10. Biscotti, M.A.; Canapa, A.; Forconi, M.; Olmo, E.; Barucca, M. Transcription of Tandemly Repetitive DNA: Functional Roles. *Chromosome Res.* **2015**, *23*, 463–477. [CrossRef]
11. Meštrović, N.; Plohl, M.; Mravinac, B.; Ugarković, D. Evolution of Satellite DNAs from the Genus Palorus—Experimental Evidence for the "library" Hypothesis. *Mol. Biol. Evol.* **1998**, *15*, 1062–1068. [CrossRef] [PubMed]
12. Bachmann, L.; Venanzetti, F.; Sbordoni, V. Characterization of a Species-Specific Satellite DNA Family of Dolichopoda Schiavazzii (Orthoptera, Rhaphidophoridae) Cave Crickets. *J. Mol. Evol.* **1994**, *39*, 274–281. [CrossRef] [PubMed]
13. Mravinac, B.; Plohl, M.; Ugarković, D. Preservation and High Sequence Conservation of Satellite DNAs Suggest Functional Constraints. *J. Mol. Evol.* **2005**, *61*, 542–550. [CrossRef] [PubMed]
14. Rodríguez, F.R.; de La Herrán, R.; Navajas-Pérez, R.; Cano-Roldán, B.; Sola-Campoy, P.J.; García-Zea, J.A.; Rejón, C.R. Centromeric Satellite DNA in Flatfish (Order Pleuronectiformes) and Its Relation to Speciation Processes. *J. Hered.* **2017**, *108*, 217–222. [CrossRef]

15. Capriglione, T.; Cardone, A.; Odierna, G.; Olmo, E. Evolution of a Centromeric Satellite DNA and Phylogeny of Lacertid Lizards. *Comp. Biochem. Physiol. Part B Comp. Biochem.* **1991**, *100*, 641–645. [CrossRef]
16. Ciobanu, D.; Grechko, V.v.; Darevsky, I.S.; Kramerov, D.A. New Satellite DNA in *Lacerta* s. str. Lizards (Sauria: Lacertidae): Evolutionary Pathways and Phylogenetic Impact. *J. Exp. Zool. B Mol. Dev. Evol.* **2004**, *302B*, 505–516. [CrossRef]
17. Giovannotti, M.; Nisi Cerioni, P.; Rojo, V.; Olmo, E.; Slimani, T.; Splendiani, A.; Caputo Barucchi, V. Characterization of a Satellite DNA in the Genera *Lacerta* and *Timon* (Reptilia, Lacertidae) and Its Role in the Differentiation of the W Chromosome. *J. Exp. Zool. B Mol. Dev. Evol.* **2018**, *330*, 83–95. [CrossRef]
18. Giovannotti, M.; S'Khifa, A.; Nisi Cerioni, P.; Splendiani, A.; Slimani, T.; Fioravanti, T.; Olmo, E.; Caputo Barucchi, V. Isolation and Characterization of Two Satellite DNAs in *Atlantolacerta andreanskyi* (Werner, 1929) (Reptilia, Lacertidae). *J. Exp. Zool. B Mol. Dev. Evol.* **2020**, *334*, 178–191. [CrossRef]
19. Rojo, V.; Martínez-Lage, A.; Giovannotti, M.; González-Tizón, A.M.; Cerioni, P.N.; Barucchi, V.C.; Galán, P.; Olmo, E.; Naveira, H. Evolutionary Dynamics of Two Satellite DNA Families in Rock Lizards of the Genus *Iberolacerta* (Squamata, Lacertidae): Different Histories but Common Traits. *Chromosome Res.* **2015**, *23*, 441–461. [CrossRef]
20. Giovannotti, M.; Nisi Cerioni, P.; Caputo, V.; Olmo, E. Characterisation of a GC-Rich Telomeric Satellite DNA in *Eumeces schneideri* Daudin (Reptilia, Scincidae). *Cytogenet. Genome Res.* **2009**, *125*, 272–278. [CrossRef]
21. Giovannotti, M.; Cerioni, P.N.; Splendiani, A.; Ruggeri, P.; Olmo, E.; Barucchi, V.C. Slow Evolving Satellite DNAs: The Case of a Centromeric Satellite in *Chalcides ocellatus* (Forskål, 1775) (Reptilia, Scincidae). *Amphib. Reptil.* **2013**, *34*, 401–411. [CrossRef]
22. Chaiprasertsri, N.; Uno, Y.; Peyachoknagul, S.; Prakhongcheep, O.; Baicharoen, S.; Charernsuk, S.; Nishida, C.; Matsuda, Y.; Koga, A.; Srikulnath, K. Highly Species-Specific Centromeric Repetitive DNA Sequences in Lizards: Molecular Cytogenetic Characterization of a Novel Family of Satellite DNA Sequences Isolated from the Water Monitor Lizard (*Varanus salvator macromaculatus*, Platynota). *J. Hered.* **2013**, *104*, 798–806. [CrossRef] [PubMed]
23. Prakhongcheep, O.; Thapana, W.; Suntronpong, A.; Singchat, W.; Pattanatanang, K.; Phatcharakullawarawat, R.; Muangmai, N.; Peyachoknagul, S.; Matsubara, K.; Ezaz, T.; et al. Lack of Satellite DNA Species-Specific Homogenization and Relationship to Chromosomal Rearrangements in Monitor Lizards (Varanidae, Squamata). *BMC Evol. Biol.* **2017**, *17*, 193. [CrossRef]
24. Yamada, K.; Nishida-Umehara, C.; Matsuda, Y. Molecular and Cytogenetic Characterization of Site-Specific Repetitive DNA Sequences in the Chinese Soft-Shelled Turtle (*Pelodiscus sinensis*, Trionychidae). *Chromosome Res.* **2005**, *13*, 33–46. [CrossRef]
25. Romanenko, S.A.; Prokopov, D.Y.; Proskuryakova, A.A.; Davletshina, G.I.; Tupikin, A.E.; Kasai, F.; Ferguson-Smith, M.A.; Trifonov, V.A. The Cytogenetic Map of the Nile Crocodile (*Crocodylus niloticus*, Crocodylidae, Reptilia) with Fluorescence In Situ Localization of Major Repetitive DNAs. *Int. J. Mol. Sci.* **2022**, *23*, 13063. [CrossRef]
26. Matsubara, K.; Uno, Y.; Srikulnath, K.; Seki, R.; Nishida, C.; Matsuda, Y. Molecular Cloning and Characterization of Satellite DNA Sequences from Constitutive Heterochromatin of the Habu Snake (*Protobothrops flavoviridis*, Viperidae) and the Burmese Python (*Python bivittatus*, Pythonidae). *Chromosoma* **2015**, *124*, 529–539. [CrossRef]
27. Thongchum, R.; Singchat, W.; Laopichienpong, N.; Tawichasri, P.; Kraichak, E.; Prakhongcheep, O.; Sillapaprayoon, S.; Muangmai, N.; Baicharoen, S.; Suntrarachun, S.; et al. Diversity of PBI-DdeI Satellite DNA in Snakes Correlates with Rapid Independent Evolution and Different Functional Roles. *Sci. Rep.* **2019**, *9*, 15459. [CrossRef] [PubMed]
28. Matsubara, K.; Tarui, H.; Toriba, M.; Yamada, K.; Nishida-Umehara, C.; Agata, K.; Matsuda, Y. Evidence for different origin of sex chromosomes in snakes, birds, and mammals and step-wise differentiation of snake sex chromosomes. *Proc. Natl. Acad. Sci. USA* **2006**, *103*, 18190–18195. [CrossRef] [PubMed]
29. Novák, P.; Robledillo, L.Á.; Koblížková, A.; Vrbová, I.; Neumann, P.; Macas, J. TAREAN: A Computational Tool for Identification and Characterization of Satellite DNA from Unassembled Short Reads. *Nucleic Acids Res.* **2017**, *45*, e111. [CrossRef] [PubMed]
30. Graphodatsky, A.S.; Yang, F.; Serdukova, N.; Perelman, P.; Zhdanova, N.S.; Ferguson-Smith, M.A. Dog Chromosome-Specific Paints Reveal Evolutionary Inter- and Intrachromosomal Rearrangements in the American Mink and Human. *Cytogenet. Genome Res.* **2000**, *90*, 275–278. [CrossRef] [PubMed]
31. Yang, F.; O'Brien, P.C.M.; Milne, B.S.; Graphodatsky, A.S.; Solanky, N.; Trifonov, V.; Rens, W.; Sargan, D.; Ferguson-Smith, M.A. A Complete Comparative Chromosome Map for the Dog, Red Fox, and Human and Its Integration with Canine Genetic Maps. *Genomics* **1999**, *62*, 189–202. [CrossRef] [PubMed]
32. Chen, S.; Zhou, Y.; Chen, Y.; Gu, J. Fastp: An Ultra-Fast All-in-One FASTQ Preprocessor. *Bioinformatics* **2018**, *34*, i884–i890. [CrossRef] [PubMed]
33. Johnson, M.; Zaretskaya, I.; Raytselis, Y.; Merezhuk, Y.; McGinnis, S.; Madden, T.L. NCBI BLAST: A Better Web Interface. *Nucleic Acids Res.* **2008**, *36*, W5–W9. [CrossRef]
34. Bao, W.; Kojima, K.K.; Kohany, O. Repbase Update, a Database of Repetitive Elements in Eukaryotic Genomes. *Mob DNA* **2015**, *6*, 11. [CrossRef]
35. Owczarzy, R.; Tataurov, A.v.; Wu, Y.; Manthey, J.A.; McQuisten, K.A.; Almabrazi, H.G.; Pedersen, K.F.; Lin, Y.; Garretson, J.; McEntaggart, N.O.; et al. IDT SciTools: A Suite for Analysis and Design of Nucleic Acid Oligomers. *Nucleic Acids Res.* **2008**, *36*, W163–W169. [CrossRef] [PubMed]
36. Biltueva, L.S.; Prokopov, D.Y.; Makunin, A.I.; Komissarov, A.S.; Kudryavtseva, A.v.; Lemskaya, N.A.; Vorobieva, N.V.; Serdyukova, N.A.; Romanenko, S.A.; Gladkikh, O.L.; et al. Genomic Organization and Physical Mapping of Tandemly Arranged Repetitive DNAs in Sterlet (*Acipenser ruthenus*). *Cytogenet. Genome Res.* **2017**, *152*, 148–157. [CrossRef]

37. Romanenko, S.A.; Biltueva, L.S.; Serdyukova, N.A.; Kulemzina, A.I.; Beklemisheva, V.R.; Gladkikh, O.L.; Lemskaya, N.A.; Interesova, E.A.; Korentovich, M.A.; Vorobieva, N.v.; et al. Segmental Paleotetraploidy Revealed in Sterlet (*Acipenser ruthenus*) Genome by Chromosome Painting. *Mol. Cytogenet.* **2015**, *8*, 90. [CrossRef]
38. Singchat, W.; Ahmad, S.F.; Sillapaprayoon, S.; Muangmai, N.; Duengkae, P.; Peyachoknagul, S.; O'Connor, R.E.; Griffin, D.K.; Srikulnath, K. Partial amniote sex chromosomal linkage homologies shared on snake W sex chromosomes support the ancestral super-sex chromosome evolution in amniotes. *Front. Genet.* **2020**, *11*, 948. [CrossRef]
39. Augstenová, B.; Mazzoleni, S.; Kratochvíl, L.; Rovatsos, M. Evolutionary dynamics of the W chromosome in caenophidian snakes. *Genes* **2017**, *9*, 5. [CrossRef]
40. Zaher, H.; Murphy, R.W.; Arredondo, J.C.; Graboski, R.; Machado-Filho, P.R.; Mahlow, K.; Montingelli, G.G.; Quadros, A.B.; Orlov, N.L.; Wilkinson, M.; et al. Large-Scale Molecular Phylogeny, Morphology, Divergence-Time Estimation, and the Fossil Record of Advanced Caenophidian Snakes (Squamata: Serpentes). *PLoS One* **2019**, *14*, e0216148. [CrossRef]
41. Ruiz-Ruano, F.J.; López-León, M.D.; Cabrero, J.; Camacho, J.P.M. High-Throughput Analysis of the Satellitome Illuminates Satellite DNA Evolution. *Sci. Rep.* **2016**, *6*, 28333. [CrossRef] [PubMed]

 *animals*

Article

Cytogenetic Analysis of the Bimodal Karyotype of the Common European Adder, *Vipera berus* (Viperidae)

Victor Spangenberg [1,2,*,†], Ilya Redekop [1,2,3,†], Sergey A. Simanovsky [2] and Oxana Kolomiets [1]

1 Vavilov Institute of General Genetics, RAS, Moscow 119991, Russia
2 Severtsov Institute of Ecology and Evolution, RAS, Moscow 119071, Russia
3 Moscow Region State Pedagogical University, Mytischi 141014, Russia
* Correspondence: vspangenberg@gmail.com
† These authors contributed equally to this work.

Simple Summary: Bimodal karyotypes, including both large chromosomes and microchromosomes, are mainly found in reptiles, birds, fish, and insects, but not mammals. Studies of microchromosomes are currently of great interest. The karyotype of the snake *Vipera berus* is a prime example of a bimodal karyotype. We conducted a comparative cytogenetic study of meiotic (synaptonemal complexes in prophase I) and mitotic chromosomes. A significant asynchrony in the assembly of meiotic bivalents and the dynamics of the appearance of the mismatch repair protein MLH1 were analyzed, and a high level of meiotic recombination was shown. Furthermore, minor species-specific markers of the *V. berus* meiotic karyotype were identified.

Abstract: *Vipera berus* is the species with the largest range of snakes on Earth and one of the largest among reptiles in general. It is also the only snake species found in the Arctic Circle. *Vipera berus* is the most involved species of the genus *Vipera* in the process of interspecific hybridization in nature. The taxonomy of the genus *Vipera* is based on molecular markers and morphology and requires clarification using SC-karyotyping. This work is a detailed comparative study of the somatic and meiotic karyotypes of *V. berus*, with special attention to DNA and protein markers associated with synaptonemal complexes. The karyotype of *V. berus* is a remarkable example of a bimodal karyotype containing both 16 large macrochromosomes and 20 microchromosomes. We traced the stages of the asynchronous assembly of both types of bivalents. The number of crossing-over sites per pachytene nucleus, the localization of the nucleolar organizer, and the unique heterochromatin block on the autosomal bivalent 6—an important marker—were determined. Our results show that the average number of crossing-over sites per pachytene nucleus is 49.5, and the number of MLH1 sites per bivalent 1 reached 11, which is comparable to several species of agamas.

Keywords: prophase I of meiosis; synaptonemal complex; crossing-over; recombination rate; nucleolar organizer; NOR; bimodal karyotype; microchromosomes; heterochromatin; chiasma

Citation: Spangenberg, V.; Redekop, I.; Simanovsky, S.A.; Kolomiets, O. Cytogenetic Analysis of the Bimodal Karyotype of the Common European Adder, *Vipera berus* (Viperidae). *Animals* **2022**, *12*, 3563. https://doi.org/10.3390/ani12243563

Academic Editor: Ettore Olmo

Received: 28 November 2022
Accepted: 15 December 2022
Published: 16 December 2022

1. Introduction

The term "viper" is often used to refer to reptiles of the subfamily Viperinae. This subfamily includes 13 genera of snakes rather diverse in terms of morphological criteria that are widespread on all continents of the Old World except for Madagascar [1]. The topographic origin of vipers is still debated, but it certainly was not Europe [2–4]. Snakes of the genus *Vipera* (Laurenti, 1768) are the most common venomous snakes in Europe and Western and Central Asia [5]. The number of species varies from 10 to 20 according to different studies, since the taxonomic status of some species and subspecies remains questionable [6–9]. These snakes inhabit extremely diverse ranges—from deserts to alpine meadows and even northern territories beyond the Arctic Circle [10].

One of the youngest evolutionary forms is the genus *Vipera*, or True vipers [11]. Particularly rapid speciation of this genus took place during the Pliocene and Pleistocene [12,13].

The most widespread species among true vipers and snakes in general, is *Vipera berus* (Linnaeus, 1758). It can be found from west to east, from the British Isles to the Sakhalin Islands, and from north to south, from Scandinavia to the Balkans [14,15]. In addition, the common viper is the only snake found in the Arctic Circle [16,17].

Numerous studies describe the ecology and variety of forms of the genus *Vipera* [18–32], including proteomic studies of the characteristics of their venom [33–36] and medical studies of their antiproliferative and cytotoxic effects in oncosuppression [10,37,38].

On the other hand, there are only few comparative molecular biological studies of the species [39–42]. Cytogenetic studies of the *Vipera* genus were carried out mainly on mitotic metaphase chromosomes or the preparation of meiocytes using light microscopy [43–48].

The most interesting problems of natural [9,49–53] or laboratory [54] hybridization between different viper species have not yet been studied using molecular cytogenetic approaches. Moreover, there is still no detailed comparative analysis of somatic and meiotic karyotypes of the key species involved in hybridization.

Bimodal karyotypes, i.e., karyotypes with a significant difference in the sizes of macro- and microchromosomes, have been described in reptiles [55,56], birds [57–59], amphibians [60], fish [61,62], and some insects [63], but surprisingly not in mammals [64]. The question of why natural selection fixed these small chromosomes in each specific lineage remains unresolved. There is great interest in such karyotypes for several reasons: they raise the question of the independent evolution of macro- and microchromosomes [55], and they present data on low heterochromatin levels in microchromosomes, higher recombination rates, a higher-mutation rate, and higher gene density in microchromosomes [64,65]. Enrichment of specific genes in microchromosomes has been revealed in snakes [64–67]. Comparative studies in several snake taxa demonstrated the conservative structure of their macro- and microchromosome sets [67]. In general, snakes could have achieved genomes characterized by higher levels of compartmentalization and smaller chromosomes, possibly resulting in an increased frequency of recombination and a greater level of speciation [68].

Synaptonemal complex (SC) karyotyping is one of the most informative methods for studying molecular markers at the chromosomal level [69].

The aim of this work was a detailed comparative cytogenetic study of the somatic and meiotic karyotypes of the most widespread Viperidae species, *V. berus*, to identify minor species-specific characteristics of the SC-karyotype which are nondetectable on the mitotic chromosomes only, to study centromere positions in microchromosomes, nucleolar organizer location, and the distribution of the crossing-over marker, namely, the MLH1 protein.

We believe that the pronounced bimodality of viper karyotypes can provide an excellent model for cytogenetic and genomic studies in Viperidae, snakes, and vertebrates as well.

2. Materials and Methods

2.1. Specimens

Three adult males and two females of *V. berus* were captured in 2019–2022 in the Tver region, Konakovsky district, and examined from May to October 2022. The manipulations with the animals followed the international rules of the Manual on Humane Use of Animals in Biomedical Research.

2.2. Mitotic Chromosome Preparation

Mitotic chromosome preparations were obtained from male and female individuals using a direct suspension technique described below. The bone marrow form ribs and the spleens were suspended in 10 mL of a 0.075 M KCl hypotonic solution and incubated for 20 min at room temperature; then, 1 mL of the freshly prepared 3:1 methanol–acetic acid fixative was added, and the cell suspension was centrifuged for 5 min at 1000 rpm. Afterwards, the supernatant was discarded, 5 mL of the fixative were added, and the cell suspension was kept at 4 °C for 15–20 min. These procedures were repeated two more times. After the third centrifugation and the elimination of the supernatant, 0.5–1.0 mL of the fixative was added, and the final cell suspension was left for storage at −20 °C. To prepare mitotic chromosome slides, several small drops of the cell suspension were released onto various sections of a slide previously maintained in distilled water at 4 °C; then, the slides were transferred to a hot plate (45 °C) for drying. The mitotic chromosome slides were stained conventionally with 4% Giemsa solution in a phosphate buffer solution at pH 6.8 for 8 min or mounted in a Vectashield antifade mounting medium with DAPI, 4′,6-diamidino-2-phenylindole (Vector Laboratories H-1200, Newark, CA, USA).

2.3. Total Preparation of SCs and Immunostaining

Seminiferous tubules were isolated and disaggregated in the phosphate-buffered saline (PBS) (PanEco, Moscow, Russia). A spread of spermatocytes I nuclei preparations were performed according to Spangenberg (2022) [69]. Poly-L-lysine-coated slides were used in all immunofluorescence studies. The slides stored in −20 °C were moved to room temperature, washed with phosphate-buffered saline (PBS) for 1 min, and incubated overnight at 4 °C with primary antibodies diluted in the antibody dilution buffer (ADB: 3% bovine serum albumin and 0.05% Triton X-100 in PBS). Axial elements of meiotic chromosomes were detected using rabbit polyclonal antibodies against the Synaptonemal complex protein 3 (SYCP3) protein (1:250; Abcam ab15093, Cambridge, UK). Centromeres were detected using the antikinetochore proteins' antibodies ACA (1:500; Antibodies Incorporated 15–234, Davis, CA, USA), known also as CREST-syndrome antisera. Antibodies against DNA mismatch repair protein MLH1 (1:250; Abcam, Cambridge, UK) were used for the detection of the late recombination sites, prospective chiasmata. Nucleolus was detected by mouse antifibrillarin monoclonal antibodies (1:250; Abcam, Cambridge, UK). After washing in PBS, the secondary antibodies diluted in Antibody Dilution Buffer (ADB) were used, namely, goat antirabbit immunoglobulin G, Alexa Fluor 488 (1:500; Abcam, Cambridge, UK), goat antimouse Alexa Fluor 555 (1:500; Invitrogen, Carlsbad, CA, USA), and goat antihuman Alexa Fluor 555 (1:500; Invitrogen, Carlsbad, CA, USA). Incubation with secondary antibodies was performed in a humid chamber at 37 °C for 2 h.

2.4. Microscopy

The synaptonemal complex slides and mitotic chromosome slides, stained with DAPI, were examined using the Leica DM microscope equipped with the Axiocam HRm CCD camera and filter sets A, I3, and N2.1, and processed with AxioVision Release 4.8. software (Carl Zeiss, Oberkochen, Germany). All preparations were mounted in a Vectashield antifade mounting medium with DAPI (Vector Laboratories H-1200, Newark, CA, USA). The mitotic chromosome slides, conventionally stained with Giemsa, were examined using an Axioplan 2 Imaging microscope (Carl Zeiss, Germany) equipped with a CV-M4+CL camera (JAI, Kanagawa, Japan) and the Ikaros software (MetaSystems, Altlussheim, Germany).

2.5. Image Analysis

Synaptonemal complex measurements were performed with ImageJ software, release 1.53k (Bethesda, AR, USA). Criteria of identification of distinct meiotic prophase I stages were used in accordance with our previous studies in reptiles [70,71]. The Origin Pro software package (OriginLab Corp., Northampton, MA, USA) was used for descriptive statistics and diagram construction.

Mitotic and synaptonemal complex karyotypes were arranged according to the centromere position following Levan et al. (1964) [72] but modified as metacentric (m), submetacentric (sm), and sub-telocentric/acrocentric (st/a). Chromosome pairs were arranged according to their size. To determine the chromosomal arm number per karyotype (fundamental number, FN), metacentrics and sub-metacentrics were considered as biarmed and sub-telocentrics/acrocentrics as monoarmed.

3. Results

3.1. Mitotic Metaphase Karyotyping and Karyotypic Formula

Both male and female mitotic karyotypes of *V. berus* have 2n = 36 and consist of 6 metacentrics (pairs 1, 3, and 4), 8 sub-metacentrics (pairs 2, 5, 7, and 8), 2 subtelocentrics/acrocentrics (pair 6), and 20 microchromosomes (pairs 9–18) (Figure 1). Pair 4 is homomorphic in the male karyotype (ZZ) and heteromorphic in the female one (ZW) (Figure 1d,f). Z and W chromosomes are metacentric. The W chromosome is similar in size to pairs 7 and 8 but differs from them in morphology (pairs 7 and 8 are sub-metacentric) (Figure 1d). DAPI staining revealed the presence of an AT-poor region on the long arm of the W chromosome (Figure 1b). The morphology of the microchromosomes is not distinguishable after mitotic metaphase karyotyping. However, analysis of synaptonemal complexes (below in the Section 3.2.2.) using SYCP3 and ACA immunostaining reveal the morphology of microchromosomes: 2 microchromosomes are sub-metacentric (pair 13) and 18 microchromosomes are sub-telocentric/acrocentric (pairs 9–12 and 14–18) (Figure 2c). Thus, the karyotypic formula for both males and females of *V. berus* is 6m + 10sm + 20st/a, FN = 52.

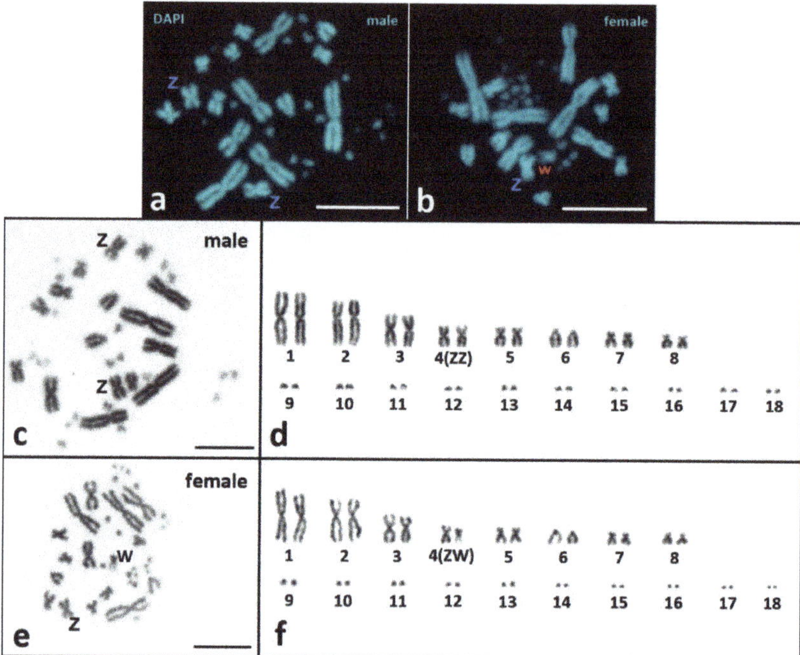

Figure 1. Mitotic karyotypes of *V. berus*. (**a**,**b**) DAPI staining and (**c**–**f**) conventional Giemsa staining. (**a**,**c**,**d**) Male karyotypes; (**b**,**e**,**f**) Female karyotypes. (**a**–**c**,**e**) Metaphase chromosome plates; (**d**,**f**) karyograms. ZZ/ZW—sex chromosomes. Scale bar—10 μm.

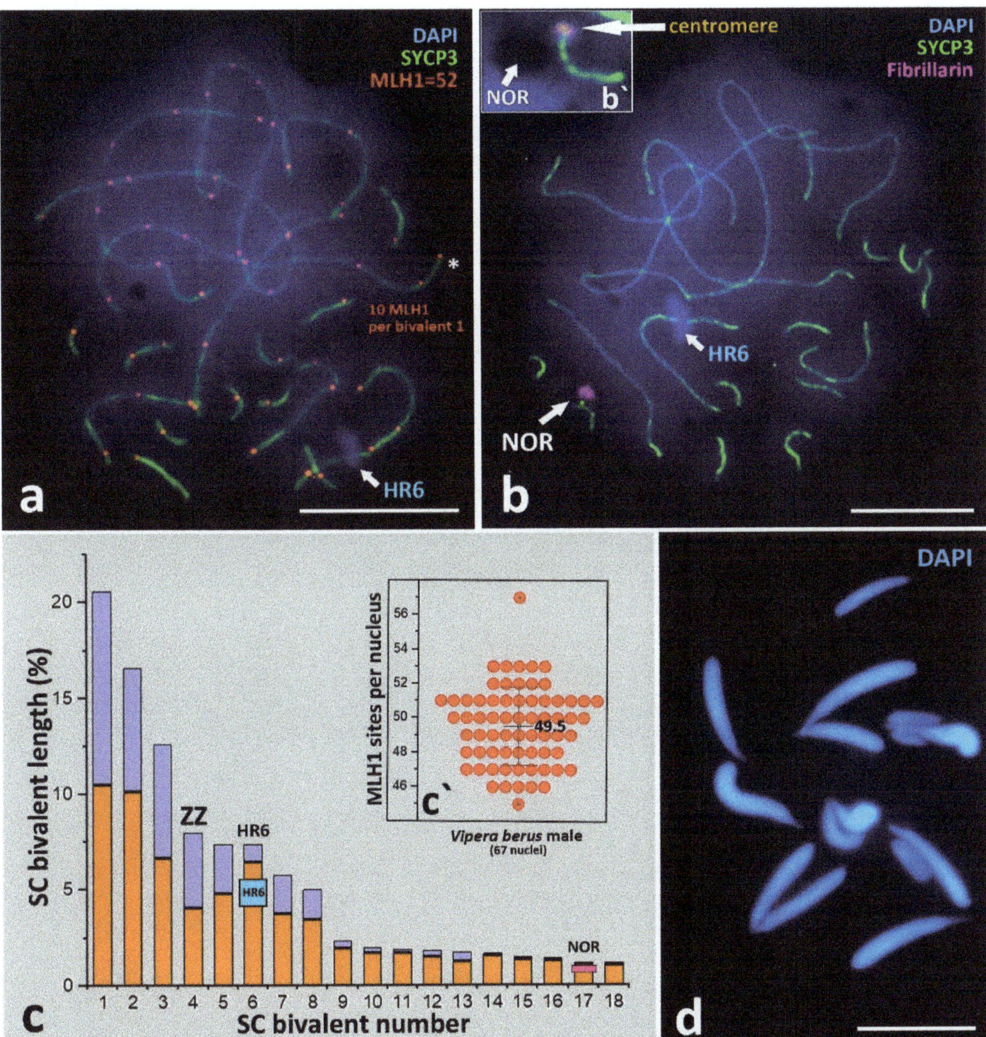

Figure 2. Markers of *V. berus* SC-karyotype, ideogram, and spermatids morphology. (**a**) Immunodetection of the crossing-over marker, MLH1 protein. A total of 52 MLH1 sites per nucleus, and 10 MLH1 sites on the bivalent 1 (asterisk). (**b**) Immunodetection of Nucleolar Organizer (NOR) on the microchromosome bivalent and (**b′**) as the DAPI-negative region. Axial elements of chromosomes are immunostained with antibodies against the SYCP3 protein (green); crossing-over sites are immunostained with antibodies against MLH1 protein (red); NOR with antibodies against Fibrillarin protein (violet). Chromatin stained with DAPI (blue). (**c**) Ideorgam of *V. berus* SC-karyotype. Blue—p arms; orange—q arms. Metacentric Z-chromosome, heterochromatin region (HR6), and NOR are indicated. Heterochromatin region on the chromosome pair 6 is indicated as HR6. (**c′**) Number of MLH1 foci per spermatocyte nucleus (mean ± SD). (**d**) Spermatids, DAPI staining (blue). Bar—10 μm.

3.2. *Immunocytochemical Analysis of Meiotic Prophase I Nuclei of V. berus spermatocytes I*

A total of 338 spermatocyte nuclei of *V. berus* at different stages of meiotic prophase I were studied. Immunostaining of protein markers allowed us to describe, in detail, all stages, which for convenience are divided into presynaptic, alignment, and postsynaptic stages.

3.2.1. Presynaptic Stages

Leptotene

The leptotene stage of *V. berus* primary spermatocytes has characteristics according to the classical description in the scientific literature as the "stage of thin long threads" or "tangled mass of threads" (Figure 3a) [73,74], which differs from those previously studied in reptiles, where the leptotene has mostly fragmented axial elements [70].

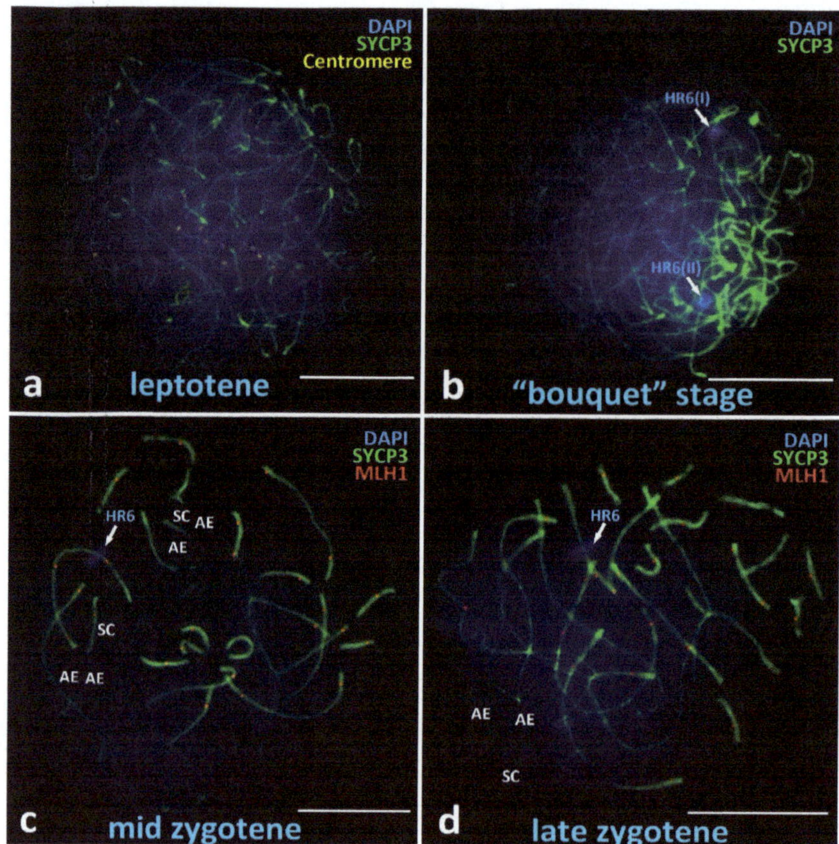

Figure 3. Presynaptic stages in *V. berus* primary spermatocytes. (**a**) Leptotene, long axial elements are completely asynapted, centromeres distributed over all spread nuclei. (**b**) Chromosomal 'bouquet' stage, U-shaped axial elements of chromosomes, clusterization of telomere ends in the local region of the spread nucleus. (**c**) Mid zygotene, asynchronous synapsis of the long chromosomes and microchromosomes. MLH1 protein loaded in bivalents of microchromosomes as well as in the regions of local synapsis of long assembling bivalents. (**d**) Late zygotene, finalization of assembly of the long bivalents. Bivalents of microchromosomes are fully assembled. Heterochromatin region on the chromosome pair 6 is indicated as HR6. AE—axial elements of asynapted chromosomes. SC—assembled regions of synaptonemal complexes. Axial elements of chromosomes are immunostained with antibodies against the SYCP3 protein (green), centromeres with antikinetochore antibodies ACA (yellow). Mismatch repair protein sites are immunostained with antibodies against MLH1 protein (red). Chromatin stained with DAPI (blue). Bar—10 μm.

Chromosomal "Bouquet" Stage

The chromosomal "bouquet" stage demonstrates clustering of the telomeric ends of all univalents and U-shaped chromosomes (Figure 3b). At this stage, we have not been able to identify clear differences between macro- and microchromosomes.

Zygotene

The zygotene stage in *V. berus* is the most remarkable meiotic prophase I stage and is characterized by asynchronous assembly of synaptonemal complexes (Figure 3c,d). In the zygotene stage, differences in the dynamics of the assembly of both types of bivalents (macro- and micro-) are clearly visible. In the middle zygotene stage (Figure 3c), all the small bivalents are already assembled in synaptonemal complexes, while all the long bivalents demonstrate significant regions of asynaptic axial elements (AE) and only short peritelomeric regions of local synapsis (SC). In the late zygotene stage, only small areas of asynapsis remain in the interstitial areas of large bivalents (Figure 3d). It is important to note the fairly early loading of the MLH1 and the mismatch repair protein (prospective chiasma sites) into bivalents already at the zygotene stage. MLH1 protein sites are found both in fully assembled short bivalents as well as in regions of partial synapsis of the long bivalents (Figure 3c,d).

3.2.2. Alignment Stage and Postsynaptic Stages

Alignment Stage

Among the identified meiotic prophase I stages in *V. berus*, a rare stage of chromosome alignment should be noted, which was found in only 5 of the 338 nuclei studied (Figure 4a). This stage, which is described in many species, including reptiles [70], is characterized by the spatial alignment of pairs of axial elements of homologous chromosomes opposite to each other, separated by about two times as far as in the assembled SCs with clearly visible space between axes, and there is no loading of MLH1 yet [74].

Pachytene

The pachytene nuclei have complete synapsis of all 18 bivalents (Figures 4b and A3), easily defined bivalent arm lengths separated by immunostained centromeres (Figure 4b), and fused NORs located on one of the microchromosome bivalents (Figure 2b). A specific problem in the majority of *V. berus* bimodal SC-spread preparations is the overlap and entanglement of the first three very long macrobivalents. To solve this problem, a large number of photographs were taken, and the most suitable were selected for measurements and analysis. Nevertheless, pachytene nuclei are most convenient for studying the basic markers of meiotic prophase I, which we describe in detail below in the section "SC-karyotyping and meiotic prophase I markers analysis".

Diplotene

The diplotene stage is characterized by desynapsis of homologues and gradual unloading of the SYCP3 protein from axial (lateral) elements (Figure 4c). In the late diplotene, the chromosome axes become diffuse (Figure 4d). At the diplotene stage, chiasmata are clearly visible, the number of which corresponds to the number of MLH1 sites in pachytene, including those for macrobivalents (Figures 2a and 4c,d).

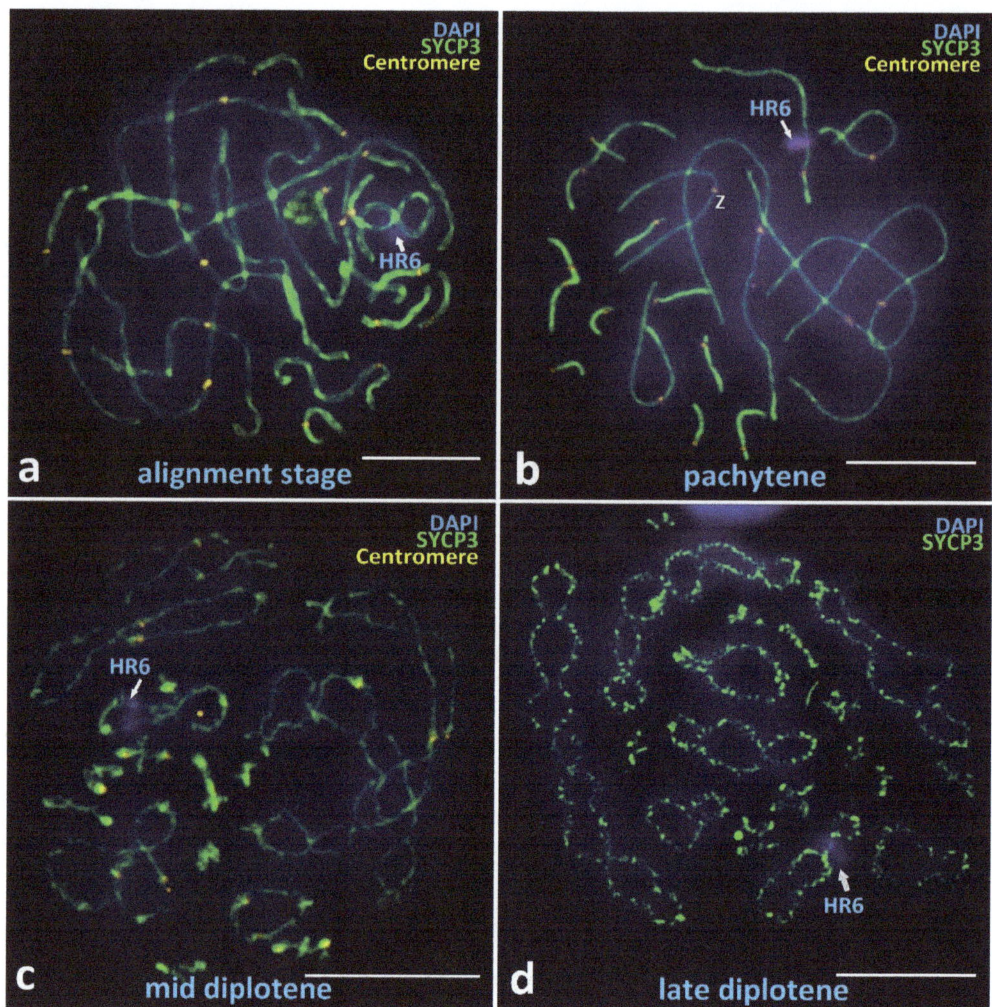

Figure 4. Alignment stage and postsynaptic stages in *V. berus* primary spermatocytes. (**a**) Alignment stage, spatial alignment of pairs of axial elements of homologous chromosomes located opposite each other. (**b**) Pachytene, complete synapsis of all 18 bivalents. (**c**) Mid diplotene, desynapsis of homologues, chiasmata. (**d**) Late diplotene, diffuse axial elements, unloading of the SYCP3, chiasmata. Heterochromatin region on the chromosome pair 6 is indicated as HR6. Axial elements of chromosomes are immunostained with antibodies against the SYCP3 protein (green), centromeres with antikinetochore antibodies ACA (yellow). Mismatch repair protein sites are immunostained with antibodies against MLH1 protein (red). Chromatin stained with DAPI (blue). Bar—10 µm.

3.2.3. SC-Karyotyping and Meiotic Prophase I Markers Analysis

Crossing-Over Marker, MLH1 Protein

Immunostaining of crossing-over associated protein MLH1 revealed a surprisingly high number of sites per pachytene nucleus. The findings revealed an average of 49.5 ± 2.27 (mean $\pm$ SD) MLH1 sites (a minimum of 45 sites and a maximum of 57) on 18 pachytene bivalents (n = 18, FN = 52) (Figures 2a and A3a). In addition, *V. berus* pachytene nuclei demonstrated a high number of MLH1 sites per one macrochromosome bivalent. Here, we revealed up to 11 MLH1 foci on the bivalent (Figure A3b). Ten microchromosomes have, on average, 11.06 ± 1.01 MLH1 sites.

Unique Heterochromatic Chromatin Region (HR6) on the Bivalent 6

One important finding is the heterochromatin region that was clearly visible on SC-spreads on the q arm of the bivalent 6 in all postzygotene nuclei we studied (Figure 3c,d and Figure 4a,d; Figures A2 and A3), indicated as HR6. Bivalent 6 was assigned to the subtelocentric/acrocentric (st/a) type. HR6 is a DAPI-intensive region, located separately from centromere but also detectable with immunostaining using ACA antibodies (Figure 4b). In addition, we confirmed the presence of HR6 before synapsis at the chromosomal "bouquet" stage as two separated heterochromatin blocks, HR6(I) and HR6(II), located separately on the yet asynapted axial elements (Figures 3b and A1).

Sex Z Chromosome Identification in the SC-Karyotype

The immunostaining of centromere proteins on the pachytene bivalents (Figure 4b) and the data of mitotic karyotyping (Figure 1) allowed us to determine Z bivalent as the only metacentric bivalent 4, which was clearly different from the bivalent 5 (sm) and bivalent 6 (st/a) of similar length. Detailed data are presented in the ideogram (Figure 2c).

Nucleolar Organizer Region (NOR)

We detected NOR location using a combination of two criteria: immunostaining with antibodies against Fibrillarin protein and the "empty" region on the DAPI staining of chromatin (NOR is an RNA- and protein-enriched region which has very weak DAPI staining). Thus, we concluded that NOR is most likely located on the micro-bivalent 17 in the proximal region to the centromere (Figure 2b,b′). However, the difference between chromosomes 17 and 18 is so small that it does not allow us to detect accurately the localization of NORs.

Spermatids

Spermatids were observed on the SC preparations using fluorescent microscopy. Spermatids of *V. berus* displayed typical characteristics of reptiles, including an elongated head and uniform chromatin staining, which corresponded to descriptions of reptile sperms in other papers [75].

4. Discussion

In many species, tiny microchromosomes are morphologically indistinguishable (dot-shaped) on mitotic metaphase plates [63]. Molecular cytogenetic techniques such as whole chromosome painting and comparative genomic hybridization are powerful methods for comparing and identifying specific microchromosomes of interest in the preparation of mitotic metaphase plates [76,77]. Basic karyological information is crucial to link cytogenetic and genomic data [78–80].

On another hand, our results indicate the applicability of comparative studies of both the somatic and meiotic chromosomes for the most complete analysis of not previously described DNA and protein markers of the bimodal karyotype.

- Our results in mitotic chromosome karyotyping are in agreement with cytogenetic data obtained previously for *V. berus*, except for minor differences in the classifications of some macrochromosomes [44]. The karyotype of *V. berus* described by us consists of 16 macrochromosomes (6m + 8sm + 2st/a) and 20 microchromosomes (2sm + 18st/a), FN = 52. For the first time, we revealed the morphology of microchromosomes in *Vipera* using high-resolution SC-karyotyping. Karyotypes with 16 macro- and 20 microchromosomes were also described previously in *V. ursinii*, *V. latastei*, and *V. seoanei* [45,81]. On the other hand, *V. aspis* and *V. ammodytes* karyotypes with 22 macro- and 20 microchromosomes were reported [45,48,81]. The presence of two variants of karyotypes in the genus *Vipera* is intriguing, and further studies on both mitotic and meiotic chromosomes are needed to understand the evolution of the karyotypes in the genus.

- Preparations of metaphase plates obtained during the study showed a structured distribution of macro- and microchromosomes: microchromosomes clustered closer to each other, forming a "microchromosome zone". This was most clearly observed on the weakly and medium-spread mitotic metaphase plates (Figure 1b,e). This partially corroborates the data of other authors, who describe a certain order of arrangement of chromosomes on metaphase plates [67]. Indeed, studies of bimodal karyotypes of birds and reptiles suggest that microchromosomes interact strongly and regularly locate together in somatic nuclei at interphase and during cell division, suggesting their functional coherence [67].
- Synaptonemal complex spread preparations provide detailed visualization of meiotic SC bivalents, which are from three- to five-times longer than mitotic metaphase chromosomes [82]. The *V. berus* karyotype is a striking example of a bimodal karyotype, combining both very large chromosomes and many microchromosomes. Thus, SC-karyotyping is a logical and very useful method in this regard, as it has allowed us to compare lengths, centromere and NOR positions, and the number of crossing-over sites in microchromosomes.

The performed immunocytochemical analysis of the stages of prophase I of meiosis makes it possible to analyze the dynamics of the synapsis of homologous pairs of macro- and microchromosomes.

It should be noted that it was almost impossible to find the pachytene stage (with completely assembled SCs) on some slides of spermatocytes I spreads. On the contrary, partial asynapsis was always found in macrobivalents. This is due to the asynchronous assembly of bivalents: microchromosomes are far ahead of long chromosomes. Therefore, an important methodological result for us was the obligatory use of tissues from different sectors of the gonad. This method, in our opinion, can be recommended for working with preparations of synaptonemal complexes of organisms with bimodal karyotypes.

The number of crossing-over sites is often considered in the context of rates of evolution [78]. Here, we immunostained MLH1 sites in the bimodal pachytene SC-karyotype of *V. berus* (Figures 2a and A3). The average number of MLH1 sites was 49.5 ± 2.27 (57 sites at maximum (Figure A3a)) on 18 pachytene bivalents (n = 18, FN = 52), which is a little lower than the crossing-over champion described in Agamidae with 69.2 MLH1 sites on 23 acrocentric bivalents (n = 23; FN = 46) [83]. Theoretically, at least one MLH1 site is needed on every bivalent for successful chiasmata formation and correct segregation in metaphase I to avoid aneuploidy of gametes. In *V. berus*, we detected an average of 11.06 ± 1.01 MLH1 sites on ten micro-bivalents, which is approximately 22.3% of the total number of MLH1 sites per nucleus (49.5 ± 2.27) (Figure 2a,c'). Of all the 18 SC bivalents in *V. berus*, these ten micro-bivalents take only 16.7% of the total SC length. Studies of bimodal karyotypes in birds and reptiles suggest that such chromosomal architecture may be connected with unequal rates of sequence evolution within one genome [64,84]. For instance, the enrichment of specific genes in microchromosomes and their intense evolution have been proposed for snakes [64,65]. On the other hand, seasonal variations in the frequency and distribution of chiasmata (terminal/intermediate) may be associated with the concentration of steroid hormones during a year [85].

The heterochromatic region we detected is a strong and easily detected marker revealed in all the nuclei under study. Our preliminary data (not shown) suggest this region is species-specific for *V. berus* and important for further comparative cytogenetic studies of closely related species of the genus *Vipera*. The fact that HR6 is detectable on anticentromere proteins immunostaining and is located separately from the centromere can be connected with the very broad specificity of these antibodies. Indeed, ACA (or CREST) antiserum is a cocktail of different anti-CENP protein family antibodies. Further studies are needed to detail the origin and specifics of the HR6 region in *V. berus* and other closely related species.

Localization of NOR on the pair of microchromosomes we revealed using antifibrillarin antibodies and DAPI staining is similar to our previous study of NOR in Lacertidae oocytes [86]. Our result is in accordance with known data on other snakes [48,87]. On the other hand, studies of several snake species using silver nitrate staining methods (AgNOR) or FISH revealed the location of NORs in two chromosome pairs [87–89]. In general, the *V. berus* karyotype shows several primitive characteristics, 2n = 36 with 16 macro- and 20 microchromosomes, fourth pair ZW heteromorphic sex chromosomes, and NORs on one microchromoosme pair described for snakes [90,91].

Further studies of Viperidae genomes are needed, such as the analysis of repetitive DNAs (rDNA, satellite DNA) and their distribution between macro- and microchromosomes [92]. Comparative SC-karyotyping of closely related species within the genus *Vipera* could help to determine the taxonomic status of many known forms and subspecies.

5. Conclusions

We performed a detailed comparative study of mitotic and meiotic karyotypes of *V. berus*. We described important protein and DNA markers, some of which are impossible to detail with mitotic metaphase karyotyping only. These markers allowed us to distinguish autosomal chromosomes of similar length, the sex chromosomes pair, and the localization of nucleolar organizer.

We traced synaptonemal complexes assembly and disassembly on the successive stages of meiotic prophase I in the bimodal karyotype of *V. berus* and visualized the highly asynchronous synapsis of macro- and microchromosomes in detail.

We revealed that the average number of sites of the crossing-over marker MLH1 in *V. berus* is 49.5 per spermatocyte's nucleus and up to 11 MLH1 sites on the largest bivalent 1. Furthermore, we detected up to 57 MLH1 sites, which is a very high rate and can be compared to the champion species from Agamidae.

The heterochromatin region HR6 that was detailed on the bivalent 6 is an important DNA marker of the *V. berus* karyotype, and it will be useful in future comparative studies.

In general, SC-karyotyping demonstrates high applicability and will complement future studies using chromosome-scale assemblies of bimodal karyotypes and genomic studies in general [56,78,79].

Author Contributions: Conceptualization, V.S. and I.R.; methodology, V.S. and S.A.S.; software, V.S., S.A.S. and I.R.; validation, O.K., V.S. and S.A.S.; resources, V.S., O.K. and S.A.S.; data curation, V.S. and S.A.S.; writing—original draft preparation, V.S. and I.R.; writing—review and editing, V.S., S.A.S. and O.K.; visualization, V.S. and S.A.S.; funding acquisition, V.S. All authors have read and agreed to the published version of the manuscript.

Funding: This work was supported by the Russian Science Foundation grant 22-14-00227.

Institutional Review Board Statement: All experimental protocols were approved by the Ethics Committee for Animal Research of the Vavilov Institute of General Genetics (protocol No. 3, 10 November 2016) in accordance with the Regulations for Laboratory Practice.

Informed Consent Statement: Not applicable, as this research did not involve humans.

Data Availability Statement: Not applicable.

Acknowledgments: The authors would like to thank Maria Folomkina, Vladislav Starkov, Konstantin Milto, Sergey Kudryavtsev, Konstantin Lotiev, Eugene Krysanov and Igor Mazheika. We thank the Genetic Polymorphism Core Facility of the Vavilov Institute of General Genetics of the Russian Academy of Sciences, Moscow.

Conflicts of Interest: The authors declare no conflict of interest.

Appendix A

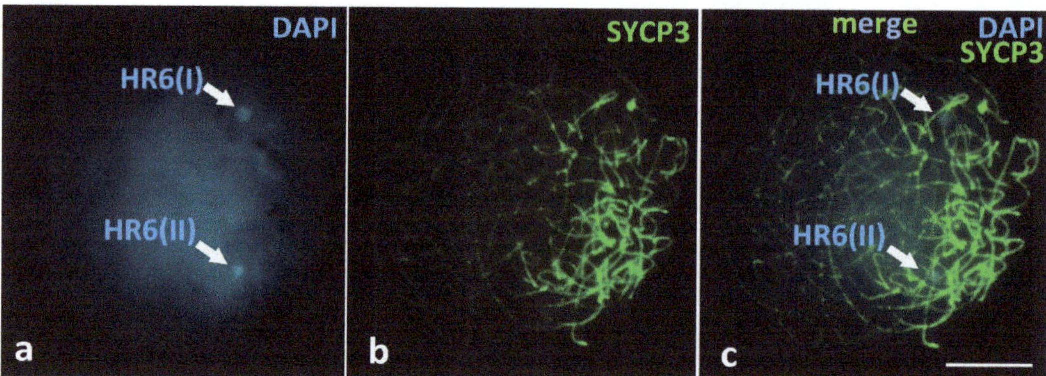

Figure A1. Chromosomal "bouquet" stage, *V. berus* spermatocyte I. Detection of the heterochromatin regions on the both chromosomes 6 before synapsis (HR6(I) and HR6(II)). (**a**) Chromatin stained with DAPI (blue), (**b**) Axial elements of chromosomes are immunostained with antibodies against the SYCP3 protein (green), and (**c**) merge. Bar—10 µm.

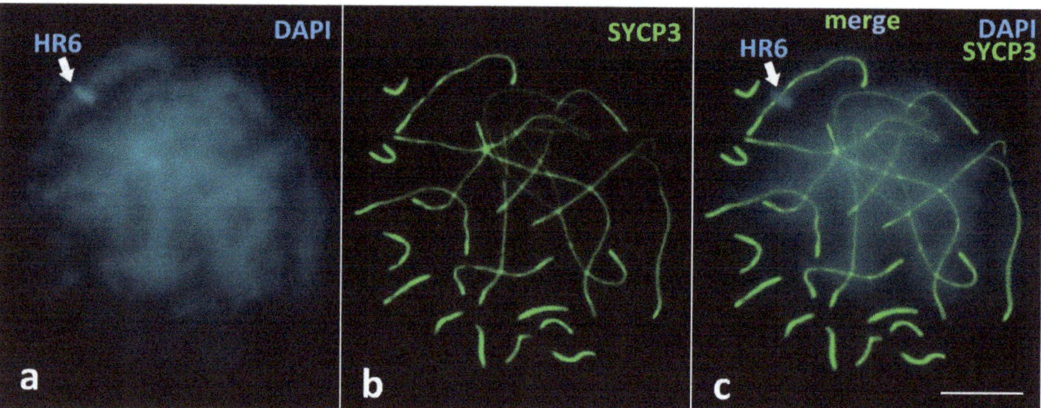

Figure A2. Late zygotene stage, *V. berus* spermatocyte I. Heterochromatin region on the bivalent 6 (HR6) after the synapsis of two homologous chromosomes 6. (**a**) Chromatin stained with DAPI (blue), (**b**) Axial elements of chromosomes are immunostained with antibodies against the SYCP3 protein (green), and (**c**) merge. Bar—10 µm.

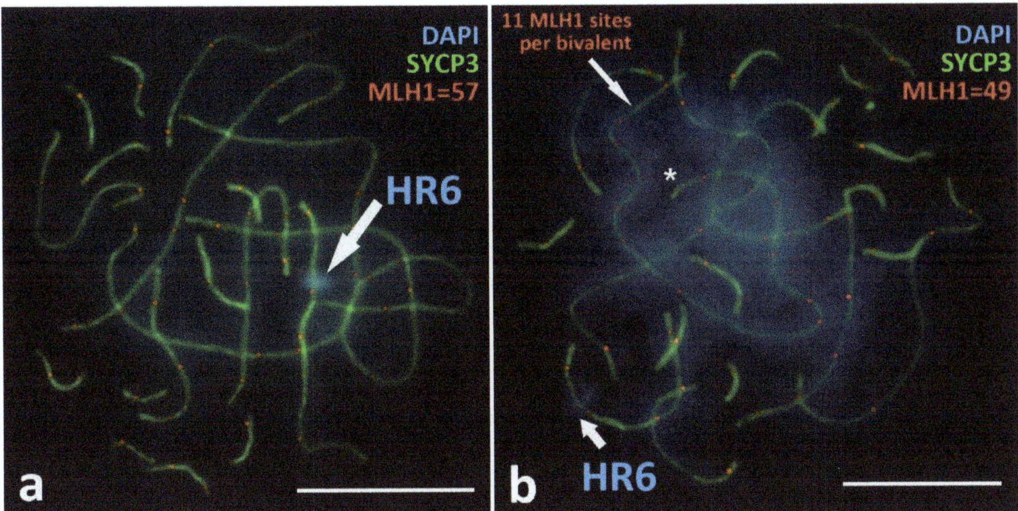

Figure A3. Pachytene stage, *V. berus* spermatocyte I. Immunodetection of crossing-over marker, MLH1 protein (red). (**a**) 57 MLH1 sites in one SC-spread. (**b**) 49 MLH1 sites in one SC-spread and 11 MLH1 sites on one macrobivalent 1 (asterisk). Heterochromatin region on the bivalent 6 (HR6) is indicated. Axial elements of chromosomes are immunostained with antibodies against the SYCP3 protein (green). Chromatin stained with DAPI (blue). Bar—10 μm.

References

1. Phelps, T. *Old World Vipers, a Natural History of the Azemiopinae and Viperinae*; Edition Chimaira: Frankfurt, Germany, 2010; 558p.
2. Wüster, W.; Peppin, L.; Pook, C.E.; Walker, D.E. A nesting of vipers: Phylogeny and historical biogeography of the Viperidae (Squamata: Serpentes). *Mol. Phylogen. Evol.* **2008**, *49*, 445–459. [CrossRef]
3. Szyndlar, Z.; Rage, J.C. Fossil Record of the True Vipers. Biology of the Vipers, Eagle Mountain Publishers 2002, pp. 419–444. Available online: https://scholar.google.com/scholar?hl=ru&as_sdt=0%2C5&q=+Fossil+record+of+the+true+vipers.+Biol.+Vipers+2002%2C+419%E2%80%93444&btnG= (accessed on 10 November 2022).
4. Gislen, T.; Kauri, H. Zoogeography of the Swedish amphibians and reptiles. With notes on their growth and ecology. *Acta Vertebr.* **1959**, *1*, 197–397.
5. Archundia, I.G.; de Roodt, A.R.; Ramos-Cerrillo, B.; Chippaux, J.P.; Olguín-Pérez, L.; Alagón, A.; Stock, R.P. Neutralization of *Vipera* and *Macrovipera* venoms by two experimental polyvalent antisera: A study of paraspecificity. *Toxicon* **2011**, *57*, 1049–1056. [CrossRef]
6. Gasc, J.P.; Cabela, A.; Crnobrnja-Isailovic, J.; Dolmen, D.; Grossenbacher, K.; Haffner, P.; Lescure, J.; Martens, H.; Martínez-Rica, J.P.; Maurin, H.; et al. *Atlas of Amphibians and Reptiles in Europe*; Societas Europaea Herpetologica Muséum National d'Histoire Naturelle (IEGB/SPN): Paris, France, 2004; 516p.
7. Steward, J.W. *The Snakes of Europe*; Fairleigh Dickinson Univ. Press: Rutherford, NJ, USA, 1971; 238p.
8. McDiarmid, R.W.; Campbell, J.A.; Touré, T. *Snake Species of the World: A Taxonomic and Geographic Reference, Version 1*; Herpetologists' League: Washington, DC, USA, 1999; 511p.
9. Bannikov, A.G.; Darevskii, I.S.; Ischenko, V.G.; Rustamov, A.K.; Scherbak, N.N. *Identification Guide to the Amphibian and Reptile Fauna of the USSR*; Moscow "Enlightenment", USSR: Moscow, Russia, 1977; p. 416.
10. Zinenko, O.; Sovic, M.; Joger, U.; Gibbs, H.L. Hybrid origin of European Vipers (*Vipera magnifica* and *Vipera orlovi*) from the Caucasus determined using genomic scale DNA markers. *BMC Evol. Biol.* **2016**, *16*, 76. [CrossRef] [PubMed]
11. Siigur, J.; Siigur, E. Biochemistry and toxicology of proteins and peptides purified from the venom of *Vipera berus berus*. *Toxicon* **2022**, *X*, 100131. [CrossRef] [PubMed]
12. Šmíd, J.; Tolley, K.A. Calibrating the tree of vipers under the fossilized birth-death model. *Sci. Rep.* **2019**, *9*, 5510. [CrossRef] [PubMed]
13. Alencar, L.R.V.; Quental, T.B.; Grazziotin, F.G.; Alfaro, M.L.; Martins, M.; Venzon, M.; Zaher, H. Diversification in vipers: Phylogenetic relationships, time of divergence and shifts in speciation rates. *Mol. Phylogenetics Evol.* **2016**, *105*, 50–62. [CrossRef] [PubMed]
14. Terhivuo, J. Provisional atlas and status of populations for the herpetofauna of Finland in 1980–92. *Ann. Zool. Fennici.* **1993**, *30*, 55–69.

15. Saint Girons, H. Biogeographie et evolution des vipers europeennes. *C. R. Soc. Biogeogr.* **1980**, *56*, 146–172.
16. Nilsson, G.; Andren, C. *Vipera berus*. In *Atlas of Amphibians and Reptiles in Europe*; Gasc, J.P., Ed.; Museum National d'Histoire Naturelle: Paris, France; Societas Europaea Herpetologica: Paris, France, 1997; pp. 388–389.
17. Elmberg, J. Grod- och kräldjurens utbredning i Norrland. *Nat. Norr* **1995**, *14*, 57–82.
18. Andersson, S. Hibernation, habitat and seasonal activity in the adder, *Vipera berus*, north of the Arctic Circle in Sweden. *Amphibia-Reptilia* **2003**, *24*, 449–457. [CrossRef]
19. Milto, K.D.; Zinenko, O.I. Distribution and morphological variability of *Vipera berus* in Eastern Europe. Herpetologia Petropolitana. In Proceedings of the 12th Ordinary General Meeting of the Societas Europaea Herpetologica, St. Petersburg, Russia, 12–16 August 2003. Available online: https://www.lacerta.de/AF/Bibliografie/BIB_5285.pdf (accessed on 10 November 2022).
20. Bakiev, A.G.; Garanin, V.I.; Gelashvili, D.B.; Gorelov, R.A.; Doronin, I.V.; Zaitseva, O.V.; Zinenko, A.I.; Klyonina, A.A.; Makarova, T.N.; Malenev, A.L.; et al. *Vipers (Reptilia: Serpentes: Viperidae: Vipera) of the Volga Basin*; Part 1; Kassandra: Togliatti, Russia, 2015; 234p.
21. Tuniyev, B.S. Rare species of shield-head vipers in the Caucasus. *Nature Conservation Research. Zapovednaya Nauka* **2016**, *1*, 11–25. [CrossRef]
22. Pavlov, A.V. On the results of the study of vipers in the republic of Tatarstan. *Mod. Herpetol.* **2000**, *1*, 47–51.
23. Shiryaev, K.A. New data on reproductive biology of Caucasian species of the genus *Vipera*. Herpetologia Petropolitana. In Proceedings of the 12th Ordinary General Meeting of the Societas Europaea Herpetologica, St. Petersburg, Russia, 12–16 August 2003. Available online: https://www.lacerta.de/AF/Bibliografie/BIB_5285.pdf (accessed on 10 November 2022).
24. Starkov, V.G.; Utkin, Y.N. New data on systematic status of the vipers in Samara Oblast. In Proceedings of the Third Regional Conference of Herpetologists of Volga Basin, Tolyatti, Russia, 5–7 February 2003; pp. 81–82.
25. Turner, R.K.; Maclean, I.M. Microclimate-driven trends in spring-emergence phenology in a temperate reptile (*Vipera berus*): Evidence for a potential "climate trap"? *Ecol. Evol.* **2022**, *12*, e8623. [CrossRef] [PubMed]
26. François, D.; Ursenbacher, S.; Boissinot, A.; Ysnel, F.; Lourdais, O. Isolation-by-distance and male-biased dispersal at a fine spatial scale: A study of the common European adder (*Vipera berus*) in a rural landscape. *Conserv. Genet.* **2021**, *22*, 823–837. [CrossRef]
27. Bauwens, D.; Claus, K. Intermittent reproduction, mortality patterns and lifetime breeding frequency of females in a population of the adder (*Vipera berus*). *PeerJ* **2019**, *7*, e6912. [CrossRef]
28. Tamagnini, D.; Stephenson, J.; Brown, R.P.; Meloro, C. Geometric morphometric analyses of sexual dimorphism and allometry in two sympatric snakes: Natrix helvetica (Natricidae) and *Vipera berus* (Viperidae). *Zoology* **2018**, *129*, 25–34. [CrossRef]
29. Nash, D.J.; Griffiths, R.A. Ranging behaviour of adders (*Vipera berus*) translocated from a development site. *Herpetol. J.* **2018**, *28*, 155–159.
30. Prestt, I. An ecological study of the viper *Vipera berus* in southern Britain. *J. Zool.* **1971**, *164*, 373–418. [CrossRef]
31. Luiselli, L.; Filippi, E.; Di Lena, E. Ecological relationships between sympatric *Vipera aspis* and *Vipera ursinii* in high-altitude habitats of central Italy. *J. Herpetol.* **2007**, *41*, 378–384. [CrossRef]
32. Ursenbacher, S.; Carlsson, M.; Helfer, V.; Tegelström, H.; Fumagalli, L. Phylogeography and Pleistocene refugia of the adder (*Vipera berus*) as inferred from mitochondrial DNA sequence data. *Mol. Ecol.* **2006**, *15*, 3425–3437. [CrossRef] [PubMed]
33. Starkov, V.G.; Utkin, Y.N. Comparison of venoms of the *Vipera* genus snakes on the data of cationic exchange chromatography. *Actual Probl. Herpetol. Toxinology* **2001**, *5*, 88–89.
34. Ramazanova, A.S.; Zavada, L.L.; Starkov, V.G.; Kovyazina, I.V.; Subbotina, T.F.; Kostyukhina, E.E.; Dementieva, I.N.; Ovchinnikova, T.V.; Utkin, Y.N. Heterodimeric neurotoxic phospholipases A2—The first proteins from venom of recently established species *Vipera nikolskii*: Implication of venom composition in viper systematics. *Toxicon* **2008**, *51*, 524–537. [CrossRef] [PubMed]
35. Dyachenko, I.A.; Murashev, A.N.; Andreeva, T.V.; Tsetlin, V.I.; Utkin, Y.N. Analysis of nociceptive effects of neurotoxic phospholipase A2 from *Vipera nikolskii* venom in mice. *J. Venom Res.* **2013**, *4*, 1–4. [PubMed]
36. Makarova, Y.V.; Kryukova, E.V.; Shelukhina, I.V.; Lebedev, D.S.; Andreeva, T.V.; Ryazantsev, D.Y.; Balandin, S.V.; Ovchinnikova, T.V.; Tsetlin, V.I.; Utkin, Y.N. The first recombinant viper three-finger toxins: Inhibition of muscle and neuronal nicotinic acetylcholine receptors. *Dokl. Biochem. Biophys.* **2018**, *479*, 127–130. [CrossRef]
37. Samel, M.; Vija, H.; Kurvet, I.; Künnis-Beres, K.; Trummal, K.; Subbi, J.; Siigur, J. Interactions of PLA2-s from *Vipera lebetina*, *Vipera berus berus* and *Naja naja oxiana* venom with platelets, bacterial and cancer cells. *Toxins* **2013**, *5*, 203–223. [CrossRef]
38. Cristina, R.T.; Kocsis, R.; Tulcan, C.; Alexa, E.; Boldura, O.M.; Hulea, C.I.; Muselin, F. Protein structure of the venom in nine species of snake: From bio-compounds to possible healing agents. *Braz. J. Med. Biol. Res.* **2020**, *53*, e9001. [CrossRef]
39. Carlsson, M.; Söderberg, L.; Tegelström, H. The genetic structure of adders (*Vipera berus*) in Fennoscandia: Congruence between different kinds of genetic markers. *Mol. Ecol.* **2004**, *13*, 3147–3152. [CrossRef]
40. Metzger, C.; Ferchaud, A.-L.; Ursenbacher, C.G.S. New polymorphic microsatellite markers of the endangered meadow viper (*Vipera ursinii*) identified by 454 high-throughput sequencing: When innovation meets conservation. *Conserv. Genet. Resour.* **2011**, *3*, 589–592. [CrossRef]
41. Ursenbacher, S.; Monney, J.C.; Fumagalli, L. Limited genetic diversity and high differentiation among the remnant adder (*Vipera berus*) populations in the Swiss and French Jura Mountains. *Conservation* **2009**, *10*, 303–315. [CrossRef]
42. Efimov, R.V.; Zav'yalov, E.V.; Velikov, V.A.; Tabachishin, V.G. Genetic divergence of *Vipera berus* and *Vipera nikolskii* (Reptilia: Viperidae, *Vipera*) populations in lower Volga and adjacent territories assessed according to the sequences of cytochrome oxidase III and 12S ribosome RNA genes. *Russ. J. Genet.* **2008**, *44*, 240–243. [CrossRef]

43. Makino, S.; Momma, E. An idiogram study of the chromosomes in some species of reptiles. *Cytologia* **1949**, *15*, 96–108. [CrossRef]
44. Kobel, H.R. Heterochromosomen bei *Vipera berus* L. (Viperidae, Serpentes). *Experientia* **1962**, *18*, 173–174. [CrossRef] [PubMed]
45. Kobel, H.R. Morphometrische Karyotypanalyse Einiger Schalangen-Arten. *Genetica* **1967**, *38*, 1–31. [CrossRef]
46. Puzachenko, A.Y.; Baklushinskaya, I.Y.; Lyapunova, E.A.; Vlasov, A.A. Karyotype of *Vipera ursini* (Reptilia, Viperidae) from Streletskaya steppe. (In Russian, English summary). *Vestn. Zool.* **1997**, *31*, 81–82.
47. Zavialov, E.V.; Kaybeleva, E.I.; Tabachishin, V.G. A comparative karyological characteristics of Forest-Steppe Viper (*Vipera (Pelias) nikolskii*) from the small river flood-lands of the Volga and Don basins. *Curr. Stud. Herpetol.* **2006**, *5/6*, 100–103.
48. Aprea, G.; Odierna, G.; Gentilli, A.; Zuffi, M. The karyology of *Vipera aspis*, *V. atra*, *V. hugyi*, and *Cerastes vipera*. *Amphibia-Reptilia* **2006**, *27*, 113–119.
49. Zinenko, O. New data about hybridization between *Vipera nikolskii* (Vedmederya, Grubant et Rudaeva, 1986) and *Vipera berus* berus (Linnaeus, 1758) and their contact zones in Ukraine. *Mertensiella* **2004**, *15*, 17–28.
50. Martínez-Freiría, F.; Lizana, M.; do Amaral, J.P.; Brito, J. Spatial and temporal segregation allows coexistence in a hybrid zone among two Mediterranean vipers (*Vipera aspis* and *V. latastei*). *Amphibia-Reptilia* **2010**, *31*, 195–212. [CrossRef]
51. Pavlov, A.V.; Zinenko, O.I.; Joger, U.; Stümpel, N.; Petrova, I.V.; Malenyov, A.L.; Zaitseva, O.V.; Shurshina, I.V.; Bakiev, A.G. Natural hybridization of the Eastern Steppe Viper *Vipera renardi* and the Common Adder *V. berus*. *Proc. Samara Sci. Cent. Russ. Acad. Sci.* **2011**, *13*, 172–178.
52. Czirják, G.Á.; Köbölkuti, L.B.; Tenk, M.; Szakács, A.; Kelemen, A.; Spînu, M. Hemorrhagic stomatitis in a natural hybrid of *Vipera ammodytes* × *Vipera berus* due to inappropriate substrate in terrarium. *J. Vet. Med. Sci.* **2015**, *77*, 701–703. [CrossRef] [PubMed]
53. Guiller, G.; Lourdais, O.; Ursenbacher, S. Hybridization between a Euro-Siberian (*Vipera berus*) and a Para-Mediterranean viper (*V. aspis*) at their contact zone in western France. *J. Zool.* **2016**, *302*, 138–147. [CrossRef]
54. Zinenko, O.I. First generation hybrids between the Nikolsky's adder, *Vipera nikolskii*, and the common adder (Reptilia, Serpentes, Viperidae). *Vestn. Zool. Kiev* **2003**, *37*, 101–104.
55. Matsubara, K.; Uno, Y.; Srikulnath, K.; Seki, R.; Nishida, C.; Matsuda, Y. Molecular cloning and characterization of satellite DNA sequences from constitutive heterochromatin of the habu snake (*Protobothrops flavoviridis*, Viperidae) and the Burmese python (*Python bivittatus*, Pythonidae). *Chromosoma* **2015**, *124*, 529–539. [CrossRef]
56. Ahmad, S.F.; Singchat, W.; Jehangir, M.; Panthum, T.; Srikulnath, K. Consequence of paradigm shift with repeat landscapes in reptiles: Powerful facilitators of chromosomal rearrangements for diversity and evolution. *Genes* **2020**, *11*, 827. [CrossRef]
57. Rodionov, A.V. Micro versus macro: A review of structure and functions of avian micro-and macrochromosomes. *Russ. J. Genet.* **1996**, *32*, 517–527.
58. Pigozzi, M.I. Distribution of MLH1 foci on the synaptonemal complexes of chicken oocytes. *Cytogenet. Genome Res.* **2001**, *95*, 129–133. [CrossRef]
59. Wilcox, J.J.; Arca-Ruibal, B.; Samour, J.; Mateuta, V.; Idaghdour, Y.; Boissinot, S. Linked-Read Sequencing of Eight Falcons Reveals a Unique Genomic Architecture in Flux. *Genome Biol. Evol.* **2022**, *14*, evac090. [CrossRef]
60. Morescalchi, A.; Odierna, G.; Olmo, E. Karyological relationships between the Cryptobranchid salamanders. *Experientia* **1977**, *33*, 1579–1581. [CrossRef]
61. Stingo, V.; Rocco, L. Chondrichthyan cytogenetics: A comparison with teleosteans. *J. Mol. Evol.* **1991**, *33*, 76–82. [CrossRef]
62. Rock, J.; Eldridge, M.; Champion, A.; Johnston, P.; Joss, J. Karyotype and nuclear DNA content of the Australian lungfish, *Neoceratodus forsteri* (Ceratodidae: Dipnoi). *Cytogenet. Genome Res.* **1996**, *73*, 187–189. [CrossRef] [PubMed]
63. Menezes, R.S.; Gazoni, T.; Costa, M.A. Cytogenetics of warrior wasps (Vespidae: *Synoeca*) reveals intense evolutionary dynamics of ribosomal DNA clusters and an unprecedented number of microchromosomes in Hymenoptera. *Biol. J. Linn. Soc.* **2019**, *126*, 925–935. [CrossRef]
64. Srikulnath, K.; Ahmad, S.F.; Singchat, W.; Panthum, T. Why do some vertebrates have microchromosomes? *Cells* **2021**, *10*, 2182. [CrossRef] [PubMed]
65. Schield, D.R.; Card, D.C.; Hales, N.R.; Perry, B.W.; Pasquesi, G.M.; Blackmon, H.; Adams, R.H.; Corbin, A.B.; Smith, C.F.; Ramesh, B.; et al. The origins and evolution of chromosomes, dosage compensation, and mechanisms underlying venom regulation in snakes. *Genome Res.* **2019**, *29*, 590–601. [CrossRef] [PubMed]
66. Li, M.; Sun, C.; Xu, N.; Bian, P.; Tian, X.; Wang, X.; Yang, N. De Novo Assembly of 20 Chicken Genomes Reveals the Undetectable Phenomenon for Thousands of Core Genes on Microchromosomes and Subtelomeric Regions. *Mol. Biol. Evol.* **2022**, *39*, msac066. [CrossRef]
67. Waters, P.D.; Patel, H.R.; Ruiz-Herrera, A.; Álvarez-González, L.; Lister, N.C.; Simakov, O.; Ezaz, T.; Kaur, P.; Frere, C.; Grützner, F.; et al. Microchromosomes are building blocks of bird, reptile, and mammal chromosomes. *Proc. Natl. Acad. Sci. USA* **2021**, *118*, e2112494118. [CrossRef]
68. Olmo, E. Rate of chromosome changes and speciation in reptiles. *Genetica* **2005**, *125*, 185–203. [CrossRef]
69. Spangenberg, V. FISH—And the Characterization of Synaptonemal Complex. In *Cytogenetics and Molecular Cytogenetics*; CRC Press: Boca Raton, FL, USA, 2022; pp. 297–305.
70. Spangenberg, V.; Arakelyan, M.; Galoyan, E.; Matveevsky, S.; Petrosyan, R.; Bogdanov, Y.; Danielyan, F.; Kolomiets, O. Reticulate evolution of the rock lizards: Meiotic chromosome dynamics and spermatogenesis in diploid and triploid males of the genus Darevskia. *Genes* **2017**, *8*, 149. [CrossRef]

71. Spangenberg, V.; Arakelyan, M.; Galoyan, E.; Pankin, M.; Petrosyan, R.; Stepanyan, I.; Grishaeva, T.; Danielyan, F.; Kolomiets, O. Extraordinary centromeres: Differences in the meiotic chromosomes of two rock lizards species *Darevskia portschinskii* and *Darevskia raddei*. *PeerJ* **2019**, *7*, e6360. [CrossRef]
72. Levan, A.; Fredga, K.; Sandberg, A.A. Nomenclature for centromeric position on chromosomes. *Hereditas* **1964**, *52*, 201–220. [CrossRef]
73. McDermott, A. Human male meiosis: Chromosome behavior at pre-meiotic and meiotic stages of spermatogenesis. *Canadian J. Genet. Cytol.* **1971**, *13*, 536–549. [CrossRef] [PubMed]
74. Zickler, D.; Kleckner, N. Meiotic chromosomes: Integrating structure and function. *Annu. Rev. Genet.* **1999**, *33*, 603. [CrossRef] [PubMed]
75. Gribbins, K.M. Reptilian spermatogenesis: A histological and ultrastructural perspective. *Spermatogenesis* **2011**, *1*, 250–269. [CrossRef]
76. Viana, P.F.; Feldberg, E.; de Bello Cioffi, M.; de Carvalho, V.T.; Menezes, S.; Vogt, R.C.; Liehr, T.; Ezaz, T. The Amazonian Red Side-Necked Turtle *Rhinemys rufipes* (Spix, 1824) (Testudines, Chelidae) Has a GSD Sex-Determining Mechanism with an Ancient XY Sex Microchromosome System. *Cells* **2020**, *9*, 2088. [CrossRef]
77. Viana, P.F.; Ezaz, T.; de Bello Cioffi, M.; Liehr, T.; Al-Rikabi Goll, L.G.; Rocha, A.M.; Feldberg, E. Landscape of snake'sex chromosomes evolution spanning 85 MYR reveals ancestry of sequences despite distinct evolutionary trajectories. *Sci. Rep.* **2020**, *10*, 12499. [CrossRef] [PubMed]
78. Deakin, J.E.; Ezaz, T. Understanding the evolution of reptile chromosomes through applications of combined cytogenetics and genomics approaches. *Cytogenet. Genome Res.* **2019**, *157*, 7–20. [CrossRef] [PubMed]
79. Iannucci, A.; Altmanová, M.; Ciofi, C.; Ferguson-Smith, M.; Pereira, J.C.; Rehák, I.; Stanyon, R.; Velenský, P.; Rovatsos, M.; Kratochvíl, L.; et al. Isolating chromosomes of the Komodo dragon: New tools for comparative mapping and sequence assembly. *Cytogenet. Genome Res.* **2019**, *157*, 123–131. [CrossRef]
80. Iannucci, A.; Makunin, A.I.; Lisachov, A.P.; Ciofi, C.; Stanyon, R.; Svartman, M.; Trifonov, V.A. Bridging the gap between vertebrate cytogenetics and genomics with Single-Chromosome Sequencing (ChromSeq). *Genes* **2021**, *12*, 124. [CrossRef]
81. Saint Girons, R.; Fons, R. Un cas de mélanisme chez *Vipera aspis* dans les Pyrénées. *Vie Milieu* **1977**, *27*, 145–146.
82. Kalikinskaya, E.I.; Kolomiets, O.L.; Shevchenko, V.A.; Bogdanov, Y.F. Chromosome aberrations in F1 from irradiated male mice studied by their synaptonemal complexes. *Mutat. Res. Lett.* **1986**, *174*, 59–65. [CrossRef]
83. Lisachov, A.P.; Solovyeva, E.N. The toad-headed agamas (*Phrynocephalus*) are champions in crossing over rate. In *Molecular cytogenetics*; Biomed Central Ltd.: London, UK, 2017; Volume 10.
84. Axelsson, E.; Webster, M.T.; Smith, N.G.; Burt, D.W.; Ellegren, H. Comparison of the chicken and turkey genomes reveals a higher rate of nucleotide divergence on microchromosomes than macrochromosomes. *Genome Res.* **2005**, *15*, 120–125. [CrossRef] [PubMed]
85. Cobror, O.; Olmo, E.; Odierna, G.; Angelini, F.; Ciarcia, G. Cyclic variation of chiasma frequency and distribution in *Podarcis sicula* (Reptilia: Lacertidae). *Genetica* **1986**, *71*, 31–37. [CrossRef]
86. Spangenberg, V.; Arakelyan, M.; Galoyan, E.; Martirosyan, I.; Bogomazova, A.; Martynova, E.; de Bello Cioffi, M.; Liehr, T.; Al-Rikabi, A.; Osipov, F.; et al. Meiotic synapsis of homeologous chromosomes and mismatch repair protein detection in the parthenogenetic rock lizard Darevskia unisexualis. *Mol. Reprod. Dev.* **2021**, *88*, 119–127. [CrossRef]
87. Oguiura, N.; Ferrarezzi, H.; Batistic, R.F. Cytogenetics and Molecular Data in Snakes: A Phylogenetic Approach. *Cytogenet. Genome Res.* **2009**, *127*, 128–142. [CrossRef]
88. Trajtengertz, I.; Beçak, M.L.; Ruiz, I.R. Ribosomal cistrons in Bothrops neuwiedi (Serpentes) subspecies from Brazil. *Genome* **1995**, *38*, 601–606. [CrossRef] [PubMed]
89. Porter, C.A.; Hamilton, M.J.; Sites, J.W., Jr.; Baker, R.J. Location of ribosomal DNA in chromosomes of squamate reptiles: Systematic and evolutionary implications. *Herpetologica* **1991**, *47*, 271–280.
90. Mezzasalma, M.; Odierna, G. Sex chromosome diversification in the smooth snake *Coronella austriaca* (Reptilia, Serpentes). *Acta Herpetol.* **2021**, *16*, 37–44. [CrossRef]
91. Cole, C.J.; Hardy, L.M. Karyotypes of six species of colubrid snakes from the Western Hemisphere, and the 140-million-year-old ancestral karyotype of Serpentes. *Am. Mus. Novit.* **2019**, *2019*, 1–14. [CrossRef]
92. Ahmad, S.F.; Singchat, W.; Panthum, T.; Srikulnath, K. Impact of repetitive DNA elements on snake genome biology and evolution. *Cells* **2021**, *10*, 1707. [CrossRef]

Review

A Brief Review of Meiotic Chromosomes in Early Spermatogenesis and Oogenesis and Mitotic Chromosomes in the Viviparous Lizard *Zootoca vivipara* (Squamata: Lacertidae) with Multiple Sex Chromosomes

Larissa Kupriyanova [1,*] and Larissa Safronova [2]

1 Zoological Institute of the Russian Academy of Sciences (ZIN), 199034 Saint Petersburg, Russia
2 Severtsov Institute of Ecology and Evolution, Russian Academy of Sciences, 119071 Moscow, Russia
* Correspondence: larissakup@zin.ru

Simple Summary: The wide-ranging Eurasian species *Zootoca vivipara* (Lichtenstein, 1823), of the family Lacertidae (Reptilia), is a rare species within the family, possessing multiple sex chromosomes (male $Z_1Z_2Z_1Z_2/Z_1Z_2W$ female). In addition, the intense reorganization of this W sex chromosome is accompanied by active subspeciation and the formation of 4–5 cryptic taxa. In the females of two cryptic forms having a similar system of multiple sex chromosomes (Z_1Z_2W) but with different morphology, the cytogenetic and specific genomic structures of the W sex chromosome's early oogenesis and meiosis have standard occurrence. Despite the ambiguous behavior of the three presumed sex chromosomes at the early stages of meiotic prophase I, variability in their number of bivalents and chromosomes and significant disturbances in chromosome segregation have not been discovered. Because in *Z. vivipara* the W sex chromosome, unlike all the other chromosomes, does not have several identified SINE-Zv and TE elements, we may assume that the specific genomic structure of this chromosome may be one of the factors ensuring meiotic stability in the cryptic taxa of the species with the multiple sex chromosomes. The question of female meiotic drive in the meiosis of the cryptic forms of the *Z. vivipara* complex is still obscure.

Abstract: This brief review is focused on the viviparous lizard *Zootoca vivipara* (Lichtenstein, 1823), of the family Lacertidae, which possesses female heterogamety and multiple sex chromosomes (male $2n = 36$, $Z_1Z_1Z_2Z_2/Z_1Z_2W$, female $2n = 35$, with variable W sex chromosome). Multiple sex chromosomes and their changes may influence meiosis and the female meiotic drive, and they may play a role in reproductive isolation. In two cryptic taxa of *Z. vivipara* with different W sex chromosomes, meiosis during early spermatogenesis and oogenesis proceeds normally, without any disturbances, with the formation of haploid spermatocytes, and in female meiosis with the formation of synaptonemal complexes (SCs) and the lampbrush chromosomes. In females, the SC number was constantly equal to 19 (according to the SC length, 16 SC autosomal bivalents plus three presumed SC sex chromosome elements). No variability in the chromosomes at the early stages of meiotic prophase I, and no significant disturbances in the chromosome segregation at the anaphase–telophase I stage, have been discovered, and haploid oocytes ($n = 17$) at the metaphase II stage have been revealed. There should be a factor/factors that maintain the multiple sex chromosomes, their equal transmission, and the course of meiosis in these cryptic forms of *Z. vivipara*.

Keywords: lizards; *Zootoca vivipara*; multiple sex chromosomes; meiosis; synaptonemal complex (SC); form and subspeciation

Citation: Kupriyanova, L.; Safronova, L. A Brief Review of Meiotic Chromosomes in Early Spermatogenesis and Oogenesis and Mitotic Chromosomes in the Viviparous Lizard *Zootoca vivipara* (Squamata: Lacertidae) with Multiple Sex Chromosomes. *Animals* **2023**, *13*, 19. https://doi.org/10.3390/ani13010019

Academic Editors: Ettore Olmo and Pietro Parma

Received: 25 October 2022
Revised: 1 December 2022
Accepted: 16 December 2022
Published: 20 December 2022

1. Introduction

Lizards are one of a few groups of reptiles whose members are characterized by temperature (TSD) and genetic (GSD) sex determination (e.g., Reference [1]). Male and female heterogamety (XY/ZW) and a variety of sex chromosome systems are also found in these animals [1–3]. About 23% of karyotyped lizard species have multiple sex chromosomes [3]. This system of sex chromosomes is usually common in XY but not in ZW groups. Presumably, as suggested by some authors [2], it is due to the different involvement of sex-specific sex chromosomes in female meiosis, and the effect of female meiotic drive. For instance, mammals possessing male heterogamety show many taxa with multiple sex chromosomes [4]. In contrast, birds (ZW), except for one species [5], do not display multiple sex chromosomes [6].

Lizards of the family Lacertidae, except for a few debated cases [7], have genetic sex determination, demonstrate female heterogamety (ZW), and only four of the karyotyped lacertid species (totaling about 115) reveal multiple sex chromosomes Z_1Z_2W [3]. During the evolution of one of those species, namely, *Zootoca vivipara* (Lichtenstein, 1823), the acrocentric (A) sex chromosome Z was involved in a rearrangement, i.e., in translocation with an autosomal acrocentric, which led to the creation of multiple sex chromosomes Z_1Z_2W and a completed W sex macrochromosome comprising the Z_1 and Z_2 chromosomes. The sex/autosome fusion has also been described among other lizard species, for example, in the Chamaeleonidae [8]. The widely distributed Eurasian lizard *Z. vivipara* is characterized by some other special features: both multiple sex chromosomes Z_1Z_2W and simple Zw system, and oviparous/viviparous reproduction in different populations. The species is polymorphic on its mitochondrial (mt) haplotype but considerably uniform morphologically; moreover, viviparous *Z. vivipara* shows considerable diversity in the morphology and structure of the W sex chromosome [9–13]. The intense reorganization of the W sex chromosome appears to be accompanied by active subspeciation and speciation, and by the formation of cryptic taxa with different W chromosomes (Table 1) [12,13]. Several chromosomal forms and subspecies of *Z. vivipara* have their distinct distribution areas (allopatric and parapatric populations) in the central and southern-central part of Europe. Some of them have a mosaic pattern of populations, and inhabit small areas, while others are relict forms [13]. All the specimens of *Z. vivipara* can be diagnosed and recognized by their mt haplotype [14,15] and by some chromosomal characteristics, in particular, by the morphology and the cytogenetical structure (the amount and distribution of heterochromatin, C-bands) of their W sex chromosome [11,13,16]. It should be noticed that the two forms, the western form and the eastern (Russian) form of *Z. vivipara*, occupy a vast territory in Europe and Asia. The western form inhabits central and western Europe, whereas the eastern (Russian) form populates eastern Europe and Asia. In addition, both of these forms of *Z. vivipara* have been discovered in the Baltic region (Figure 1, map) [17,18].

Thus, it is clear that karyological differentiation in the *Z. vivipara* complex is high, and in particular in the morphology and the cytogenetical structure (the amount and distribution of heterochromatin, C-bands) of their W sex chromosome. As is well known, chromosome reorganization may play an important role in sex chromosome differentiation in different lizard groups [3]. The role of variable sex chromosomes in the evolution of the cryptic *Z. vivipara* complex is still poorly studied, as is the connection between sex chromosomes and reproductive isolation. According to King [19,20], simple or multiple sex chromosomes may play a role in meiosis (as a reproductive meiotic barrier) and the speciation of different groups. Moreover, reproductive isolation seems to evolve faster among species with heteromorphic sex chromosomes [21,22]. Alterations in some patterns of multiple sex chromosomes, among other things, may reinforce their isolation effect.

Furthermore, genomic composition is a factor favoring the fixation of mutant karyotypes [23]. In addition to the variable multiple sex chromosomes in the karyotype of *Z. vivipara*, several new molecular markers, namely, some different short interspersed elements (SINEs) and transposable elements (TE), have recently been detected and identified in its genome [24]. SINE elements, as is well known, often have preferred sites in the genome and may also influence the process of meiosis, speciation rate, etc.

All the mentioned characteristics offer a rare possibility to use *Z. vivipara* as a model for studying some general evolutionary problems; for instance, sex chromosome evolution and its impact on subspeciation and form formation. Given the above, the karyotype, especially the characteristics of sex chromosomes, meiosis, and behavior of chromosomes during the meiosis of the described cryptic taxa with variable multiple sex chromosomes, are of particular interest.

In this brief review, we consider mainly the features of karyotype and sex chromosomes, and the course of spermato- and oogenesis, meiosis, and the behavior of chromosomes in the early stages of prophase I of meiosis, in two closely related cryptic chromosomal forms of *Z. vivipara* with multiple sex chromosomes Z_1Z_2W: the eastern (Russian) form and the western form from the Baltic region of Russia (Figure 1, map).

Table 1. Karyotype of *Zootoca vivipara*, subspecies, and chromosomal forms with characteristics of sex chromosomes system, morphology of w/W chromosomes, reproductive modality and distribution.

№	2n ♂/♀	System of Sex Chromosomes ♂/♀	Morphology of w/W Sex Chromosomes	Mode of Reproduction, O/V (Ovi-/ Viviparous)	Localities	Species, Subspecies, Chromosomal Forms
			The first group of karyotype			
1.	36A/36A	ZZ/Zw	M	O	Central, southwestern Europe	*Z. vivipara*, now *Z. carniolica*
2.	36A/36A	ZZ/Zw	M	V	Central Europe	*Z. vivipara*, now *Z. vivipara* Hungarian form
			The second group of karyotype			
3.	36A/35 (34A + 1 A/ST)	$Z_1Z_1Z_2Z_2/Z_1Z_2W$	A, ST	O	Western Europe	*Z. vivipara*, now *Z. v. louislantzi* Pyrenean form
4.	36A/35 (34 A + 1 A/ST)	$Z_1Z_1Z_2Z_2/Z_1Z_2W$	A/ST	V	Central Europe	*Z. vivipara*, now *Z. vivipara* Austrian form?
5.	36A/35 (34A + 1 SV)	$Z_1Z_1Z_2Z_2/Z_1Z_2W$	SV/ST	V	Central Europe	*Z. vivipara*, now *Z. vivipara* Romanian form
6.	36A/35 (34A + 1 A/ST)	$Z_1Z_1Z_2Z_2/Z_1Z_2W$	A/ST	V	Eastern Europe, eastern Baltic region, Asia	*Z. vivipara*, now *Z. vivipara* Eastern (Russian) form
7.	36A/35 (34A + 1 SV)	$Z_1Z_1Z_2Z_2/Z_1Z_2W$	SV	V	Western, Central Europe, Baltic region	*Z. vivipara*, now *Z. vivipara* Western form

Characteristics of *Z. vivipara:* karyotype and system of sex chromosomes: ZZ/Zw—simple system; ZZ_1ZZ_2/Z_1Z_2W—multiple sex chromosome system; shape and morphology of w/W sex chromosomes: w—microchromosome, W—macrochromosome; A—acrocentric, ST—subtelocentric, SV—submetacentric; reproductive modality: O—oviparous; V—viviparous; distribution area.

They have the viviparous mode of reproduction and are diverse in their karyotypes, namely, the morphology and cytogenetical structure of the W sex chromosomes [17,25,26]. Several samples of the *Z. vivipara* under study have already been used in other studies. The species is not included in the national Red Data Book and lists of protected taxa. It is not included in international agreements. The specimens were treated by ether according to ethical practices and were deposited in the collection of ZISP, chromosomal collection, accession numbers №№ 9261–9263; 9448–9450. Chromosomal material and the preparations from oocytes were stored in a freezer (minus 22–25 °C) and some of them were used in this work for the first time. A total of six males and females of *Z. vivipara* were collected in the Leningrad and Kaliningrad areas (Baltic region, Russia). Chromosomal preparations had been obtained by the scraping and air-drying method from intestine, gonads, and germinal lamina cells, and then they were stained with Giemsa. C-banding was carried out according to Summer's method [27], and fluorochrome AT staining (DAPI) using the method of Drs. M. Schmid and M. Guttembach [28]. Meiotic preparations were obtained by using the method of total oocyte nuclei spreading developed by M. Dresser and M. Moses [29]. Chromosome preparations were stained with Giemsa, and for the visualization of synaptonemal complexes (SCs), total preparations were stained with silver nitrate and DAPI. Fluorescent analysis with the help of incubation with primary and secondary antibodies SYPC3 (the protein of synaptonemal complexes (SC) of central elements), and fluorochrome AT DAPI staining was performed on the preparations. The lengths of the SCs of bivalents were measured using Leica Application Suite V3 on the digital microphotographs. The SCs of the bivalents in a karyotype were numbered in the decreasing order of their linear sizes. Analysis of the photos and SC karyotyping were conducted on the basis of the measurements of the SC by the relative lengths of each individual SC.

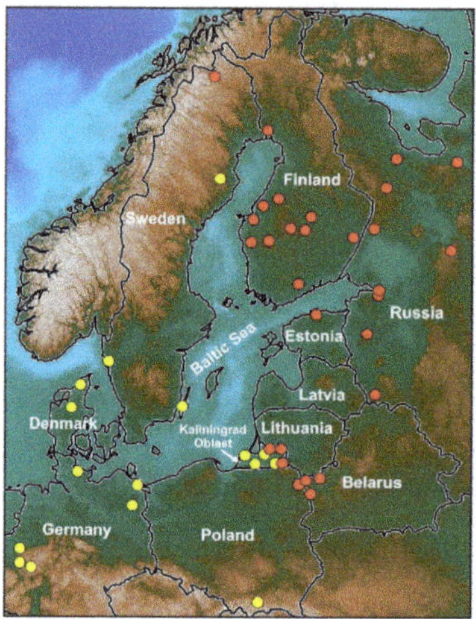

Figure 1. Map showing the distribution of eastern (Russian) (•) and western (•) forms of *Zootoca vivipara* in the Baltic Sea basin based mainly on their karyotypes. Topography is adapted from the GEBCO world map 2014. The points of distribution are from reference [18] with additions.

2. Characteristics of Karyotype

In the karyotype markers, these forms of *Z. vivipara* are characterized by different diploid chromosome numbers (male $2n$ = 36 acrocentrics (A) and female $2n$ = 35) and different numbers of sex chromosomes (in male $Z_1Z_1Z_2Z_2$ and Z_1Z_2W in female), with different morphology of the W sex chromosome (the eastern (Russian) cryptic form W is acro/subtelocentric (A/ST) and the western cryptic form W is submetacentric (SV)) [11,12,17,18]. Thus, the male karyotype is $2n$ = 36 A: 32A + $Z_1Z_1Z_2Z_2$, while the female karyotype is $2n$ = 35: 32 A+ Z_1Z_2W, where W is (A/ST) or (SV). Further comparative staining analyses of C-banding/CMA_3/DAPI have also shown the different cytogenetic structure (the distribution of conspicuous centromeric and telomeric C-bands), and the presence of an additional interstitial C-positive heterochromatin block, by staining with an AT-specific fluorochrome (DAPI) in the W sex chromosome [12,30]. All the karyotype markers of the two cryptic forms considered allow us to make the suggestion that the submetacentric W sex chromosome resulted from a pericentric inversion of the acrocentric W sex chromosome [12]. Thus, we can see that, in the evolution of *Z. vivipara*, the formation of the viviparous cryptic western form (Z_1Z_2W, W-SV) has been accompanied by the changing of the W sex chromosome. It should also be noted that the mechanisms and steps of chromosomal changes in W sex chromosomes for all the described cryptic forms and subspecies of the *Z. vivipara* complex include heterochromatinization event, deletion, tandem fusion, and inversion. These and other mechanisms have also been described in other lizard groups [1,3].

As indicated earlier, alterations in sex chromosomes (and especially in multiple sex chromosomes) are important for evolution, and they may influence the process of meiosis and play a role in isolation and a female meiotic drive (unequal transmission of Z and W chromosomes). Moreover, nonrandom segregation of chromosomes of different morphology (acrocentric versus metacentric) during female meiosis has been documented in birds and mammals [31,32].

Therefore, we reviewed some characteristics of the spermatogenesis and of the oogenesis and early meiosis of two cryptic chromosomal forms of the *Z. vivipara* complex (the eastern (Russian) form and the western form) that have a similar system ($Z_1Z_1Z_2Z_2/Z_1Z_2W$) but different morphology, and different cytogenetic and genomic structure in the W sex chromosomes [25,26,30,33,34].

3. Characteristics of Meiosis in Spermatogenesis and Early Oogenesis

The male diploid karyotype of these forms of *Z. vivipara* is 2n = 36 A: 32 A + $Z_1Z_1Z_2Z_2$ (pairs 5 or 6 Z_1 and 13 Z_2), and the haploid number is equal to 18 (n = 18). During their early meiosis, the synaptonemal complex (SC) bivalents at the prophase I meiosis (the late zygotene–middle pachytene stage and the middle pachytene stage) were found. All SCs did not form asymmetric configuration, and they appeared to be successfully synaptic, including SC $Z_1Z_1Z_2Z_2$ sex chromosomes according to the lengths of SC bivalents. However, a wave-shaped morphology of the sex bivalent SC (fifth to sixth in length in a karyotype) was noted [25]. At the diakinesis stage of prophase I meiosis, 18 bivalents were also discovered, including the sex bivalents, without any disruptions in chromosome conjugation (Figure 2A,B). All bivalents were represented by cross-shaped, ring-shaped, or baculiform figures. Their regular segregation with the formation of haploid spermatocytes, 18 chromosomes, at the metaphase II stage of meiosis was constantly revealed [25,34]. The obtained results demonstrated the standard course of meiosis, with formation of constant-haploid-number chromosomes (n = 18) in spermatocytes. No clear disturbances in the segregation of chromosomes were detected and the results suggest the stability of their male meiosis.

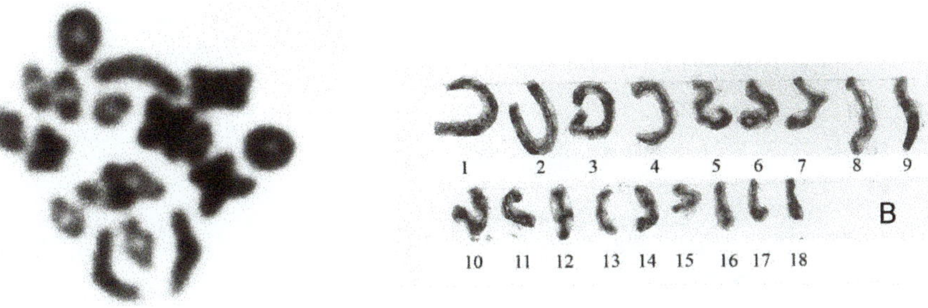

Figure 2. Cells of male specimens of eastern (Russian) cryptic form of viviparous lizard *Zootoca vivipara*: (**A**)—meiotic testis cell at the diakinesis stage; bivalents, n = 18; (**B**)—synaptonemal complex (SC) karyotype of spermatocytes, n = 18. (**B**) from reference [30].

The female diploid karyotype of both cryptic forms of *Z. vivipara* is 35, 2n = 35: 32 A + Z_1Z_2W, but in the eastern (Russian) form the W sex chromosome has an acro/subtelocentric (A/ST) shape, whereas in the western form the W sex chromosome has a submetacentric (SV) shape (as a result of a pericentric inversion) (Figure 3A,B) [12]. The ovarian lumen germinal vesicles (oocytes), as well as germinal lamina cells of the females (of the eastern (Russian) form) were examined. During oogenesis, primary follicles enter the early stages of the meiotic prophase I and some characteristics of early oocytes during the early stages of prophase I meiosis (from leptotene to diplotene), synaptonemal complexes (SCs), and lampbrush chromosomes were revealed (Figure 3C,D) [30]. The obtained results demonstrated the standard course of early oogenesis and early meiosis (with formation of a constant number of SC configurations.)

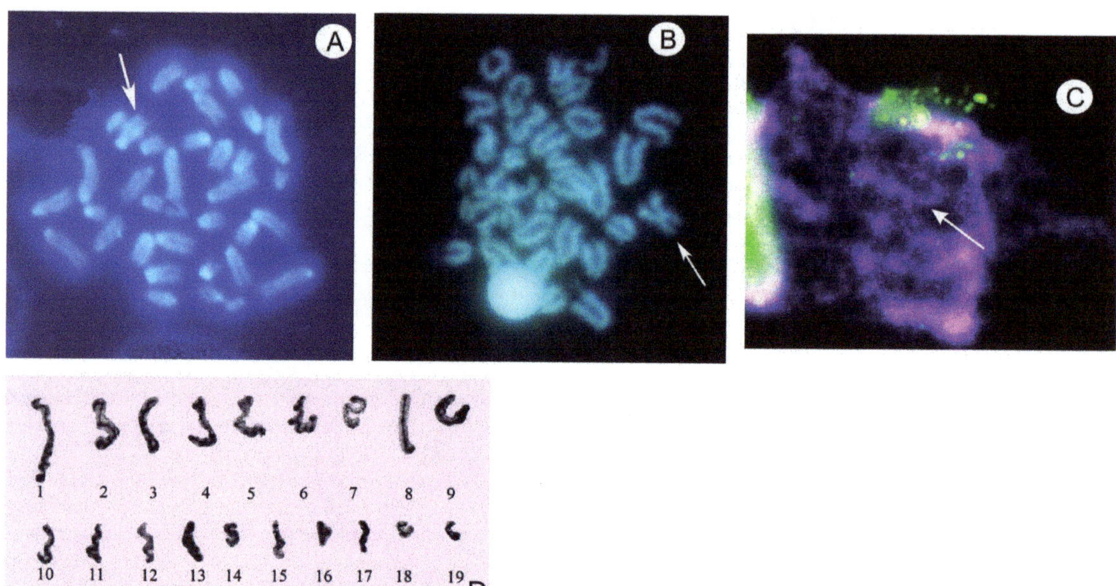

Figure 3. Cells of female specimens of eastern (Russian) form and of western form of viviparous lizard *Zootoca vivipara*: (**A**)—metaphase plate of eastern (Russian) form, specific DAPI stained, $2n$ = 35: 32 A + Z_1Z_2W. Arrow points to centromeric and interstitial DAPI blocks of acrocentric W sex chromosome. (**A**) from reference [30]; (**B**)—metaphase plate of western form, specific DAPI stained, $2n$ =35: 32 A + Z_1Z_2W. Arrow points to centromeric and weak interstitial DAPI blocks of submetacentric W sex chromosome; (**C**)—the spread oocyte nuclei of female of eastern (Russian) form at pachytene–diplotene stages. Incubation with antibody (SYPC3) and after incubation specific fluorochrome AT DAPI stained. Arrow points to the lampbrush chromosomes. (**C**) from reference [33]; (**D**)—SC karyotype of female eastern (Russian) form, n = 16 autosomal bivalents and 3 SC elements of presumed Z_1Z_2W sex chromosomes. (**D**) from reference [30].

It should be stressed that in meiosis of a female with a diploid chromosome number equal to 35, with multiple sex chromosomes and with the indicated cytogenetical chromosome structure ($2n$ = 35: 32 A + 3 sex Z_1Z_2W chromosomes), at meiotic prophase I, 16 autobivalents and a complex trivalent of sex chromosomes or complex bivalent and univalent or univalents could be expected. On the basis of light microscopic analysis of the oocyte SCs, and taking into account their length in a female, SC analysis showed that oocytes of these females contained 19 fully synaptic SC elements. The results of previous studies [25] supported the correlation between the morphometric characteristics of relative SC lengths in meiotic prophase I and the metaphase chromosome lengths in somatic cells. Because of this, the female 19 SC elements were assembled and numbered according to their length in descending order (Figure 3C) [26]. The sex chromosome W, on the basis of its size and cytogenetical structure (distribution of C-bands, C/DAPI/CMA3 structure), was attributed to the chromosome pair 5–6 [11]. It should be stressed that SCs were visualized at the stages of late zygotene–middle pachytene. Neither asymmetric configuration nor complex units during the female meiotic prophase I were noticed, only successful synaptic SC elements without any asymmetric configuration [26]. In the eastern (Russian) form, the SC number was constantly equal to 19 (16 SC autosomal bivalents plus 3 SC elements). According to the SC lengths, three SC elements might be univalent of three sex chromosomes Z_1Z_2W, or one SC bivalent of W and Z_1 sex chromosomes and univalent of Z_2 chromosome and B chromosome univalent. As a result, during meiosis in these female viviparous lizards, 19 SC elements might be formed [26].

During oogenesis and meiosis in the female of the cryptic western form (with reorganized morphological and different cytogenetic structure (C/DAPI/CMA3) of the submetacentric W sex chromosome), primary follicles also entered the early stages of the prophase I of meiosis (stages from leptotene to diplotene) and SC bivalents and lampbrush chromosomes were formed. Again, at these early stages of meiosis (the late zygotene–middle pachytene), neither asynaptic SC configurations nor complex configurations were identified, and the exact SC number has been difficult to count thus far. Nevertheless, at the stages of anaphase–telophase I of meiosis, only rare cells (2 out of 20) with some disturbance in the segregation of bivalents have been revealed [34]. Moreover, in the metaphase II oocytes, haploid numbers equal to 17 with the W sex chromosome (SV) (n = 17) were previously determined [35].

Thus, in the females of two cryptic forms of the *Z. vivipara* complex that have a similar system of multiple sex chromosomes (Z_1Z_2W) but different morphology, differences also in cytogenetical structure (distribution of heterochromatin, of C/DAPI/CMA3 elements) of the W sex chromosome, resulting from reorganization, early oogenesis and meiosis, have standard occurrence. At the same time, the ambiguous behavior of the three presumed sex chromosomes in the eastern (Russian) form and the lack of variability in their number has been indicated, and in both forms there is a lack of significant disturbances in the chromosome segregation. As is known, the studied populations of *Z. vivipara* in nature do not show the disturbances in sex ratio caused probably by unequal transmission in female multiple sex chromosomes. The obtained results suggest that, along with the shape and cytogenetic structure, there should be a factor (or several factors) that maintains the multiple sex chromosomes and their equal transmission.

As mentioned above, two different short interspersed elements (SINEs) and transposable elements (TE), described in the genome of *Z. vivipara*, have been identified (SINE-Zv 700 and SINE-Zv 300) [24]. SINE-Zv 700 appears to be restricted to *Z. vivipara*, whereas SINE-Zv 300 (including the Gypsy-like fragment) appears to be conserved in many different Squamata species. The active role of the SINEs and the Gypsy-like element in the genomic evolution and differentiation of the *Z. vivipara* complex has been suggested [24]. It is well known that the effects of TEs on the origin of new species are widely discussed in the literature. The activity of TEs might lead to genomic changes, and genetic and phenotypic diversity, often due to new specific gene regulations [36]. TEs are considered by some researchers as potential causes of reproductive isolation across a diversity of taxa [37]. This assumption is associated with the suggestion of some researchers [38] that in reptiles the evolution of sex chromosomes seems to be also explained by some molecular mechanisms, such as gene regulatory mechanisms and others.

It should also be highlighted that fluorescence in situ hybridizations showed a preferential localization of SINE-Zv sequences in the peritelomeric regions of almost all chromosomes, except for the W sex chromosome [24]. The centromere and telomere regions, as is known, are often of key importance for the spatial orientation of chromosomes in the nucleus and are very important for the coincidence of the sites of communication in the hybrid or reorganized (rearrangement) chromosomes with the nuclear envelope, as well as for the conjugation and segregation of chromosomes during meiosis [39]. Apart from this, both centromere and telomere regions play an important role in female meiotic drive [40]. It may be assumed that specific cytogenetic and genomic composition of the W sex chromosome, and the SINE-Zv sequences in the peritelomeric region of chromosomes, might play a role in the meiotic process and the behavior of the sex chromosomes of *Z. vivipara*.

4. Conclusions

It becomes clear that in the *Z. vivipara* complex, two closely related viviparous cryptic chromosomal forms (the Eastern (Russian) and the Western), with similar karyotype (male 36/35 female) and system of multiple sex chromosomes ($Z_1Z_1Z_2Z_2/Z_1Z_2W$) but with different morphology and cytogenetic and specific molecular structure (genomic composition) of their W sex chromosome, demonstrate the standard course of early oogenesis and of female and male meiosis.

In male meiosis, no clear disturbances in the segregation of chromosomes were observed, and the results suggest the stability of male meiosis with formation of a constant number of haploid spermatocytes ($n = 18$).

In female meiosis, during the early stages of prophase I meiosis, the synapted SC bivalents and the lampbrush chromosomes were formed. The SC elements appeared to be fully synapted at the pachytene stage, and no asynaptic SC configurations were observed. The number of SC elements was equal to 19 (in the eastern (Russian) form); however, no significant disturbances, including chromosomal segregation at the anaphase–telophase I stage, were revealed, and haploid oocytes with 17 chromosomes ($n = 17$) were found.

The characteristics of early oogenesis and early meiosis in these two forms of *Z. vivipara* show them maintaining the course of their meiosis and the segregation of multiple sex chromosomes. The question of female meiotic drive in the meiosis of the cryptic forms of the *Z. vivipara* complex is still obscure.

Future studies on genome and karyotype, meiosis, behavior and segregation of multiple sex chromosomes, and their molecular composition, in particular of the centromere/telomere regions, may help to clarify the factors behind the plasticity and the preservation of stability, and the maintenance of high genetic diversity and sex ratio, in the cryptic *Z. vivipara* complex with multiple sex chromosomes.

Author Contributions: All authors contributed to the obtaining of chromosomal preparations, the discussing of the results obtained and final version of the manuscript. All authors have read and agreed to the published version of the manuscript.

Funding: This study was supported by grant no. 1021051302397-6.

Institutional Review Board Statement: Not applicable.

Informed Consent Statement: Not applicable.

Data Availability Statement: The data considered in this work can be found in the manuscript, in Table 1.

Acknowledgments: We are grateful to the editor, Ettore Olmo, for the kind invitation to present our paper. We thank the three anonymous reviewers for constructive reviews of the manuscript that led to substantial revisions and improvements.

Conflicts of Interest: The authors declare no conflict of interest.

References

1. Mezzasalma, M.; Guarino, F.; Odierna, G. Lizards as model organisms of sex Chromosome evolution: What we really know from a systematic distribution of available data? *Genes* **2021**, *12*, 1341. [CrossRef] [PubMed]
2. Pokorná, M.; Almanová, M.; Kratochvil, L. Multiple sex chromosomes in the light of female meiotic drive in amniote vertebrates. Cytogenet. *Genome Res.* **2014**, *22*, 35–44. [CrossRef]
3. Ezaz, T.; Sarre, S.; Meally, D.; Graves, J.M.M.; Georges, A. Sex chromosome evolution in lizards: Independent origins and rapid transitions. *Cytogenet. Genome Res.* **2009**, *127*, 249–260. [CrossRef] [PubMed]
4. Yoshida, K.; Kitano, J. The contribution of female meiotic drive to the evolution of neo-sex chromosomes. *Evolution* **2012**, *66*, 3198–3208. [CrossRef] [PubMed]
5. Kretschmer, R.; Ferguson-Smith, M.A.; De Oliveira, E.H.C. Karyotype evolution in birds: From conventional staining to chromosome painting. *Genes* **2018**, *9*, 181. [CrossRef]
6. Pala, I.; Naurin, S.; Stervander, M.; Hasselquist, D.; Bensch, S.; Hansson, B. Evidence of a neo-sex chromosome in birds. *Heredity* **2011**, *108*, 264–272. [CrossRef]

7. Rovatsos, M.; Vukic, J.; Mrugala, A.; Suwala, C.; Lymberakis, P.; Kratochvil, L. Little evidence for switches to environmental sex determination and turnover of sex chromosomes in lacertid lizards. *Sci. Rep.* **2019**, *9*, 7832. [CrossRef]
8. Rovatsos, M.; Almatová, M.; Augstenová, B.; Mazzoleni, S.; Velenský, P.; Kratochvil, L. ZZ/ZW sex determination with multiple neo-sex chromosomes in common in madagascan chameleons of the genus Furcifer (Reptilia: Chamaeleonidae). *Genes* **2019**, *10*, 1020. [CrossRef]
9. Olmo, E.; Odierna, G.; Capriglione, T.; Cardone, A. DNA and chromosome evolution in lacertid lizards. In *Book Cytogenetics of Amphibians and Reptiles, ALS*; Olmo, E., Ed.; Birkhäuser Verlag: Basel, Switzerland; Boston, MA, USA; Berlin, Germany, 1990; pp. 181–204.
10. Kupriyanova, L.; Rudi, E. Comparative karyological analysis of Lacerta vivipara (Lacertidae, Sauria) populations. *Zool. J.* **1990**, *69*, 93–101. (In Russian)
11. Odierna, G.; Aprea, G.; Capriglione, T.; Arribas, O.; Kupriyanova, L.; Olmo, E. Progressive differentiation of the W sex chromosome between viviparous ndoviparous populations of Zootoca vivipara (Reptilia, Lacertidae). *Ital. J. Zool.* **1998**, *65*, 295–302. [CrossRef]
12. Kupriyanova, L.; Odierna, G.; Caproglione, T.; Olmo, E.; Aprea, G. Chromosomal changes and form-formation, subspeciation in the wideranged Europasian species Zootoca vivipara (evolution, biogeography). Herpetologia Petropolitana. In Proceedings of the 12th Ordinary General Meeting of the Societas Europaea Herpetologica, St. Petersburg, Russia, 12–16 August 2003; Ananjeva, N., Tsinenko, O., Eds.; Folium: St. Petersburg, Russia, 2003; Volume 12, pp. 47–52.
13. Kupriyanova, L.; Böhme, W. A review of the cryptic diversity of the viviparous lizard, Zootoca vivipara (Lichtenstein, 1823) (Squamata: Lacertidae) in Central Europe and its postglacial re-colonization out of the Carpathian basin: Chromosomal and molecular data. In *Book Europe: Environmental, Political and Social Issues*; Thygesen, E., Ed.; Nova Science Publisher: New York, NY, USA, 2020; Chapter 2; pp. 25–43.
14. Surget-Groba, Y.; Heulin, B.; Guillaume, C.-P.; Puky, M.; Semenov, D.; Orlova, V.; Kupriyanova, L.; Ghira, I.; Smajda, B. Multiple origins of viviparity, or reversal from viviparity to oviparity and the evolution of parity. *Biol. J. Linn. Soc.* **2006**, *87*, 1–11. [CrossRef]
15. Velekei, B.; Lakatos, F.; Biro, P.; Acs, E.; Puky, M. The genetic structure of Zootoca vivipara (Lichtenstein, 1823) populations did not support the existence of a north-south corridor of the VB haplogroup in eastern Hungary. *North-West. J. Zool.* **2014**, *10*, 187–189.
16. Kupriyanova, L.; Kuksin, A.; Odierna, G. Karyotype, chromosome structure, reproductive modalities of three souuuuthern Eurasia populations of the common lacertid lizards, Zootoca vivipara (Jacquin, 1787). *Acta Herpetol.* **2008**, *3*, 99–106.
17. Kupriyanova, L.; Melashchenko, O. The common Eurasian lizard Zootoca vivipara (Jacqouin, 1787) from Russia: Sex chromosomes, subspeciation, and colonization. *Russ. J. Herpetol.* **2011**, *18*, 99–104.
18. Kupriyanova, L.; Kirschey, T.; Böhme, W. Distribution of the common or viviparous lizard, Zootoca vivipara (Lichtenstein, 1823) (Squamata: Lacertidae) in Central Europe and re-colonization of the Baltic Sea basin: New karyological evidence. *Russ. J. Herpetol.* **2017**, *24*, 311–317. [CrossRef]
19. King, M. The evolution of sex chromosomes in lizards. In *Evolution and Reproduction*; Australian Academy of Science: Canberra, Australia, 1977; pp. 55–60.
20. King, M. Species Evolution. The Role of Chromosome Change. Cambridge University Press: Cambridge, UK, 1993; 336p.
21. Payseur, B.; Presgraves, D.; Filatov, D. Sex chromosomes and speciation. *Mol. Ecol.* **2018**, *27*, 3745–3748. [CrossRef]
22. Pennell, M.W.; Mank, J.E.; Peichel, C.L. Transition in sex determination and sex chromosomes across vertebrate species. *Mol. Ecol.* **2018**, *27*, 3950–3963. [CrossRef]
23. Olmo, E. Rate of chromosome changes and speciation in reptiles. *Genetica* **2005**, *125*, 185–203. [CrossRef]
24. Petraccioli, A.; Guarino, F.; Kupriyanova, L.; Mezzasalma, M.; Odierna, G.; Picariello, O.; Capriglione, T. Isolation and characterization of interspersed repeated sequences in the European common lizard, Zootoca vivipara, and their conservation in Squamata. *Cytogenet. Genome Res.* **2019**, *157*, 65–76. [CrossRef]
25. Safronova, L.D.; Kupriyanova, L.A. Metaphase and meiotic chromosomes, Synaptonemal complexes (SC) of the lizard Zootoca vivipara. *Russ. J. Genet.* **2016**, *52*, 1184–1191, ISSN 1022-7954. [CrossRef]
26. Kupriyanova, L.A.; Safronova, L.D.; Chekunova, A.I. Meiotic chromosomes, Synaptonemal complexes (SC) in a female viviparoys lizard Zootoca vivipara. *Russ. J. Genet.* **2019**, *55*, 774–778, ISSN 1022-7954. [CrossRef]
27. Summer, A.T. A simple technique for demonstrating centromeric heterochromatin. *Exp. Cell Res.* **1972**, *75*, 304–306. [CrossRef] [PubMed]
28. Schmid, M.; Guttembach, M. Evolutionary diversity of reverse (R) fluorescent chromosome bands in vertebrates. *Chromosoma* **1988**, *97*, 101–114. [CrossRef] [PubMed]
29. Dresser, M.; Moses, M. Synaptonemal complex karyotyping in spermatocytes of the Chinesehamster (Cricetulus gricus).iY. Light and electron microscopy of synapsis and nucleolar development by silver staining. *Chromosoma* **1980**, *76*, 1–22. [CrossRef]
30. Kupriyanova, L.A.; Safronova, L.D. Results and perspectives of cyto- and genetic studying of "cryptic" group of the Lacertidae. *Trudyzin* **2020**, *324*, 100–107. [CrossRef]
31. Dinkel, B.J.; O'Laughlin-Phillips, E.A.; Fechheimer, N.S.; Jaap, R.G. Gametic products transmitted by chickens heterozygous for chromosomal rearrangements. *Cytogenet. Cell. Genet.* **1979**, *23*, 124–136. [CrossRef]
32. De Villena, F.P.-M.; Sapienza, C. Female meiosis drives karyotypic evolution in mammals. *Genetics* **2001**, *159*, 1179–1189. [CrossRef]

33. Kupriyanova, L.; Safronova, L.; Chekunova, A.; Osipov, F. The study of oocytes in the early oogenesis of the viviparous lizard (*Zootoca vivipara*) in prophase of its meiosis. *Bull. Mosc. Soc. Nat.* **2021**, *126*, 3–11.
34. Kupriyanova, L.; Safronova, L.; Kirschey, T.; Böhme, W. Meiotic chromosomes, Synaptonemal complexex (SC) in the cryptic forms of the euroasian complex Zootoca vivipara with multiple sex chromosomes. In *Book Abstract: The 12st European Congress of herpetology, Belgrade, Serbia 5–9 September 2022*; Crnobrnja-Isailović, J., Vukov, T., Vučić, T., Tomović, L., Eds.; Institute for Biological Research "Siniša Stanković", National Institute of Republicnof Serbia, University of Belgrade: Belgrade, Serbia, 2022; Volume 82, p. 261.
35. Capriglione, T.; Olmo, E.; Odierna, G.; Kupriyanova, L. Mechanisms of differentiation in the sex chromosomes of some Lacertidae. *Amphibia-Reptilia* **1994**, *15*, 1–8.
36. Chalopin, D.; Naville, M.; Plard, F.; Galiana, D.; Volff, J.N. Comparative analysis of transposable elements highlights mobilome diversity and evolution in vertebrates. *Genome Biol. Evol.* **2015**, *7*, 567–580. [CrossRef]
37. Serrato-Capuchina, A.; Matute, D. The role of transposable elements in speciation. *Genes* **2018**, *9*, 254. [CrossRef] [PubMed]
38. Alam, S.M.I.; Sarre, S.; Gleeson, D.; Georges, A.; Ezaz, T. Did lizards follow unique pathways in sex chromosome evolution? *Genes* **2018**, *9*, 239. [CrossRef] [PubMed]
39. Bakloushinskaya, I. Chromosomal rearrangements, reorganization of genome, and Speciation. *Zool. J.* **2016**, *95*, 376–393. [CrossRef]
40. Axelsson, E.; Alrechtsen, A.; van A., P.; Li, L.; Megens, H.J.; Vereijken, A.L.J.; Crooijmans, R.P.M.A.; Groenen, M.A.M.; Ellegren, H.; Wierslev, E.; et al. Segregation distortion in chicken and the evolutionary consequences of female meiotic drive in birds. *Heredity* **2010**, *105*, 290–298. [CrossRef] [PubMed]

Article

Temperature Incubation Influences Gonadal Gene Expression during Leopard Gecko Development

Maria Michela Pallotta [1,2,*], Chiara Fogliano [2,*] and Rosa Carotenuto [2]

[1] Institute for Biomedical Technologies, National Research Council, 56124 Pisa, Italy
[2] Department of Biology, University of Naples Federico II, Via Cinthia 26, 80126 Naples, Italy
* Correspondence: mariamichela.pallotta@irgb.cnr.it (M.M.P.); chiara.fogliano@unina.it (C.F.)

Simple Summary: Environmental sex determination is a modality of sex determination related to external factors and that has implicated determinants such as climatic conditions, which act on the embryo after fertilization and deposition of the egg. For reptiles, the temperature is the main element for sex determination; this factor affects laid eggs in different ways. Details remain to be elucidated concerning the temporal gene expression and the functions of their protein products. Therefore, the aim of the present work was to determine the genetic determinants differentially represented during the embryonic development of a model species already known in temperature-dependent sex determination, the leopard gecko *Eublepharis macularius*. Following this investigation, new data were acquired on genes expressed in the sexual differentiation of *E. macularius*. In addition, new genes potentially involved in the mechanisms of tissue and metabolic sexual differentiation of the embryo of this species have been identified. This study could bring new useful information in order to correctly interpret the regulatory pathway underlying the determination of sex in vertebrates.

Abstract: During development, sexual differentiation results in physiological, anatomical and metabolic differences that implicate not only the gonads but also other body structures. Sex in Leopard geckos is determined by egg incubation temperature. Based on the premise that the developmental decision of gender does not depend on a single gene, we performed an analysis on *E. macularius* to gain insights into the genes that may be involved in gonads' sexual differentiation during the thermosensitive period. All the genes were identified as differentially expressed at stage 30 during the labile phase of sex differentiation. In this way, the expression of genes known to be involved in gonadal sexual differentiation, such as *WNT4, SOX9, DMRT1, Erα, Erβ, GnRH, P450 aromatase, PRL and PRL-R*, was investigated. Other genes putatively involved in sex differentiation were sought by differential display. Our findings indicate that embryo exposure to a sex-determining temperature induces differential expression of several genes that are involved not only in gonadal differentiation, but also in several biological pathways (*ALDOC, FREM1, BBIP1, CA5A, NADH5, L1 non-LTR retrotransposons, PKM*). Our data perfectly fit within the new studies conducted in developmental biology, which indicate that in the developing embryo, in addition to gonadal differentiation, sex-specific tissue and metabolic polarization take place in all organisms.

Keywords: lizard; sex determination; incubation temperature; gonadal differentiation

Citation: Pallotta, M.M.; Fogliano, C.; Carotenuto, R. Temperature Incubation Influences Gonadal Gene Expression during Leopard Gecko Development. *Animals* **2022**, *12*, 3186. https://doi.org/10.3390/ani12223186

Academic Editor: Ettore Olmo

Received: 5 October 2022
Accepted: 16 November 2022
Published: 17 November 2022

1. Introduction

In many organisms, sex is determined by the presence of heteromorphic chromosomes and by factors encoded by them, which establish balances between specific regulatory patterns. In mammals, for example, the SRY factor linked to the sex chromosome Y is decisive for testicular differentiation [1]: in these cases, we speak of genotyping sex determination (GSD). Several cases of GSD exist: one of the best known is that of *D. melanogaster* [2], in which sex determination depends on the ratio of the number of X sex chromosomes to the number of autosomes A. Sexual development in mammals, on the other hand, is a more complex process and is independent of the ratio of the number of sex/autosomal chromosomes present in the individual's genome but originates from the presence/absence of the Y chromosome [3], from which derives a transcript, Sex-determining region Y protein (*SRY*), which, by interacting with factors such as SRY-box containing gene 9 (*Sox9*), *Sf1* and *Dmrt1* (dsx and mab-3-related transcription factor 1) [4] controls the differentiation of the bipotential gonad in the male direction in embryos with XY chromosomes. There is another possibility of sex determination, which is dependence on environmental factors (environmental sex determination—ESD). The latter is related to factors external to the organism's genome and sees implicated determinants, such as climatic conditions, which act on the embryo after fertilization and deposition of the egg [5]. Reptiles, unlike some birds and mammals, exhibit sex determination that is both dependent on sex chromosomes and dependent on external factors [6]. The modes of GSD include at least three different conditions of heterogametes. When we talk about environmental determinants, we refer, for reptiles, to temperature as the main one; this factor affects laid eggs in different ways. The prerequisite for the persistence of ESD in reptiles seems to be related to the thermally heterogeneous environment, in which the natural and different incubation temperatures of the different microenvironments in which eggs are laid ensure the determination and development of both gonads and the physiological characterization of the male and female sexes. Granting that temperature can act as a motive for the activation of specific pathways that induce sex differentiation [7], it is possible to assume the existence of a period during egg incubation when the embryo is temperature sensitive (TSP) [8]. Although several genetic factors are important for sex determination and are regulated by temperature, details remain to be elucidated in regard to temporal gene expression and the functions of their protein products [9].

Reptiles are among the organisms that tolerate temperature variations worst [10,11]. To date, however, information on the gene and molecular network that would guide gonadogenesis in these species remains limited, both regarding the different components and regarding their respective functions. Even in mammals, the lack of data on all genes expressed during the early stages of gonadal development has limited the ability to delineate the complete pathway of genes that would regulate the early stages of ovary development [12,13]. The aim of the present work was mainly to individuate which are the genetic determinants differentially represented during the embryonic development of a model species, the leopard gecko (*Eublepharis macularius*), a well-known model in the study of the mechanisms of embryonic development and temperature-dependent sex determination [14–16]. The possible variation in expression of genes involved in gonad differentiation and in other pathways was then evaluated. All the genes were identified as differentially expressed at stage 30 during the labile phase of sex differentiation [17]. For our aims, we used differential display (DDRT-PCR), and the results obtained were validated by means of real-time PCR. Since the use of model species has proved useful for identifying new genetic factors and understanding their mechanisms of action, the study of reptiles could bring new useful information in order to correctly interpret the regulatory pathway underlying the determination of the sex in vertebrates.

2. Materials and Methods

2.1. Animals

Two females and one male specimen of *E. macularius* were housed in a terrarium at the Department of Biology of Università di Napoli "Federico II", according to the institution's Animal Welfare Office guidelines and policies and to international rules and to the recommendations of the Guide for the Care and Use of Laboratory Animals of the American National Institutes of Health and of the Italian Health Ministry. The experimental protocol was approved by the institutional Animal Experiments Ethics Committee (Centro Servizi Veterinari) (permit number: 2014/0017970). Fertilization occurred naturally. Every month (May, June and July), after fertilization, each female deposed two eggs. Each experiment was technically replicated three times. Each pair of eggs (6 pairs, 12 eggs in total) were collected and immediately placed in two precision incubators (±0.1 °C) set to a constant temperature of 26 °C (FPT, six eggs) or 32.5 °C (MPT, six eggs), for 7 days, roughly corresponding to stage 30 [17]. Temperature and moisture were monitored daily using HOBO temperature loggers (Onset Computer Corporation, Pocassett, MA, USA).

2.2. RNAs

Each embryo (*N* = 12; 6 from MPT and 6 from FPT derived from two females and deposed in different months) were eviscerated, and the area strictly adjacent to the not fully formed gonads was taken. Total RNA was extracted from each single embryo according to the TRI-Reagent protocol (Sigma Aldrich, Saint Louis, MO, USA). The concentration and purity of RNA samples were determined by UV absorbance spectrophotometry; RNA integrity was checked by 2.0% agarose gel electrophoresis. RNA extracted from the 12 embryos at stage 30, the stage in which the undifferentiated gonad is sensible to the temperature [17], was subdivided between qRT-PCR (*N* = 6; 3 from MPT and 3 from FPT) and DDRT-PCR (*N* = 6; 3 from MPT and 3 from FPT). First-strand cDNA, used for all amplification reactions, was synthesized from singularly extracted RNA from each MPT and FPT embryo, then utilized to obtain 1 µg of total RNA using Super Script III Reverse Transcriptase (Invitrogen, Waltham, MA, USA) and used for the two screening protocols [16].

2.3. Expression Analysis of Genes Involved in Gonadal Sex Differentiation

Differential expression analysis of nine genes involved in gonadal sex differentiation, estrogen receptor α (*Erα*), estrogen receptor β (*Erβ*), gonadotropin-releasing hormone (*GnRH*), P450 aromatase, prolactin (*PRL*), prolactin receptor (*PRL-R*), Wnt family member 4 (*WNT4*), sex determining region Y-Box 9 (*SOX9*) and doublesex and Mab-3 related transcription factor 1 (*DMRT1*) was carried out by quantitative RT (qRT) PCR using primers designed on vertebrate sequences found in GenBank. Sequences of interest were aligned by using a Multiple Sequence Alignment free software (http://www.genome.jp/tools-bin/clustalw, accessed on 10 November 2022) and primers designed on the sequence regions with the highest degree of identity by means of Primer 3Plus software (http://www.bioinformatics.nl/cgibin/primer3plus/primer3plus.cgi, accessed on 10 November 2022) (Table 1) [16].

Table 1. Primers used in the qRT-PCR analysis performed to validate the expression profiles of the studied genes.

Gene	Oligo Forward Sequence (5′-3′)	Oligo Reverse Sequence (5′-3′)
Erα	CACCCTGGAAAGCTGTTGTT	TTCGGAATCGAGTAGCAGTG
Erβ	ATCCCGGCAAGCTAATCTTT	CAGCTCTCGAAACCTTGAAGT
GnRH	GTCTTGCTGGCCTCTCCTC	GTGGTCTCCTGCCAGTGTTC
P450 aromatase	TGAACACCCTCAGTGTGGAA	TCAGVTTTGGCATGTCTTCA
PRL	AAGGCCATGGAGATTGAGG	GGAGGCCTGACCAAGTAGAA
PRL-R	ATGGAGGTCTCCCCACTAAT	AACAGGAATTGGGTCCTCCT
WNT4	CTGCAACAAGACCTCCAAGG	AGCAGCACCGTGGAATTG
SOX9	GGGCAAGCTTTGGAGGTTAC	TGGGCTGGTACTTATAGTCTGGA
DMRT1	GCAGGGGATCCTACCAAAGT	AGAAGGCAGCAAGCTCAAGA
ALDOC	TACCATGGTGTTGTGCAAGC	CTTCACGCTGCATTTTCTCA
FREM1	GGAATGTCAACCAAGATGTGG	CAGGGGAGATCAGAACCACT
BBIP1	CGTGAGCTGTAGCTTTGCAG	CTGCCTTACCCACAGCACTT
CA5A	TTGCAAAGTTATGGGGAGGA	TCAAGCAGGGTTTATTTCTCATC
NADH5	GCTACAGGTAAATCCGCTCAA	AGTAGGGCAGAAACGGGAGT
L1 non-LTR retrotransposons	ATCATCGTGGGCCTCTTTGC	AGCAGCACCGTGGAATTG
PKM	AGAGCTGCTTGTACGCCTGT	CCAGATTTCCAAAGGACAGTG

2.4. DDRT-PCR

Differential display allows one to compare and identify changes in gene expression at the mRNA level between two or more cell populations. Briefly, RNA was reverse transcribed using anchored oligo-dT primers designed to specifically bind to the 5′ ends of the poly-A tails. Successively, cDNAs were amplified by using the anchored oligo-dT primers in combination with a series of arbitrary 5′ primers and amplification products, then separated and visualized by electrophoresis. For our purpose, RT-PCR and PCR were performed using the RNA spectra kit and fluorescent mRNA Differential Display System (GenHunter®Corporation, Nashville, TN, United States). RNA extracted from each MPT embryo (N = 3) was then utilized to obtain 1 µg of total RNA, and the same was performed for the FPT embryo (N = 3). After extraction, total RNA was reverse-transcribed in two 20 µL reaction mixtures at 37 °C for 60 min using MMLV reverse transcriptase and a set of three one-base anchored oligo(dT) primers (H-T11A/C/G). MPT and FPT cDNA fragments were amplified using combinations of the anchored H-T11 primers from the reverse transcription step and eight different AP upstream primers (Table 2).

Table 2. Sequences of the 3 oligo (dT) primers and 5 arbitrary primers (H-AP) used in differential display.

Primers	Sequence
3′ oligo (dT) H-T11G	AAGCTTTTTTTTTTTG
3′ oligo (dT) H-T11A	AAGCTTTTTTTTTTTA
3′ oligo (dT) H-T11C	AAGCTTTTTTTTTTTC
H-AP1	AAGCTTGATTGCC
H-AP2	AAGCTTCGACTGT
H-AP3	AAGCTTTGGTCAG
H-AP4	AAGCTTCTCAACG
H-AP5	AAGCTTAGTAGGC
H-AP6	AAGCTTGCACCAT
H-AP7	AAGCTTAACGAGG
H-AP8	AAGCTTTTACCGC

The DNA fragments differentially amplified by DDRT-PCR were purified from agarose gel using WIZARDR SV Gel and the PCR Clean-Up System (Promega, Milano, Italy). The purified fragments were T/A inserted into Vector pCRR 4-TOPOR and cloned into *Escherichia coli* DH5α using a TOPOR TA CloningR Kit for Sequencing (Invitrogen, Waltham, MA, USA) according to the manufacturer's recommendations. Plasmids were extracted by Fast Plasmid Mini Kit (Eppendorf, Hamburg, Germany). Sequencing was performed by Primmbiotech srl (Milan, Italy). Sequences were queried against the NCBI database using Nucleotide BLAST tool and related to known proteins using the tBLASTX algorithm and gene ontology hierarchy [16].

2.5. *Confirmation of Differential Gene Expression: Real-Time PCR*

qRT-PCR analysis was performed on all the genes of interest, using the same RNA samples employed for the experiments, as previously described. All the primers used for qRT-PCR (Table 1) were designed using the software Primer 3Plus (http://www.bioinformatics.nl/cgiin/primer3plus/primer3plus.cgi, accessed on 10 November 2022). qRT-PCR reactions were carried out using iTaqTM Universal SYBR Green Supermix kit (Bio-Rad, Hercules, CA, USA) in a final reaction volume of 20 µL. For transcript quantification, samples were normalized to the expression level of the endogenous reference gene (GAPDH) to take into account possible differences in cDNA quantity and quality. The amplification protocol involved one cycle at 95 °C for 10 min, to activate Taq DNA polymerase, and 40 cycles consisting of a denaturation step at 95 °C for 15 s and annealing and extension steps at 60 °C for 1 min [16]. Reactions were conducted in an iCycler iQ5 system. The magnitudes of change in gene expression relative to males were determined by the $2^{-\Delta\Delta Ct}$ method of Livak and Schmittgen [18]. Statistical significance was determined using a t-test analysis with the Holm–Sidak correction for multiple comparison method using GraphPad Prism 6.0.7 software.

3. Results

3.1. Genes Involved in Gonadal Sex Differentiation

Nine transcript fragments of *E. macularius* genes, which are critical for SD in mammals and other vertebrates (*Erα, Erβ, GnRH, P450 aromatase, PRL, PRL-R, WNT4, SOX9 and DMRT1*), were amplified by qRT-PCR from RNA of leopard-gecko embryos incubated at sex-specific temperatures and sacrificed at 7 (stage 30) days. In gonads of stage 30, *PRL-R* appeared more expressed in embryos incubated at 26 °C (FPT); *WNT4, SOX9* and *DMRT1* were more expressed at 32.5 °C (MPT). *Erα, Erβ, GnRH* and *P450 aromatase* did not exhibit any statistically significant differential expression (Figure 1). At this stage, we failed to detect expression of *PRL* by qRT-PCR.

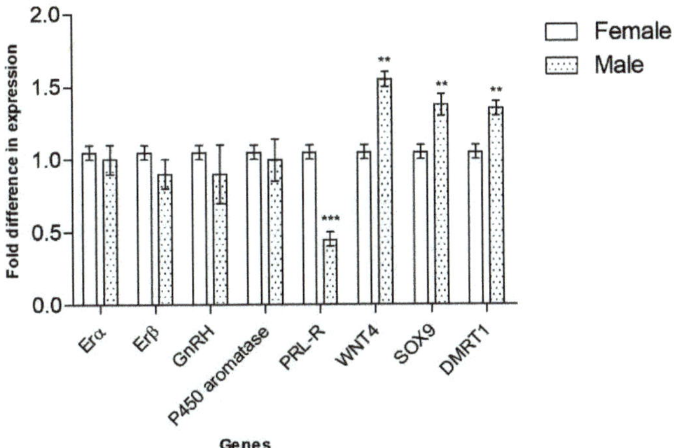

Figure 1. Expression analysis of genes canonically involved in sexual differentiation by qRT-PCR. mRNA levels of FPT embryos are related to MPT embryos. Data are presented as mean with SD. Statistical significance was determined using *t*-tests with Holm–Sidak correction for multiple comparison. ** $p < 0.01$, *** $p < 0.001$.

3.2. Identification and Expression Analysis of Seven New Transcripts by DDRT-PCR

In analysis of the arbitrary primers provided in the kit, only four (H-AP2, H-AP5, H-AP6, and H-AP7) yielded expression profiles containing bands that were differently expressed (Figure 2). The sequences of these cloned fragments (Supplementary Table S1) were compared to those found in Genbank and Embl using BLASTN and TBLASTX. Some correspondence with: *Anolis carolinensis* pyruvate kinase muscle isozyme-like, *Gekko japonicus* aldolase fructose-bisphosphate C (*ALDOC*), *Anolis carolinensis FRAS1-related extracellular matrix protein 1-like, Anolis carolinensis BBSome-interacting protein 1-like, Sphaerodactylus townsendi* carbonic anhydrase 5A (*CA5A*), *NADH dehydrogenase subunit 5* (mitochondrion) of *Hemitheconyx caudicinctus* and *L1 non-LTR retrotransposons* of *A. carolinensis,* were found (Table 3).

3.3. qRT-PCR Expression Analysis of Genes Identified by DDRT-PCR

All the data collected by DDRT-PCR were validated using qRT-PCR, which partially confirmed the results obtained with the method (Table 4). At stage 30, differential expression was confirmed for six of the seven genes identified: *CA5A* and *L1 non-LTR* expression were stronger in embryos incubated at the FPT, whereas *PKM, FREM1, BBIP1* and *ALDOC* appeared to be expressed more strongly in MPT. Equal expression levels of *NADH5* were found in male and female embryonic gonads (Figure 3).

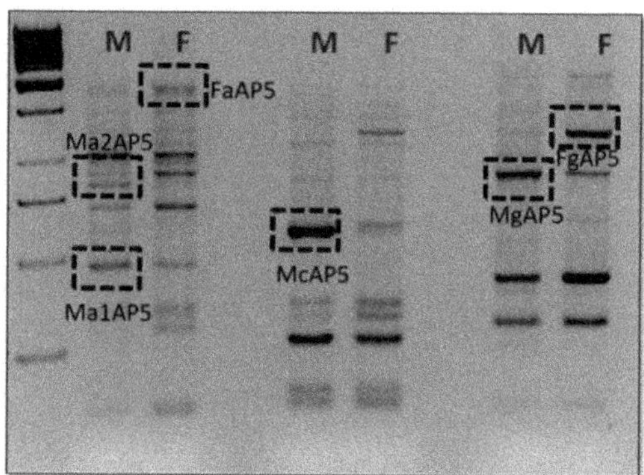

Figure 2. Representative gel image of DDRT-PCR band pattern. The image shows amplifications of cDNA from gonad embryos performed using a 5′ arbitrary primer (H-AP5) in combination with 3′ oligo (dT) H-T11A primer on male (Ma) and female (Fa) embryos.

Table 3. Sequence analysis of differentially expressed mRNAs isolated by DDRT-PCR.

Clone	E-Value	Length (BP)	Identity (%)	Results
McAP2	e−121	450	98%	*Gekko japonicus, aldolase fructose-bisphosphate C (ALDOC)*
Ma1AP5	2e−41	528	83%	*Anolis carolinensis, pyruvate kinase muscle isozyme-like (PKM)*
Ma2AP5	2e−60	710	95%	*Anolis carolinensis, FRAS1 related extracellular matrix 1 (FREM1)*
FaAP5	4e−148	1011	73%	*Hemitheconyx caudicinctus, NADH dehydrogenase subunit 5 (mitochondrion) (NADH-CoQ reduttasii)*
McAP5	1e−144	592	83%	*Anolis carolinensis, BBSome-interacting protein 1-like (BBIP1)*
Fg1AP6	9e−15	137	77%	*Sphaerodactylus townsendi, carbonic anhydrase 5A (CA5A)*
FcAP7	1e−14	202	76%	*Aanolis carolinensis, L1 non-LTR retrotransposons*

Table 4. Summary of the differentially expressed genes in *E. macularius* embryo at stage 30, listed according to the two different methodologies used. MPT stands for genes more expressed at a male-producing temperature and FPT stands for genes more expressed at a female-producing temperature.

Gene	Identification	Embryo Gonads Stage 30
Erα	qRT-PCR	No differential expression
Erβ	qRT-PCR	No differential expression
GnRH	qRT-PCR	No differential expression
P450 aromatase	qRT-PCR	No differential expression
PRL	qRT-PCR	Not detected
PRL-R	qRT-PCR	FPT
WNT4	qRT-PCR	MPT
SOX9	qRT-PCR	MPT
DMRT1	qRT-PCR	MPT
ALDOC	DDRT-PCR	MPT
FREM1	DDRT-PCR	MPT
BBIP1	DDRT-PCR	MPT
CA5A	DDRT-PCR	FPT
NADH5	DDRT-PCR	No differential expression
L1 non-LTR retrotransposons	DDRT-PCR	FPT
PKM	DDRT-PCR	MPT

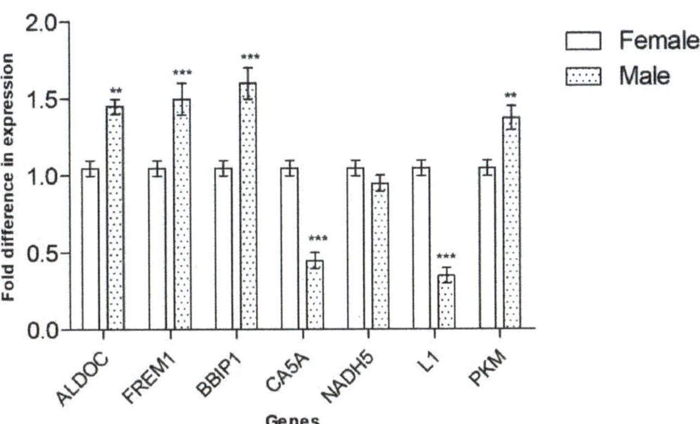

Figure 3. Validation of DDRT-PCR by qRT-PCR. Data are presented as mean with SD. Statistical significance was determined using *t*-tests with Holm–Sidak correction for multiple comparisons. ** $p < 0.01$, *** $p < 0.001$.

4. Discussion

In the animal world, it is usually the genome that provides adequate instructions for the embryonic development of morphological structures; however, in some cases, such as in reptiles, the environment can drive morphogenesis events by modulating the expression of specific genes. Temperature-dependent sex determination (TSD) makes some reptile species ideal models for acquiring information on the space-time path of gene activation. Up to now, the information on the genes that trigger the gene and molecular network at the basis of vertebrate gonadogenesis is still incomplete. Even in mammals, the lack of data on genes expressed during the early stages of gonadal development has limited the possibility of drawing a definitive framework for the genes that would regulate the early stages of ovarian development [12]. It is known that, in the early stages of embryonic development, some important genes involved in sexual differentiation are expressed in a sex-specific way.

Sox9, which is part of the family of transcription factors with the HMG box, equivalent to the *SRY* of mammals, is expressed in *E. macularius* only a few days after deposition, at stage 28–30, which, in this species, presumably corresponds to the beginning of the temperature-sensitive period (TSP). *Sox9* could, therefore, have a role in this species in determining the initiation of gonadal differentiation, and its expression found in both gonadal sketches not yet differentiated could therefore trigger this process. Subsequently, its expression decreases in the female embryonic gonad, remaining constant in the male one of some species, until the end of the temperature-sensitive period, or increasing, in others, during the testicular morphogenesis phase [19]. The high degree of identity of the *Sox9* sequence in mammals, birds, fish and reptiles, even in geckos, suggests, however, that there is conservation of its function in all vertebrates.

Wnt4, which acts in antagonism with *Sox9*, seems to guide the differentiation pattern of the ovary. Surprisingly, in the analyzed samples, in stages 28 and 29 it was more expressed in the embryo at MDT than in the one held at FDT. In *Trachemis scripta*, the species that gave the greatest amount of information related to the pathway underlying the reptilian TSD, similar expression levels in males and females for *Wnt4* were found at the beginning of the TSP phase, stages 16–19 [20]. In this species, *Wnt4* appears to be over-expressed in females only during ovarian differentiation.

The expression of *DAX* is also highly variable. In organisms with TSD that have been studied, *Dax* shows a species-specific trend. It is initially expressed at similar levels in MSD and FSD in all species. It was localized, by WISH, both from Muller's duct and from Wolff's duct. It then decreases dramatically during embryonic development in

T. scripta [21] and *L. olivacea* [18]; it increases slightly in the alligator, *A. mississipiensis* [22], and in *C. picta* [23]; but it remains constant in *C. serpentina* [20].

Such a variable trend suggests that Dax has different functions in the various organisms, or that, in these functions, it may have activation timing that does not necessarily correspond in the various species. In the leopard gecko, it was not possible to detect the presence of *DAX* in the early stage we studied; this could agree with an actual involvement of this gene in later stages of gonadal formation. For *Dmrt1*, which is involved in determining the formation of the testicle, the results obtained confirmed what is known in the literature: even if only in minimal quantities, its expression is detectable in males from the beginning of embryonic development [21]. In fact, in situ hybridizations of *Dmrt1* on embryonic gonads of *Podarcis sicula* at 7 days from fertilization show no signals, but the gene is clearly expressed both in the ovary and in the embryonic testis at a later stage, at about 15 days, and then expression is localized only in the testicle until hatching [24]. *Dmrt1* has been the subject of in-depth analysis, as it is considered one of the oldest genes in the sex determination of vertebrates. In fact, genes belonging to the DM family are considered phylogenetically close to *dS* of *D. melanogaster* and *Mab3* of *C. elegans*. The expression of *Dmrt1* is considered not only important for guiding the correct development of the testicle in the embryo but also for maintaining correct testicular function in the adult. Additionally, the expression of the estrogen receptors *Er-b* is clearly present in *E. macularius* embryos from stage 28–30, in both MSD and FSD, without variations. Since these genes are also involved in determining the proper development of the central nervous system and in morphogenesis in general, their early activation may be required to perform these additional functions. During the LP phase (stage 30), the temperature seems to affect aromatase activity and synthesis of estrogens. These data were found not only for the gonads but also for other body structures, such as the brain [17]. Our work is in accordance with previous studies that by utilizing differential display, highlighted how temperatures induced differential expression of several genes involved not only in gonadal differentiation but also, for example, in neural differentiation, in basal metabolic processes or in cell proliferation and differentiation [16]. In our case, we found a group of genes not strictly related to sexual differentiation that also displayed differential expression. One of the differentially expressed sequences, FcAP7, aligned with a portion belonging to the 3′ UTR region of a transposable element present in the genome of a reptile. The data, also validated by analysis for real-time PCR, are interesting because they confirm the presence of repeated elements transcribed during moments of cell differentiation, once again suggesting their probable role in the functionality of the genome. For the Ma1AP5 sequence, on the other hand, we found 83% similarity to the *Anolis carolinensis* pyruvate kinase muscle isozyme-like (*PKM*) sequence, and analysis using the EMBL database revealed similarity with ovarian and testis cDNA libraries of *Anolis carolinensis*. *PKM* is a glycolytic isozyme that catalyzes the transfer of a phosphoryl group from phosphoenolpyruvate to ADP, generating ATP [25]. Different enzymatic forms of it are known [26]; considering their implications in the phenomena of cell growth and proliferation and their involvement in some tumor pathologies [27], it could be hypothesized that the isolated enzyme form is an isozyme expressed only in the phase of embryonic development that could have a role in the induction of cell growth. The *FRAS1-related extracellular matrix 1 transcript*, with which the Ma2AP5 sequence aligns with 95% identity, is associated with craniofacial and renal embryonic formation and development, and its mutation leads precisely to renal agenesis in mice [28]. Given that the gonad and the kidney have the same embryonic origin, one could hypothesize a role of this transcript in the differentiation of the renal portion with respect to the induction of the differentiation of the primordial gonad. The sequence of the transcript of *Anolis carolinensis BBSome-interacting protein 1-like* was instead found following interrogation in the EMBL database (tBLASTx) with the McAP5 sequence (83% identity). We found that the query sequence shows high similarity to *Anolis carolinensis* cDNA libraries derived from transcripts present in the testes and primordial kidneys. The BBSome complex, which contains several isoforms of the BBSome-interacting protein, also forms a protein complex involved in cell trafficking, ciliogenesis and microtubular stability [29]. In Zhang et al. [30], the mutation of this protein is responsible for male infertility in mice due to defects in the formation of the sperm flagellum. Having observed, in the present study, greater expression of the transcript in the male embryo than the female embryo, one could hypothesize a role of *BBSome-interacting protein 1-like* in the induction of male differentiation of the primordial gonad and a default role in both embryos for cell communication.

Aldolase C fructose-bisphosphate (ALDOC, or ALDC) is an enzyme that, in humans, is encoded by the *ALDOC* gene on chromosome 17. This gene encodes a member of the class I fructose-bisphosphate aldolase gene family [31]. For the first time, this gene was found differentially expressed in male and female embryo gonads. This suggests that there are genes not yet studied during sexual development. *CA5A* (carbonic anhydrase 5A) is a protein-coding gene. Diseases associated with CA5A include carbonic anhydrase Va deficiency, hyperammonemia and carbonic anhydrase Va deficiency. Among its related pathways are metabolism and reversible hydration of carbon dioxide. *CA5A* was shown to be expressed in the ovaries of the Pelibuey breed of sheep; the gene was upregulated in a subset of ewes that gave birth to two lambs compared to uniparous animals [32]. The level of expression led the authors to conclude that *CA5A* is heritable and potentially an imprinted gene [33]. That result is in agreement with our findings. In fact, *CA5A* was more expressed in female embryo gonads.

5. Conclusions

Through the present study, new data have been acquired on genes expressed in the early stages of development and sexual differentiation in *E. macularius*. We demonstrated that not only genes related to sexual differentiation, but also genes involved in different developmental pathways, modify their expression in relation to breeding temperature. Our data perfectly fit within the new studies conducted in developmental biology, which indicate that in the developing embryo, in addition to gonadal differentiation, sex-specific tissue and metabolic polarization take place in all organisms. Further investigations will be necessary on embryos at later stages of embryonic development, in order to test the roles of traced transcripts in the determination of gonads and tissues, define any progressive variations in their levels of expression, identify other genes differentially expressed in the later stages of development and analyze their behavior during the reproductive life of the organism.

Supplementary Materials: The following supporting information can be downloaded at: https://www.mdpi.com/article/10.3390/ani12223186/s1. Table S1. Nucleotide sequences identified by DDRT-PCR.

Author Contributions: Conceptualization, M.M.P. and R.C.; methodology, M.M.P. and C.F.; software, M.M.P.; validation, M.M.P., C.F. and R.C.; formal analysis, M.M.P. and C.F.; investigation, M.M.P. and C.F.; resources, R.C.; data curation, M.M.P., C.F. and R.C.; writing—original draft preparation, M.M.P. and C.F.; writing—review and editing, M.M.P., C.F. and R.C.; visualization, R.C.; supervision, R.C.; project administration, R.C. All authors have read and agreed to the published version of the manuscript.

Funding: This research received no external funding.

Institutional Review Board Statement: For this study, we used samples already collected for other previously published studies with the approval of institutional committees, and no further sampling was performed.

Data Availability Statement: The newly generated sequences are available in the Supplementary Materials.

Acknowledgments: This contribution is dedicated to the memory of our friend and colleague Teresa Capriglione. We will cherish her memory forever.

Conflicts of Interest: The authors declare no conflict of interest.

References

1. Quinn, A.; Koopman, P. The Molecular Genetics of Sex Determination and Sex Reversal in Mammals. *Semin. Reprod. Med.* **2012**, *30*, 351–363. [CrossRef] [PubMed]
2. Mullon, C.; Pomiankowski, A.; Reuter, M. Molecular evolution of Drosophila Sex-lethal and related sex determining genes. *BMC Evol. Biol.* **2012**, *12*, 5. [CrossRef] [PubMed]
3. Uguz, C.; Iscan, M.; Togan, I. Developmental genetics and physiology of sex differentiation in vertabrates. *Environ. Toxicol. Pharmacol.* **2003**, *14*, 9–16. [CrossRef]
4. Barrionuevo, F.J.; Burgos, M.; Scherer, G.; Jiménez, R. Genes promoting and disturbing testis development. *Histol. Histopathol.* **2012**, *27*, 1361–1383. [CrossRef]
5. Ciofi, C.; Swingland, I.R. Environmental sex determination in reptiles. *Appl. Anim. Behav. Sci.* **1997**, *51*, 251–265. [CrossRef]
6. Janzen, F.J.; Paukstis, G.L. A preliminary test of the adaptive significance of environmental sex determination in reptiles. *Evolution* **1991**, *45*, 435–440. [CrossRef]
7. Miller, D.; Summers, J.; Silber, S. Environmental versus genetic sex determination: A possible factor in dinosaur extinction? *Fertil. Steril.* **2004**, *81*, 954–964. [CrossRef]
8. Girondot, M.; Ben Hassine, S.; Sellos, C.; Godfrey, M.; Guillon, J.-M. Modeling Thermal Influence on Animal Growth and Sex Determination in Reptiles: Being Closer to the Target Gives New Views. *Sex. Dev.* **2010**, *4*, 29–38. [CrossRef]
9. Modi, W.S.; Crews, D. Sex chromosomes and sex determination in reptiles: Commentary. *Curr. Opin. Genet. Dev.* **2005**, *15*, 660–665. [CrossRef]
10. Valenzuela, N. Multivariate Expression Analysis of the Gene Network Underlying Sexual Development in Turtle Embryos with Temperature-Dependent and Genotypic Sex Determination. *Sex. Dev.* **2010**, *4*, 39–49. [CrossRef]
11. Du, W.-G.; Shine, R.; Ma, L.; Sun, B.-J. Adaptive responses of the embryos of birds and reptiles to spatial and temporal variations in nest temperatures. *Proc. R. Soc. Boil. Sci.* **2019**, *286*, 20192078. [CrossRef] [PubMed]
12. Piprek, R.P. Molecular Mechanisms Underlying Female Sex Determination—Antagonism Between Female and Male Pathway. *Folia Biol.* **2009**, *57*, 105–113. [CrossRef] [PubMed]
13. Yildirim, E.; Aksoy, S.; Onel, T.; Yaba, A. Gonadal development and sex determination in mouse. *Reprod. Biol.* **2020**, *20*, 115–126. [CrossRef] [PubMed]
14. Endo, D.; Kanaho, Y.-I.; Park, M.K. Expression of sex steroid hormone-related genes in the embryo of the leopard gecko. *Gen. Comp. Endocrinol.* **2008**, *155*, 70–78. [CrossRef]
15. Valleley, E.M.; Cartwright, E.J.; Croft, N.J.; Markham, A.F.; Coletta, P.L. Characterisation and expression ofSox9 in the Leopard gecko, *Eublepharis macularius*. *J. Exp. Zool.* **2001**, *291*, 85–91. [CrossRef]
16. Pallotta, M.M.; Turano, M.; Ronca, R.; Mezzasalma, M.; Petraccioli, A.; Odierna, G.; Capriglione, T. Brain Gene Expression is Influenced by Incubation Temperature During Leopard Gecko (*Eublepharis macularius*) Development. *J. Exp. Zool. Mol. Dev. Evol.* **2017**, *328*, 360–370. [CrossRef]
17. Wise, P.A.; Vickaryous, M.K.; Russell, A.P. An Embryonic Staging Table for in ovo Development of *Eublepharis macularius*, the Leopard Gecko. *Anat. Rec.* **2009**, *292*, 1198–1212. [CrossRef]
18. Livak, K.J.; Schmittgen, T.D. Analysis of relative gene expression data using real-time quantitative PCR and the $2^{-\Delta\Delta CT}$ Method. *Methods* **2001**, *25*, 402–408. [CrossRef]
19. Maldonado, L.T.; Piedra, A.L.; Mendoza, N.M.; Valencia, A.M.; Martínez, A.M.; Larios, H.M. Expression profiles of Dax1, Dmrt1, and Sox9 during temperature sex determination in gonads of the sea turtle Lepidochelys olivacea. *Gen. Comp. Endocrinol.* **2002**, *129*, 20–26. [CrossRef]
20. Shoemaker, C.M.; Crews, D. Analyzing the coordinated gene network underlying temperature-dependent sex determination in reptiles. *Semin. Cell Dev. Biol.* **2009**, *20*, 293–303. [CrossRef]
21. Shoemaker, C.M.; Queen, J.; Crews, D. Response of Candidate Sex-Determining Genes to Changes in Temperature Reveals Their Involvement in the Molecular Network Underlying Temperature-Dependent Sex Determination. *Mol. Endocrinol.* **2007**, *21*, 2750–2763. [CrossRef] [PubMed]
22. Western, P.S.; Harry, J.L.; Graves, J.A.; Sinclair, A.H. Temperature-dependent sex determination in the American alligator: Expression of SF1, WT1 and DAX1 during gonadogenesis. *Gene* **1999**, *241*, 223–232. [CrossRef]
23. Valenzuela, N. Evolution of the gene network underlying gonadogenesis in turtles with temperature-dependent and genotypic sex determination. *Integr. Comp. Biol.* **2008**, *48*, 476–485. [CrossRef] [PubMed]
24. Capriglione, T.; Vaccaro, M.; Morescalchi, M.; Tammaro, S.; De Iorio, S. Differential DMRT1 Expression in the Gonads of Podarcis sicula (Reptilia: Lacertidae). *Sex. Dev.* **2010**, *4*, 104–109. [CrossRef]
25. Ikeda, Y.; Noguchi, T. Allosteric Regulation of Pyruvate Kinase M2 Isozyme Involves a Cysteine Residue in the Intersubunit Contact. *J. Biol. Chem.* **1998**, *273*, 12227–12233. [CrossRef] [PubMed]
26. Tsutsumi, H.; Tani, K.; Fujii, H.; Miwa, S. Expression of L- and M-type pyruvate kinase in human tissues. *Genomics* **1988**, *2*, 86–89. [CrossRef]
27. Yang, W.; Xia, Y.; Ji, H.; Zheng, Y.; Liang, J.; Huang, W.; Gao, X.; Aldape, K.; Lu, Z. Nuclear PKM2 regulates β-catenin transactivation upon EGFR activation. *Nature* **2011**, *480*, 118–122. [CrossRef]
28. Vrontou, S.; Petrou, P.; I Meyer, B.; Galanopoulos, V.K.; Imai, K.; Yanagi, M.; Chowdhury, K.; Scambler, P.; Chalepakis, G. Fras1 deficiency results in cryptophthalmos, renal agenesis and blebbed phenotype in mice. *Nat. Genet.* **2003**, *34*, 209–214. [CrossRef]

29. Loktev, A.V.; Zhang, Q.; Beck, J.S.; Searby, C.C.; Scheetz, T.E.; Bazan, J.F.; Slusarski, D.C.; Sheffield, V.C.; Jackson, P.K.; Nachury, M.V. A BBSome Subunit Links Ciliogenesis, Microtubule Stability, and Acetylation. *Dev. Cell* **2008**, *15*, 854–865. [CrossRef]
30. Zhang, Q.; Nishimura, D.; Vogel, T.; Shao, J.; Swiderski, R.; Yin, T.; Searby, C.; Carter, C.C.; Kim, G.; Bugge, K.; et al. BBS7 is required for BBSome formation and its absence in mice results in Bardet-Biedl syndrome phenotypes and selective abnormalities in membrane protein trafficking. *J. Cell Sci.* **2013**, *126*, 2372–2380. [CrossRef]
31. Juszczak, G.R.; Stankiewicz, A.M. Glucocorticoids, genes and brain function. *Prog. Neuro-Psychopharmacol. Biol. Psychiatry* **2018**, *82*, 136–168. [CrossRef] [PubMed]
32. Hernández-Montiel, W.; Collí-Dula, R.C.; Ramón-Ugalde, J.P.; Martínez-Núñez, M.A.; Zamora-Bustillos, R. RNA-seq Transcriptome Analysis in Ovarian Tissue of Pelibuey Breed to Explore the Regulation of Prolificacy. *Genes* **2019**, *10*, 358. [CrossRef] [PubMed]
33. Pokharel, K.; Peippo, J.; Weldenegodguad, M.; Honkatukia, M.; Li, M.-H.; Kantanen, J. Gene Expression Profiling of Corpus luteum Reveals Important Insights about Early Pregnancy in Domestic Sheep. *Genes* **2020**, *11*, 415. [CrossRef] [PubMed]

Article

A Cautionary Tale of Sexing by Methylation: Hybrid Bisulfite-Conversion Sequencing of Immunoprecipitated Methylated DNA in *Chrysemys picta* Turtles with Temperature-Dependent Sex Determination Reveals Contrasting Patterns of Somatic and Gonadal Methylation, but No Unobtrusive Sex Diagnostic

Beatriz A. Mizoguchi * and Nicole Valenzuela

Department of Ecology, Evolution and Organismal Biology, Iowa State University, Ames, IA 50011, USA
* Correspondence: biakemi@iastate.edu

Simple Summary: Identifying the sex of turtle hatchlings is important to assess the sex ratio of populations, which is important to study their ecology and evolution and for conservation programs. However, turtle hatchlings rarely display morphological differences detectable to the naked eye, and existing sexing techniques are either harmful, lethal, or non-viable for turtles with temperature-dependent sex determination. We investigated two methodologies that rely on differences in DNA methylation, a modification that occurs naturally in the DNA without changing its sequence, but that affects the expression of genes. As DNA methylation is known to differ in the gonads of male and female painted turtle hatchlings, we investigated whether the same is true in their tails We found that the painted turtle displays differential DNA methylation in the gonads, but not in the tails. We conclude that DNA methylation is tissue-specific in the painted turtle and that this epigenetic modification plays an important role in sexual development in this species but not in the somatic tissue of the tails.

Abstract: Background: The gonads of *Chrysemys picta*, a turtle with temperature-dependent sex determination (TSD), exhibit differential DNA methylation between males and females, but whether the same is true in somatic tissues remains unknown. Such differential DNA methylation in the soma would provide a non-lethal sex diagnostic for TSD turtle hatchings who lack visually detectable sexual dimorphism when young. **Methods:** Here, we tested multiple approaches to study DNA methylation in tail clips of *Chrysemys picta* hatchlings, to identify differentially methylated candidate regions/sites that could serve as molecular sex markers To detect global differential methylation in the tails we used methylation-sensitive ELISA, and to test for differential local methylation we developed a novel hybrid method by sequencing immunoprecipitated and bisulfite converted DNA (MeDIP-BS-seq) followed by PCR validation of candidate regions/sites after digestion with a methylation-sensitive restriction enzyme. **Results:** We detected no global differences in methylation between males and females via ELISA. While we detected inter-individual variation in DNA methylation in the tails, this variation was not sexually dimorphic, in contrast with hatchling gonads. **Conclusions:** Results highlight that differential DNA methylation is tissue-specific and plays a key role in gonadal formation (primary sexual development) and maintenance post-hatching, but not in the somatic tail tissue.

Keywords: epigenetic DNA methylation; temperature-dependent sex determination—TSD; bisulfite conversion of immunoprecipitated DNA MeDIP-BS-seq; methylation-sensitive ELISA; methylation-sensitive PCR; methylation-sensitive restriction enzyme; vertebrate reptilian turtle; somatic versus gonadal tissue; sexing diagnosis; conservation ecology

Citation: Mizoguchi, B.A.; Valenzuela, N. A Cautionary Tale of Sexing by Methylation: Hybrid Bisulfite-Conversion Sequencing of Immunoprecipitated Methylated DNA in *Chrysemys picta* Turtles with Temperature-Dependent Sex Determination Reveals Contrasting Patterns of Somatic and Gonadal Methylation, but No Unobtrusive Sex Diagnostic. *Animals* **2023**, *13*, 117. https://doi.org/10.3390/ani13010117

Academic Editor: Ettore Olmo

Received: 6 December 2022
Revised: 22 December 2022
Accepted: 23 December 2022
Published: 28 December 2022

1. Introduction

Epigenetic modifications mark DNA nucleotides chemically without altering their sequence in response to normal environmental signals (e.g., nutrition and temperature fluctuations) [1] or to environmental stressors during development, including sexual development [2]. DNA methylation is the most commonly studied epigenetic modification. It is a trait characterized by the replacement of the carbon 5′ of a deoxycytidine next to guanine (CpG) by a methyl group which alters the conformation of the major groove of the DNA, which in turn, affects the interaction of the DNA with the transcriptional machinery [3,4]. Hence, DNA methylation changes are tightly related to gene regulation [5]. As methylated cytosines undergo spontaneous deamination resulting in C to T (thymine) mutations, the abundance of CpG dinucleotides is reduced over evolutionary time from the expectation based on the frequency of Cs and Gs in the genome [6,7]. Various studies demonstrated that DNA methylation is sexually dimorphic in the developing or post-hatching gonads of vertebrates with temperature-dependent sex determination (TSD), including turtles [6,8–11] and alligator [12], and in fish with a mixed system of genotypic-sex determination susceptible to thermal effects (GSD + TE) [13–15]. These observations raise the possibility that DNA methylation, if it were sexually dimorphic in somatic tissues, could be used as a non-lethal sex diagnostic.

Sex diagnosis has important implications for basic and applied biology, as it is necessary to study a myriad of sexually dimorphic traits [16], as well as to monitor sex ratios to study population dynamics or to evaluate conservation efforts [17–21]. Sexing individuals is also important for research on sex determination in turtles to understand the effects of environmental factors (or lack thereof) on sexual development and its evolutionary consequences (e.g., [22–26]). As hatchling turtles usually display little sexual dimorphism that could be easily discerned by external observation, the development of sexing techniques is necessary.

Unfortunately, some earlier sexing techniques used in turtles are either lethal, as they rely on the gonadal inspection and/or gonadal tissue collection [27,28], while others require special training or equipment, such as laparoscopy/endoscopy of live animals, radioimmunoassay (RIA) of circulating hormone levels, or immunohistochemistry [29–33]. Non-invasive geometric morphometric techniques were also developed for a variety of species [34–37], but fast and simple field techniques remain elusive for young turtles. In recent years, alternative non-lethal sexing methods were reported, such as penile stimulation with vibrators and penis eversion by hind limb and neck stimulation [38,39], that are applicable in the field without harming the animal. Regrettably, results obtained with these last two methods are affected by the stress level of the animal post-capture and thus, are not reliable [38,39].

Molecular sexing methods were developed to identify the sex of turtles by detecting sex-specific genetic sequences or gene dosage in species with sex chromosome systems of GSD [reviewed in [40]]. For instance, gene dosage was detected with sexing primers for *Apalone spinifera*, *Glyptemys insculpta* and *Glyptemys muhlenbergii* [41], whereas quantification of the sex-specific abundance of rRNA repeats using qPCR was used to sex *A. spinifera* and other trionychid species such as *Pelodiscus sinensis* and *Chitra indica* [40,42]. In contrast, molecular sexing of turtles with temperature-dependent sex determination (TSD), who lack sex chromosomes or any consistent genotypic differences between the sexes [43], has been accomplished by measuring circulating testosterone levels after a hormonal challenge [17], and more recently, by sex-specific circulating proteins in neonate blood [44]. However, no study has explored the use of epigenetic markers for non-lethal sex diagnosis in any turtle. Any sexually dimorphic DNA methylation present in easily sampled somatic tissues (such as tail clips) could be used as a non-lethal sex diagnostic.

DNA methylation can be measured globally (genome-wide) by DNA methylation-sensitive ELISA (Enzyme-Linked Immunoassay) [45], or by high-throughput sequencing of immunoprecipitated methylated DNA (MeDIP-seq) [6]. DNA methylation can also be assessed locally (gene-by-gene or region by-region) by MeDIP-seq, by sequencing bisulfite-converted DNA (BS-seq) (which reveals the methylation status of individual nucleotides) [45], or by PCR after DNA digestion with a methylation-sensitive restriction enzyme [46]. The latter is the simplest method and was applied to sex chickens [46], a GSD species whose CpG-rich region on the Z chromosome, called MHM region (Male Hyper-Methylated region), constitutes an ideal male-specific molecular marker. This technique was also successful to identify differential methylation in the gene *Fezf2* in TSD turtle gonads [6], yet it is unknown if somatic tissues display the same pattern.

Here, we investigated the global and local DNA methylation in a somatic tissue (the tail) of *Chrysemys picta* hatchlings using a multi-pronged approach, to test the hypothesis that differential DNA methylation exists in somatic tissue and can be used as a sexing technique for TSD turtles. We chose tails because their shape is sexually dimorphic in turtles of the family Emydidae to which *C. picta* belongs [47], and because tail clips can be easily collected in the field without sacrificing the individual. Our analysis included methylation-sensitive ELISA (global), plus a novel hybrid method we developed to provide global and local methylation information by combining MeDIP-seq with BS-seq (MeDIP-BS-seq). Our novel MeDIP-BS-seq offers an alternative to quantify methylation in genes or regions while also providing base-by-base methylation information simultaneously. Additionally, we assembled the methylomes of tails of male and female *C. picta* hatchlings to identify candidate molecular sex markers in TSD turtles for PCR detection after methylation-sensitive DNA digestion.

2. Materials and Methods

2.1. Tissue Collection and DNA Extraction

Freshly laid eggs were collected from an Iowa turtle farm and transported in moist vermiculite to the laboratory for incubation following standard protocols [48]. Specifically, eggs were cleaned from excess mud, marked with a unique ID, randomly assigned to boxes with moist sand (30 eggs per box), and placed in incubators at 26 °C (Male Producing Temperature—MPT), 28 °C (Pivotal Temperature—PivT, which produces males and females in equal numbers) and 31 °C (Female Producing Temperature—FPT). Boxes were rotated daily in a clockwise fashion to control for potential temperature gradients within the incubators. Moisture inside the egg boxes was maintained constant by replacing evaporated water weekly. We obtained 20 hatchlings from the 26 °C treatment, 20 hatchlings from the 28 °C treatment and 23 hatchlings from the 31 °C treatment. Hatchlings were assigned a unique ID according to their incubation treatment and order of hatching and were notched at their carapace scutes for identification [49]. Hatchlings were raised for 3 months in water tanks at 26 °C, fed ad libitum and cleaned daily. All animals were euthanized by a lethal injection of propofol and sex was determined by visual gonadal inspection. We collected tail clips from each hatchling and preserved the tissue in RNA later at −20 °C until further processing. All procedures were approved by Iowa State University IACUC.

DNA was extracted separately from the tail of each individual collected as described above, using Gentra Puregene DNA extraction kits (Gentra), and following the manufacturer's instructions. DNA quality and quantity were assessed by Nanodrop Spectrophotometer and 1% agarose gel electrophoresis. DNA was diluted to 200 ng/uL and stored at −20 °C until processing.

2.2. Global Methylation Analysis via ELISA

We measured global methylation levels in 20 randomly selected individuals per temperature (26 °C, 28 °C and 31 °C) via methylation-specific ELISA using the MethylFlashTM Global DNA Methylation (5-mC) ELISA Easy Kit (Epigentek), following the manufacturer's instructions. DNA samples were diluted to 10 ng/uL for ELISA. Reactions were run in a Chromate 4300 machine at the Iowa State University proteomics facility. All plates included a positive control standard curve from the kit plus a turtle-specific standard curve of eight standards obtained by serially diluting (1:1) a sample of pooled DNA from all individuals.

The normality of the absorbance values was tested using QQ plots in RRPP [50], and results indicated that no data transformation was necessary (Supplementary Figure S1). We converted the absorbance values to methylation percentage, following the equation [51]:

$$5\text{mC}\% = \frac{\text{SampleOD} - \text{NCOD}}{\text{Slope} \times \text{S}} \times 100\% \quad (1)$$

where 5 mC% = percentage of 5-methylcytosines, OD = optimal density, NC = negative control, Slope = standard curve slope, S = input DNA in ng. For the statistical analysis of these values, we evaluated first if the standard curves of positive control (provided with the kit) and our turtle standard curves were linear using a generalized linear model (GLM). Then, we performed an ANCOVA to compare slopes between these two types of standard curves. Next, we tested for differences in global methylation using ANOVA. Tests were applied first to the calculated methylated percentages that represent the total 5-mC fraction in the sample accounting for the kit's specificity in detecting DNA methylation, given that it is calculated proportionally to the OD intensity measured [51]. Second, ANOVA was applied to the absorbance values, following a traditional ELISA data analysis [52]. As the interaction between temperature and sex was not significant for the analysis of 5 mC% in the full factorial ANOVA ($p > 0.05$), we performed a reduced ANOVA that excluded the interaction term. Additionally, because the temperature and sex terms were not significant in the reduced model, we then tested for differences combining samples by sex (26 °C male + 28 °C male and 31 °C female + 28 °C female), and by temperature (31 °C, 26 °C and 28 °C). On the other hand, because the sex and temperature interaction was significant for the absorbance values ($p < 0.05$) we did a pairwise comparison of all temperature by sex combinations.

2.3. MeDIP-BS-Seq Library Construction and Sequencing

Twenty random samples of DNA per incubation treatment (26 °C and 31 °C) were divided into two groups to obtain two biological replicates of pooled DNA per temperature (10 samples per pool). DNA was processed by EpiGentek using a hybrid approach we developed to detect methylated regions. Specifically, methylated DNA was immunoprecipitated first (5 mC MeDIP) and then subjected to bisulfite conversion, after which Illumina NextSeq 500 libraries were prepared and sequenced at Duke University sequencing facility (75 bp PE sequencing). This hybrid sequencing approach was designed to quantify methylation via MeDIP, which targets mostly CpG-rich regions and to assess the base-by-base cytosine methylation status of the immunoprecipitated DNA from the bisulfite-conversion.

2.4. Methylome Assembly and Analysis

The quality of the library reads was assessed by FASTQC [53] followed by an adaptor trimming step using trimgalore [54]. Trimmed reads were quality controlled in an additional step via FASTQC to check for the adaptor removal. We used the *Chrysemys picta* 3.0.3 genome assembly [55] as a reference to map the MeDIP-seq+ BS-seq reads using Bismark [56]. We applied the genome preparation step, by which the software converts the reference genome into a 3-base genome (cytosines are converted to thymines and adenines are converted to guanines), followed by single- and non-directional read mapping, with a score of −120. Alignments were sorted using Samtools [57], and then imported into RStudio [58] using the process BismarkAln from methylkit [59] for further analysis. The conversion rate was calculated by methylkit as the number of thymines divided by coverage for each non-CpG cytosine. Non-methylated cytosines are converted to uracils during the bisulfite conversion, which in turn are converted to thymines during PCR amplification. Coverage is calculated by the number of reads per base, with a minimum of 10 reads per base to ensure the high quality of the data and methylation percentage [59].

We tiled the genome in windows of 1000 bp for differential methylation analysis, as recommended by methylkit. This tiling process allows methylkit to summarize methylation information using these windows rather than individual bases. Following the window-tiling process, we calculated differential methylation using the methylkit function "calculateDiffMeth", with the q-value set at 0.01. We used Bedtools [60] to obtain region coordinates of exons, introns, intergenic regions and promoter regions (i.e., 500 bp, 1000 bp and 3500 bp upstream of exon 1) from the *C. picta* genome ver 3.0.3 [55] from NCBI. Regions with $p < 0.05$ were annotated using genomation [61] to identify methylation present in promoters, introns, exons and intergenic regions. Differential methylation analysis was also performed at individual nucleotides, using the same functions of methylkit and q-values as for the analysis by windows.

In an alternative approach, we run a coverage-based analysis on edgeR [62], to identify differentially methylated regions using 500 bp windows, following [6]. For this, we created a count table (Data S1) from our alignments, using "bedtools coverage" and imported it into RStudio for edgeR analysis. We used the quasi-likelihood F-test (QLF) to calculate differential methylation, and methylation levels were measured as the natural log of the Counts Per Million (logCPM) [63]. Scripts used in this study can be found in Material S1.

2.5. DNA Digestion and Methylome Validation by PCR

Twenty regions that showed significant differential methylation ($p < 0.05$) between males and females in the previous analyses were selected as candidate regions for a role in sex diagnostics and were inspected visually in Geneious [64]. Regions were selected for validation according to the difference in methylation between treatments (26 °C and 31 °C), and to the presence of the restriction site (CCGG) recognized by the methylation-sensitive restriction enzyme HpaII. To validate regions, we designed primers (Table S1) according to the location of the highest coverage peak within the region (i.e., the peak location enriched with aligned reads which indicates a reliable methylation call) and the location of the restriction site. To validate methylation at specific bases, we designed primers according to the differential methylated site location, given that it is the only area where there is a difference in methylation that would be detected by the restriction enzyme (Figure 1).

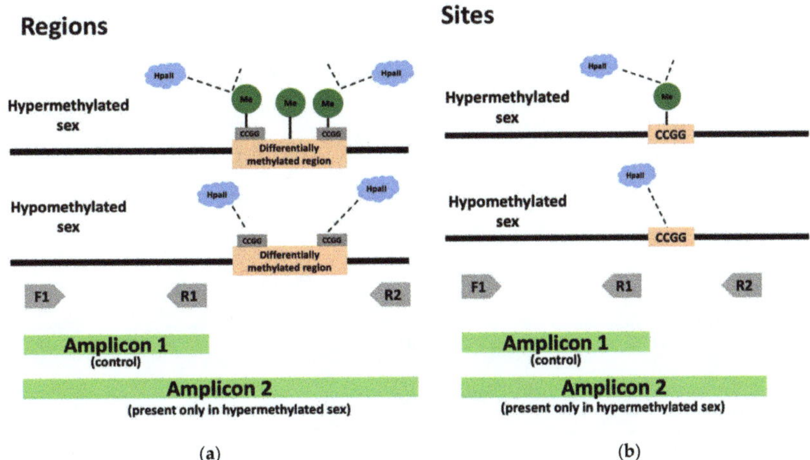

Figure 1. Methylation-sensitive PCR to validate the candidate regions and sites for tail sexing. In regions (**a**), the hypermethylated sex has one or more restriction cleavage sites for the methylation-sensitive restriction enzyme HpaII (CCGG) surrounded by methyl groups. In candidate sites (**b**), a single differentially methylated cytosine is located within the restriction cleavage site. In both these cases (regions and sites), methylation within the restriction sequence prevents DNA digestion by HpaII, while hypomethylation at this site permits DNA digestion. Primers F1 and R1 produce an amplicon irrespective of DNA methylation at the candidate region or site, and thus serve as a control, whereas amplicon 2 is only produced by primers F1 and R2 in the presence of methylation at the restriction site (when DNA digestion is prevented).

DNA (100 ng per reaction) was digested with HpaII (Thermo Fisher, Waltham, MA USA), following the manufacturer's instructions, and digestion was verified visually by comparing digested and undigested DNA in 1% agarose gels stained with ethidium bromide (EtBr) against a 1 kb plus ladder (Invitrogen). Undigested DNA should concentrate above the 25 kb standard whereas digested DNA (unmethylated) produces a smear in the gel between 1.5 Kb and 12 kb. PCR amplification used 10 ng of digested and undigested DNA (control) in 15 µL reactions containing 1× Tag buffer, 1.5 mM $MgCl_2$, 0.2 mM dNTPs, 0.4 U Taq polymerase, 10.5 µL water, and a 0.4 µM primer cocktail containing the three primers in equimolar concentrations. PCR conditions included an initial denaturing step at 94 °C for 3 min, followed by 35 cycles of denaturing at 94 °C for 30 s, annealing at 58 °C for 30 s, and extension at 72 °C for 90 s. Amplicons were visualized in EtBr-stained 1% agarose gel, and their size estimated against a 1 kb plus ladder (Invitrogen, Waltham, MA USA).

3. Results

3.1. Global Methylation by ELISA

In order to investigate whether overall differential methylation in *C. picta* tails is present such that it would be a good indicator of the individual's sex, we performed an ELISA assay to detect global methylation differences between temperature treatments (26 °C, 28 °C, and 31 °C). Standard curves were linear and displayed an R^2 = 0.95 and R^2 = 0.99 for the kit positive control (PC) and the turtle standard curve (TC), respectively. An ANCOVA revealed no significant difference between the slopes of the PC and TC standard curves (p = 0.6). Our ANOVA analysis between temperature and sex groups uncovered no significant difference in global methylation percentage or absorbance between males and females that could be used as a sexing marker (p > 0.05) (Figure 2, Table 1), although a permutation procedure [50] detected significantly higher within-group variance in percent methylation of 31 °C females than 26 °C males (but not among other groups).

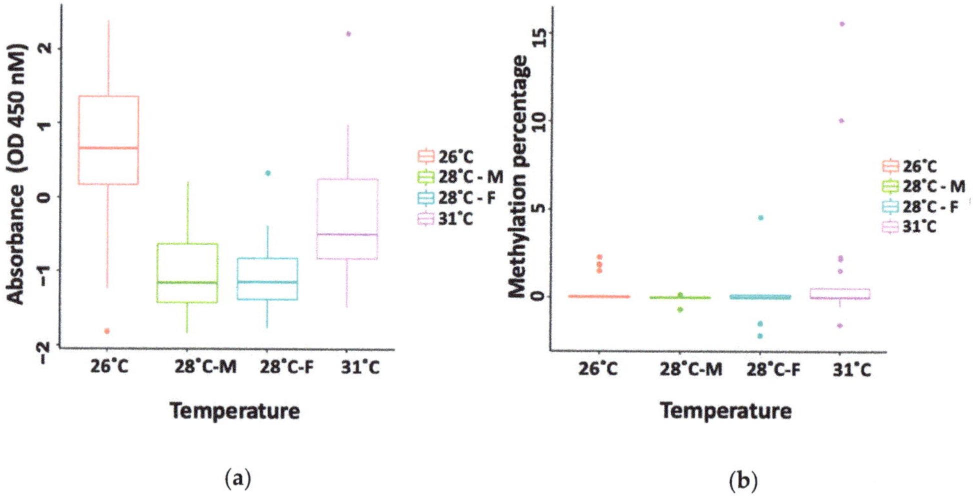

Figure 2. (**a**) Absorbance and (**b**) percent 5-mC DNA methylation values by temperature and sex measured by ELISA in the tail of *Chrysemys picta* hatchling tail tissue.

3.2. MeDIP-BS-Seq Methylome Assembly and Analysis

A reference methylome was assembled using pooled reads from the 26 °C and 31 °C temperature treatments, which in *Chrysemys picta* (TSD) produce exclusively males and females, respectively. All reverse reads (R2) from the paired-end RNA-sequencing were comprised of guanines (Gs), an artifact later found to be commonly caused by the two-color Illumina Nextseq chemistry which over-calls no-signal N bases as high confidence

Gs. Therefore, reverse reads were discarded from further analysis, and the methylome assembly and analysis were based on single (forward) reads only, which are typically used for bisulfite sequencing [65–67]. We obtained a mapping efficiency of reads to the CPI 3.0.3 genome between 60% and 70% and a bisulfite conversion rate for all samples between 83–92% (83.15% and 90.41% for females, and 91.56%, and 91.81% for males). Similar to the ELISA results, Bismark detected no differences between the sexes in global methylation levels for cytosines in CpG context (78.60% and 84.00% for female, and 81.00% and 81.30% in males). In contrast, cytosines in non-CpG context (CHH) exhibited lower methylation in females than in males in these CpG-rich regions pulled down during MeDIP (4.80% and 4.30% for females, but 6.90% and 8.60% for males) despite female samples experiencing lower conversion efficiency which would have caused the overestimation of their methylation level compared to males. However, having duplicates does not provide enough power to detect the significance of this difference using a *t*-test ($p > 0.05$).

Table 1. ANOVA results for absorbance and percent 5-mC DNA methylation values measured by ELISA in the tail of *Chrysemys picta* hatchling tail tissue.

	ANOVA—Methylation Percentage (Full Factorial Model)			
	Sum of Squares	**Mean Square**	**F**	***p*-Value**
Temperature	14.4	14.376	2.62	0.138
Sex	0.0	0.015	0.002	0.961
Temperature:Sex	8.4	8.39	1.320	0.255
Residuals	356	6.357		
	ANOVA—methylation percentage (reduced model)			
	Sum of Squares	**Mean Square**	**F**	***p*-value**
Temperature	14.4	14.376	2.249	0.139
Sex	0.0	0.015	0.002	0.961
Residuals	364.4	6.393		
	ANOVA—methylation percentage by sex			
	Sum of Squares	**Mean Square**	**F**	***p*-value**
Sex	9.5	9.511	1.494	0.227
Residuals	369.3	6.367		
	ANOVA—methylation percentage by temperature			
	Sum of Squares	**Mean Square**	**F**	***p*-value**
Temperature	14.4	14.376	2.288	0.136
Residuals	364.4	6.283		
	ANOVA—absorbance (full factorial model)			
	Sum of Squares	**Mean Square**	**F**	***p*-value**
Temperature	4.26	4.262	5.509	0.5942
Sex	0.22	0.222	0.287	**2.33e−0.6**
Temperature:Sex	21.41	21.414	27.679	
Residuals	43.33	0.774		
Pairwise comparisons of absorbance. Mean absorbance per temperature by sex combination (diagonal, underlined), mean squares (below diagonal, italics), *p*-values (above diagonal, significant values are denoted in bold)				
	26 °C—Male	28 °C—Male	31 °C—Female	28 °C—Female
26 °C—Male	<u>0.5999</u>	**0.0008**	0.126	**0.005**
28 °C—Male	*32.56*	<u>−1.0218</u>	**0.035**	0.771
31 °C—Female	*7.507*	*10.722*	<u>−0.2172</u>	**0.067**
28 °C—Female	*24.402*	*0.084*	*8.803*	<u>−0.9854</u>

At the 1000 bp window level (using methylkit), we detected 164 differentially methylated regions while at the 500 bp window level (using edgeR) we identified 761,800 differentially methylated regions. No region exhibited a presence/absence pattern, that is, showing 100% methylation at one temperature (26 °C or 31 °C) and 0% methylated in the other temperature, i.e., the ideal scenario for primer design and DNA digestion by methylation-sensitive restriction enzymes, perhaps due to the lower conversion efficiency. Therefore, we selected the top three regions (Table 2) that displayed the greatest differences in fold-change of methylation levels between the 26 °C and 31 °C treatments for downstream methylation-sensitive PCR. At the site-by-site level, we detected 34 individual sites that exhibited the 100–0% methylation pattern between temperatures. Of those 34 sites, only two were located at a restriction site that would be amenable for methylation-sensitive PCR, and both were selected for further tests (Table 2).

Table 2. Differentially methylated 500 bp regions and differentially methylated cytosines in the tail of *Chrysemys picta* selected for methylation-sensitive PCR. Intergenic regions are located outside any specific gene.

Regions			
CPI 3.0.3. Scaffold	**Start Position**	**End Position**	**Gene**
NW_007359905.1	3,604,500	3,605,000	Intergenic
NC_024218.1	55,635,611	55,636,011	Intergenic
NW_007281443.1	1,473,502	1,474,001	FAM170B
Sites			
CPI 3.0.3. Scaffold	**Position**		**Gene**
NC_024218.1	18,209,143		CASQ2
NC_024220.1	23,784,593		FOSL2

3.3. Methylation-Sensitive PCR

We used DNA from 10 individuals incubated at 26 °C and 10 at 31 °C, sexed by gonadal inspection, to test for differential methylation by the PCR assay in three candidate regions and two candidate sites (Table 1). Amplification patterns did not differ between the sexes. Indeed, amplicons of the size expected if methylation was absent were observed between undigested and digested samples in both sexes. Visual inspection of the DNA template before PCR in an agarose gel showed a smear for the digested DNA sample but not for the undigested DNA. Thus, the PCR results revealed that the restriction enzyme did not digest the DNA at the candidate regions at a significant level in either sex. Therefore, the regions selected because they exhibited the greatest differential methylation in the methylome analysis exhibited negligible methylation in both sexes when tested by PCR, and thus, cannot be used as a sex diagnostic.

4. Discussion

4.1. A Novel Complexity-Reduction and Site-by-Site Approach for DNA Methylation Analysis

DNA methylation is an important biological process, and several methods have been designed to discern DNA methylation patterns in ecological, evolutionary and medical research [68–70]. In this study, we tested a new approach to study DNA methylation, by combining MeDIP and BS-Seq. We applied this new hybrid technique to study the DNA methylation of somatic tissue (tail clips) of *C. picta* hatchlings with the goal of identifying molecular markers that could serve as a sex diagnostic tool. To our knowledge, this is the first time a MeDIP + BS-Seq hybrid approach has been used in a DNA methylation study. This complexity reduction procedure allowed us to enrich the genomic DNA samples to those regions with higher DNA methylation via MeDIP, and to obtain information on the individual base methylation status from the BS-Seq. Other methods used to study CpG density at one or a few regions of interest, generally combine restriction enzymes and fluorescence along with bisulfite conversion and sequencing. These methods include either tagging methylated CpG dinucleotides, or labeling S-adenosylmethionine—(SAM—a methyl donor) to incorporate methyl groups to bisulfite converted PCR amplicons, in order to identify all the CpGs that are methylated in a fragment [69,71,72]. These methods focus on exploring regional CpG density in targeted genomic areas, while our approach allows an unbiased genome-wide DNA methylation profiling [73]. Furthermore, these other methods have limited data acquisition compared to our hybrid method, as they are highly dependent on the quantity and location of the restriction sites, on the quality of the digested DNA, and because they are also restricted to genomic regions of known sequence [69,73]. In contrast, our hybrid method is able to capture genome-wide regions that are enriched in methylation through the MeDIP-Seq step and to provide additional site-specific information thanks to the bisulfite-sequencing step. We achieved satisfactory mapping results (60–70%) using Bismark compared to the 56% and 77% mapping efficiency attained for human blood samples [65,74], and 24.6% using Bismark on mice [75]. This was true despite the fact that our mapping efficiency was reduced compared to other studies of DNA methylation in turtles that used exclusively MeDIP [6], partly because we used single-end reads, and partly because no software exists specifically designed to handle the combined MeDIP + bisulfite data. In addition, we note that the calculated bisulfite conversion efficiency obtained in this study (83–92%) is lower compared to what others have obtained (99%) [76]. This is expected to increase false positives, as some unmethylated cytosines were not converted to uracils and would have been counted as methylated cytosines during the analysis (i.e., misinterpreted as being protected from deamination by the presence of a methylation mark). Thus, subtle but biologically significant differential methylation between the sexes may have been obscured (particularly higher male than female methylation, since conversion efficiency was lower in female than male tail samples).

Further optimization of the MeDIP-BS-seq protocol is warranted as well as improvements in bioinformatics pipelines to handle hybrid data of this kind to improve results, and we hope that our work will foster new developments in this area. Importantly, because our novel hybrid method was successful in providing a genome-wide assessment of the DNA methylation status in hatchling tails and thus, it should be applicable at a broad taxonomic scale to other DNA methylation analyses of somatic or gonadal tissue, and particularly useful to reduce the complexity of samples for the study of large genomes.

4.2. Inter-Individual Variation in DNA Methylation Exists in Both Males and Females

Despite identifying multiple differentially methylated regions and sites in the tails among individuals, and overall higher CHH methylation in females than males in the regions pulled down by MeDIP (which are CpG-enriched), no reliable sexing marker was detected by methylation-sensitive PCR using any of the top candidate regions. Indeed, although differential methylation was identified with the MeDIP-BS-Seq data between pooled samples of males and females, the differences turned out to not be dichotomous enough for the restriction enzyme to yield sexually dimorphic methylation-sensitive PCR markers. Further, the difference in overall CHH methylation levels detected here does not provide a cheap diagnostic tool for sex TSD individuals at the scale needed for population-level analysis. Thus, the current method for sexing TSD turtles with relative reliability includes the recent immunoassay of circulating AMH (Anti-Mullerian Hormone) protein in the blood, which was tested in two species (*Trachemys scripta* and *Caretta caretta*) and was 100% accurate in neonates, although it is less accurate at an older age (accuracy dropped to 90% in 2.7–6 mo old juveniles) [44] and is expected to drop in accuracy once the Mullerian ducts are fully resorbed in males. While sampling blood is minimally invasive (albeit not always easy), this immunoassay is simpler than the radioimmunoassay of circulating testosterone after the FSH challenge, which has been applied to sea and freshwater turtle hatchlings [29,30,77]. Importantly, the only non-invasive sexing method to date is able to discern very subtle external sexual dimorphism using landmark-based geometric morphometrics, which was 90%-98% accurate to diagnose the sex of *Podocnemis expansa* and *C. picta* hatchlings [24,34], and was applied to sex hatchlings of *Podocnemis lewyana* and *Chelydra serpentina* [22,78].

4.3. Contrasting Patterns of DNA Methylation between Somatic and Gonadal Tissues

Our results concur with earlier reports of extensive methylation in the genome of *C. picta* hatchlings [6], yet, the response of DNA methylation to incubation temperature differ drastically between gonadal and somatic tissue (i.e., tails). Indeed, previous studies uncovered sexually dimorphic DNA methylation in the gonadal tissue of TSD turtles (*C. picta* and *Lepidochelys olivacea*) [6,10], whereas we observed mostly monomorphic DNA methylation in tails. This monomorphism was evident both at the global (genome-wide, determined by ELISA) and local (at small regions and at individual sites, determined by MeDIP-BS-seq) levels. Gonads are sexually dimorphic tissues by definition, and in *C. picta*, they display differential DNA methylation in hatchlings and differential gene expression patterns since early development [6,79–84]. Turtle tails exhibit morphological differences between the sexes that are relevant for mating, such as contrasting size, texture, or the relative position of the cloaca [85]. For instance, tails are sexually dimorphic in *Trachemys scripta* turtles, an emydid close relative of *C. picta* [47,86], yet we did not observe sexually dimorphic DNA methylation in the hatchling tails, perhaps because the tail dimorphism has not yet developed in painted turtle hatchlings [47] or because its development is not controlled epigenetically via DNA methylation. Further, the painted turtle does not rely on male combat or forced insemination, two mating strategies linked to male-specific body size and shape [85,86]. On the contrary, *C. picta* mating relies on female choice where pre-coital male behavior and display structures such as foreclaws and coloration are relevant [85]. Nonetheless, the discrepancy between our results and those previously reported for gonadal tissue underscores the importance of DNA methylation for the sex-specific maintenance and/or function of the gonads but not for some somatic tissues, such as the tail. Future studies should explore the sexual dimorphism of DNA methylation in other somatic tissues such as blood in turtles, which can be collected non-lethally. Blood exhibits sexually dimorphic DNA methylation in humans linked to other traits [87] and serves as a biomarker for multiple medical purposes, such as detecting aging [88] or cancer types [68,89,90].

5. Conclusions

In conclusion, here we build the somatic methylome of a TSD turtle with the goal to identify molecular sex markers. We found substantial differences in tail methylation among individuals, but no consistent sex-specific pattern that could be used to diagnose the sex of hatchlings accurately using this somatic tissue, in contrast with the sexually dimorphic gonadal methylation previously reported [6]. Our results underscore the importance of DNA methylation in primary sexual development and gonadal maintenance post-hatching and highlight that sexually dimorphic methylation is not ubiquitous in the soma. Our study led to the development of a new hybrid method that combines MeDIP-Seq and bisulfite-sequencing which provides greater insight to profile the genome-wide methylation status of large genomes with relative ease, and thus should be widely applicable, but whose further optimization is warranted.

Supplementary Materials: The following supporting information can be downloaded at: https://www.mdpi.com/article/10.3390/ani13010117/s1, Data S1: Count tables produced as input to edgeR analysis; Material S1: Scripts used in differential methylation analysis; Table S1: Primer sequences used for PCR assay.

Author Contributions: Conceptualization, B.A.M. and N.V.; methodology, B.A.M. and N.V.; software, B.A.M.; validation, B.A.M.; formal analysis, B.A.M.; investigation, B.A.M. and N.V.; resources, N.V.; data curation, B.A.M.; writing—original draft preparation, B.A.M.; writing—review and editing, B.A.M. and N.V.; visualization, B.A.M.; supervision, N.V.; project administration, N.V.; funding acquisition, N.V. All authors have read and agreed to the published version of the manuscript.

Funding: This work was supported in part by the National Science Foundation of USA grant IOS-1555999 to N.V. B.A.M. was supported in part by a fellowship from Science without Borders (CAPES—Brazil).

Institutional Review Board Statement: The animal study protocol was approved by the IOWA STATE UNIVERSITY IACUC (protocol number 6-07-6383-J approved 4/7/16).

Informed Consent Statement: Not applicable.

Data Availability Statement: All files are available in Genbank BioProject accession number PRJNA681606.

Conflicts of Interest: The authors declare no conflict of interest.

References

1. Vandegehuchte, M.B.; Janssen, C.R. Epigenetics and its implications for ecotoxicology. *Ecotoxicology* **2011**, *20*, 607–624. [CrossRef] [PubMed]
2. Mizoguchi, B.A.; Valenzuela, N. Ecotoxicological Perspectives of Sex Determination. *Sex. Dev.* **2016**, *10*, 45–57. [CrossRef] [PubMed]
3. Senner, C.E. The role of DNA methylation in mammalian development. *Reprod. Biomed. Online* **2011**, *22*, 529–535. [CrossRef] [PubMed]
4. Piferrer, F. Epigenetics of sex determination and gonadogenesis. *Dev. Dyn.* **2013**, *242*, 360–370. [CrossRef]
5. Cedar, H. DNA methylation and gene activity. *Cell* **1988**, *53*, 3–4. [CrossRef] [PubMed]
6. Radhakrishnan, S.; Literman, R.; Mizoguchi, B.; Valenzuela, N. MeDIP-seq and nCpG analyses illuminate sexually dimorphic methylation of gonadal development genes with high historic methylation in turtle hatchlings with temperature-dependent sex determination. *Epigenetics Chromatin* **2017**, *10*, 28. [CrossRef] [PubMed]
7. Elango, N.; Hunt, B.G.; Goodisman, M.A.D.; Yi, S.V. DNA methylation is widespread and associated with differential gene expression in castes of the honeybee, *Apis mellifera*. *Proc. Natl. Acad. Sci. USA* **2009**, *106*, 11206–11211. [CrossRef] [PubMed]
8. Matsumoto, Y.; Buemio, A.; Chu, R.; Vafaee, M.; Crews, D. Epigenetic Control of Gonadal Aromatase (cyp19a1) in Temperature-Dependent Sex Determination of Red-Eared Slider Turtles. *PLoS ONE* **2013**, *8*, e63599. [CrossRef]
9. Matsumoto, Y.; Hannigan, B.; Crews, D. Temperature Shift Alters DNA Methylation and Histone Modification Patterns in Gonadal Aromatase (cyp19a1) Gene in Species with Temperature-Dependent Sex Determination. *PLoS ONE* **2016**, *11*, e0167362. [CrossRef]
10. Venegas, D.; Marmolejo-Valencia, A.; Valdes-Quezada, C.; Govenzensky, T.; Recillas-Targa, F.; Merchant-Larios, H. Dimorphic DNA methylation during temperature-dependent sex determination in the sea turtle Lepidochelys olivacea. *Gen. Comp. Endocrinol.* **2016**, *236*, 35–41. [CrossRef]
11. Ge, C.; Ye, J.; Zhang, H.; Zhang, Y.; Sun, W.; Sang, Y.; Capel, B.; Qian, G. *Dmrt1* induces the male pathway in a turtle with temperature-dependent sex determination. *Development* **2017**, *144*, 2222–2233. [CrossRef] [PubMed]
12. Parrott, B.B.; Kohno, S.; Cloy-McCoy, J.A.; Guillette, L.J. Differential Incubation Temperatures Result in Dimorphic DNA Methylation Patterning of the SOX9 and Aromatase Promoters in Gonads of Alligator (Alligator mississippiensis) Embryos1. *Biol. Reprod.* **2014**, *90*, 2. [CrossRef] [PubMed]
13. Sun, L.-X.; Wang, Y.-Y.; Zhao, Y.; Wang, H.; Li, N.; Ji, X.S. Global DNA Methylation Changes in Nile Tilapia Gonads during High Temperature-Induced Masculinization. *PLoS ONE* **2016**, *11*, e0158483. [CrossRef]
14. Navarro-Martín, L.; Viñas, J.; Ribas, L.; Díaz, N.; Gutiérrez, A.; Di Croce, L.; Piferrer, F. DNA Methylation of the Gonadal Aromatase (cyp19a) Promoter Is Involved in Temperature-Dependent Sex Ratio Shifts in the European Sea Bass. *PLOS Genet.* **2011**, *7*, e1002447. [CrossRef]
15. Valenzuela, N.; Adams, D.C.; Janzen, F.J. Pattern Does Not Equal Process: Exactly When Is Sex Environmentally Determined? *Am. Nat.* **2003**, *161*, 676–683. [CrossRef] [PubMed]
16. Andersson, M. *Sexual Selection*, 1st ed.; Princeton University Press: Princeton, NJ, USA, 1994.
17. Valenzuela, N.; Lance, V. *Temperature-Dependent Sex Determination in Vertebrates*, 1st ed.; Smithsonian Books: Washington, DC, USA, 2004.
18. Mrosovsky, N. Sex ratio bias in hatchling sea turtles from artificially incubated eggs. *Biol. Conserv.* **1982**, *23*, 309–314. [CrossRef]
19. Valenzuela, N.; Literman, R.; Neuwald, J.L.; Mizoguchi, B.; Iverson, J.B.; Riley, J.L.; Litzgus, J.D. Extreme thermal fluctuations from climate change unexpectedly accelerate demographic collapse of vertebrates with temperature-dependent sex determination. *Sci. Rep.* **2019**, *9*, 1–11. [CrossRef] [PubMed]
20. Holleley, C.E.; O'Meally, D.; Sarre, S.D.; Graves, J.A.M.; Ezaz, T.; Matsubara, K.; Azad, B.; Zhang, X.; Georges, A. Sex reversal triggers the rapid transition from genetic to temperature-dependent sex. *Nature* **2015**, *523*, 79–82. [CrossRef]
21. Jensen, M.P.; Allen, C.D.; Eguchi, T.; Bell, I.P.; LaCasella, E.L.; Hilton, W.A.; Hof, C.A.; Dutton, P.H. Environmental Warming and Feminization of One of the Largest Sea Turtle Populations in the World. *Curr. Biol.* **2018**, *28*, 154–159.e4. [CrossRef]
22. Ceballos, C.P.; Valenzuela, N. The Role of Sex-specific Plasticity in Shaping Sexual Dimorphism in a Long-lived Vertebrate, the Snapping Turtle Chelydra serpentina. *Evol. Biol.* **2011**, *38*, 163–181. [CrossRef]
23. Ceballos, C.P.; Adams, D.C.; Iverson, J.B.; Valenzuela, N. Phylogenetic Patterns of Sexual Size Dimorphism in Turtles and Their Implications for Rensch's Rule. *Evol. Biol.* **2012**, *40*, 194–208. [CrossRef]

24. Ceballos, C.P.; Hernández, O.E.; Valenzuela, N. Divergent Sex-Specific Plasticity in Long-Lived Vertebrates with Contrasting Sexual Dimorphism. *Evol. Biol.* **2013**, *41*, 81–98. [CrossRef]
25. Valenzuela, N. Sexual Development and the Evolution of Sex Determination. *Sex. Dev.* **2008**, *2*, 64–72. [CrossRef] [PubMed]
26. Valenzuela, N. Causes and consequences of evolutionary transitions in the level of phenotypic plasticity of reptilian sex determination. In *Transitions Between Sexual Systems: Understanding the Evolution of Hermaphroditism and Dieocy, Hermaphroditism and Other Sexual Systems*; Leonard, J.L., Ed.; Springer Nature: Cham, Switzerland, 2018; pp. 345–363.
27. American Society of Ichthyologists and Herpetologists. *Nature* **1937**, *139*, 1014. [CrossRef]
28. Van Der Heiden, A.M.; Briseno-Duenas, R.; Rios-Olmeda, D. A Simplified Method for Determining Sex in Hatchling Sea Turtles. *Copeia* **1985**, *1985*, 779. [CrossRef]
29. Owens, D.W.; Hendrickson, J.R.; Lance, V.A.; Callard, I.P. A Technique for Determining Sex of Immature Chelonia mydas Using a Radioimmunoassay. *Herpetologica.* **1978**, *34*, 270–273. Available online: https://www.jstor.org/stable/3891551 (accessed on 1 December 2022).
30. Rostal, D.C.; Grumbles, J.S.; Lance, V.A.; Spotila, J.R. Non-Lethal Sexing Techniques for Hatchling and Immature Desert Tortoises (*Gopherus agassizii*). *Herpetol. Monogr.* **1994**, *8*, 83. [CrossRef]
31. Tezak, B.M.; Guthrie, K.; Wyneken, J. An Immunohistochemical Approach to Identify the Sex of Young Marine Turtles. *Anat. Rec.* **2017**, *300*, 1512–1518. [CrossRef] [PubMed]
32. Kuchling, G.; Kitimasak, W. Endoscopic sexing of juvenile softshell turtles, Amyda cartilaginea. *Nat. Hist. Chulalongkorn Univ.* **2009**, *9*, 91–93.
33. Miller, W.J.; Wagner, F.H. Sexing Mature Columbiformes by Cloacal Characters. *Ornithology* **1955**, *72*, 279–285. [CrossRef]
34. Valenzuela, N.; Adams, D.C.; Bowden, R.M.; Gauger, A.C. Geometric Morphometric Sex Estimation for Hatchling Turtles: A Powerful Alternative for Detecting Subtle Sexual Shape Dimorphism. *Copeia* **2004**, *2004*, 735–742. [CrossRef]
35. Sönmez, B.; Turan, C.; Özdilek, Ş.Y.; Turan, F. Sex determination of green sea turtle (*Chelonia mydas*) hatchlings on the bases of morphological characters. *J. Black Sea/Mediterr Environ.* **2016**, *22*, 93–102.
36. Boone, J.L.; Holt, E.A. Sexing Young, Free-ranging Desert Tortoises (*Gopherus agassizii*) Using External Morphology. *Chel. Conserv. Biol.* **2001**, *4*, 28–33.
37. Michel-Morfin, J.E.; Munoz, V.M.G.; Rodriguez, C.N. Morphometric Model for Sex Assessment in Hatchling Olive Ridley Sea Turtles. *Chelon. Conserv. Biol.* **2001**, *4*, 53–58.
38. McKnight, D.T.; Howell, H.J.; Hollender, E.C.; Ligon, D.B. Good vibrations: A novel method for sexing turtles. *Acta Herpetol.* **2017**, *13*, 13–19.
39. Rodrigues, J.; Soares, D.; Silva, J. Sexing freshwater turtles: Penile eversion in Phrynops tuberosus (Testudines: Chelidae). *Acta Herpetol.* **2014**, *9*, 259–263. [CrossRef]
40. Literman, R.; Badenhorst, D.; Valenzuela, N. qPCR-based molecular sexing by copy number variation in rRNA genes and its utility for sex identification in soft-shell turtles. *Methods Ecol. Evol.* **2014**, *5*, 872–880. [CrossRef]
41. Literman, R.; Radhakrishnan, S.; Tamplin, J.; Burke, R.; Dresser, C.; Valenzuela, N. Development of sexing primers in Glyptemys insculpta and Apalone spinifera turtles uncovers an XX/XY sex-determining system in the critically-endangered bog turtle Glyptemys muhlenbergii. *Conserv. Genet. Resour.* **2017**, *9*, 651–658. [CrossRef]
42. Rovatsos, M.; Praschag, P.; Fritz, U.; Kratochvšl, L. Stable Cretaceous sex chromosomes enable molecular sexing in softshell turtles (Testudines: Trionychidae). *Sci. Rep.* **2017**, *7*, srep42150. [CrossRef]
43. Valenzuela, N.; Badenhorst, D.; Montiel, E.E.; Literman, R. Molecular Cytogenetic Search for Cryptic Sex Chromosomes in Painted Turtles *Chrysemys picta*. *Cytogenet. Genome Res.* **2014**, *144*, 39–46. [CrossRef]
44. Tezak, B.; Sifuentes-Romero, I.; Milton, S.; Wyneken, J. Identifying Sex of Neonate Turtles with Temperature-dependent Sex Determination via Small Blood Samples. *Sci. Rep.* **2020**, *10*, 1–8. [CrossRef] [PubMed]
45. Kurdyukov, S.; Bullock, M. DNA Methylation Analysis: Choosing the Right Method. *Biology* **2016**, *5*, 3. [CrossRef] [PubMed]
46. Caetano, L.; Gennaro, F.; Coelho, K.; Araújo, F.; Vila, R.; Araújo, A.; Bernardo, A.D.M.; Marcondes, C.; Lopes, S.C.D.S.; Ramos, E. Differential expression of the MHM region and of sex-determining-related genes during gonadal development in chicken embryos. *Genet. Mol. Res.* **2014**, *13*, 838–849. [CrossRef] [PubMed]
47. Readel, A.M.; Dreslik, M.J.; Warner, J.K.; Banning, W.J.; Phillips, C.A. A Quantitative Method for Sex Identification in Emydid Turtles Using Secondary Sexual Characters. *Copeia* **2008**, *2008*, 643–647. [CrossRef]
48. Valenzuela, N. Egg Incubation and Collection of Painted Turtle Embryos. *Cold Spring Harb. Protoc.* **2009**, *2009*. [CrossRef] [PubMed]
49. Cagle, F.R. A System of Marking Turtles for Future Identification. *Copeia* **1939**, *1939*, 170. [CrossRef]
50. Collyer, M.L.; Adams, D.C. RRPP: An r package for fitting linear models to high-dimensional data using residual randomization. *Methods Ecol. Evol.* **2018**, *9*, 1772–1779. [CrossRef]
51. Epigentek. EpiGentek MethylFlash ™ Global DNA Methylation. 2019, pp. 1–13. Available online: https://www.epigentek.com/docs/P-1030.pdf (accessed on 1 December 2022).
52. Canellas, P.F.; Karu, A.E. Statistical package for analysis of competition ELISA results. *J. Immunol. Methods* **1981**, *47*, 375–385. [CrossRef]
53. Andrews, S. FastQC: A Quality Control Tool for High Throughput Sequence Data. Babraham Bioinformatics. 2010. Available online: https://www.bioinformatics.babraham.ac.uk/projects/fastqc/ (accessed on 1 December 2022).

54. Krueger, F.; Trim galore. A Wrapper Tool Around Cutadapt and FastQC to Consistently Apply Quality and Adapter Trimming to FastQ Files. *Babraham Bioinformatics.* 2017. Available online: https://www.bioinformatics.babraham.ac.uk/projects/trim_galore/ (accessed on 1 December 2022).
55. Badenhorst, D.; Hillier, L.W.; Literman, R.; Montiel, E.E.; Radhakrishnan, S.; Shen, Y.; Minx, P.; Janes, D.E.; Warren, W.C.; Edwards, S.V.; et al. Physical Mapping and Refinement of the Painted Turtle Genome (*Chrysemys picta*) Inform Amniote Genome Evolution and Challenge Turtle-Bird Chromosomal Conservation. *Genome Biol. Evol.* **2015**, *7*, 2038–2050. [CrossRef]
56. Krueger, F.; Andrews, S.R. Bismark: A flexible aligner and methylation caller for Bisulfite-Seq applications. *Bioinformatics* **2011**, *27*, 1571–1572. [CrossRef]
57. Li, H.; Handsaker, B.; Wysoker, A.; Fennell, T.; Ruan, J.; Homer, N.; Marth, G.; Abecasis, G.; Durbin, R. 1000 Genome Project Data Processing Subgroup. The Sequence Alignment/Map format and SAMtools. *Bioinformatics* **2009**, *25*, 2078–2079. [CrossRef] [PubMed]
58. Allaire, J.J. RStudio: Integrated Development Environment for R. 2012. Available online: https://www.r-project.org/conferences/useR-2011/abstracts/180111-allairejj.pdf (accessed on 1 December 2022).
59. Akalin, A.; Kormaksson, M.; Li, S.; Garrett-Bakelman, F.E.; Figueroa, M.E.; Melnick, A.; E Mason, C. methylKit: A comprehensive R package for the analysis of genome-wide DNA methylation profiles. *Genome Biol.* **2012**, *13*, R87. [CrossRef] [PubMed]
60. Quinlan, A.R.; Hall, I.M. BEDTools: A flexible suite of utilities for comparing genomic features. *Bioinformatics* **2010**, *26*, 841–842. [CrossRef] [PubMed]
61. Akalin, A.; Franke, V.; Vlahoviček, K.; Mason, C.E.; Schübeler, D. Genomation: A toolkit to summarize, annotate and visualize genomic intervals. *Bioinformatics* **2014**, *31*, 1127–1129. [CrossRef] [PubMed]
62. Robinson, M.D.; McCarthy, D.J.; Smyth, G.K. EdgeR: A Bioconductor package for differential expression analysis of digital gene expression data. *Bioinformatics* **2010**, *26*, 139–140. [CrossRef]
63. Liu, Y.; Zhou, J.; White, K.P. RNA-seq differential expression studies: More sequence or more replication? *Bioinformatics* **2013**, *30*, 301–304. [CrossRef] [PubMed]
64. Kearse, M.; Moir, R.; Wilson, A.; Stones-Havas, S.; Cheung, M.; Sturrock, S.; Buxton, S.; Cooper, A.; Markowitz, S.; Duran, C.; et al. Geneious Basic: An integrated and extendable desktop software platform for the organization and analysis of sequence data. *Bioinformatics* **2012**, *28*, 1647–1649. [CrossRef]
65. Tran, H.; Porter, J.; Sun, M.; Xie, H.; Zhang, L. Objective and Comprehensive Evaluation of Bisulfite Short Read Mapping Tools. *Adv. Bioinform.* **2014**, *2014*, 1–11. [CrossRef]
66. Gu, H.; Bock, C.; Mikkelsen, T.S.; Jäger, N.; Smith, Z.D.; Tomazou, E.; Gnirke, A.; Lander, E.S.; Meissner, A. Genome-scale DNA methylation mapping of clinical samples at single-nucleotide resolution. *Nat. Methods* **2010**, *7*, 133–136. [CrossRef]
67. Chatterjee, A.; Ozaki, Y.; A Stockwell, P.; Horsfield, J.A.; Morison, I.M.; Nakagawa, S. Mapping the zebrafish brain methylome using reduced representation bisulfite sequencing. *Epigenetics* **2013**, *8*, 979–989. [CrossRef]
68. Hu, C.; Hart, S.N.; Bamlet, W.R.; Moore, R.M.; Nandakumar, K.; Eckloff, B.W.; Lee, Y.K.; Petersen, G.M.; McWilliams, R.R.; Couch, F.J. Prevalence of pathogenic mutations in cancer predisposition genes among pancreatic cancer patients. *Cancer Epidemiol. Biomark. Prev.* **2016**, *25*, 207–211. [CrossRef] [PubMed]
69. Fraga, M.F.; Esteller, M. DNA Methylation: A Profile of Methods and Applications. *BioTechniques* **2002**, *33*, 632–649. [CrossRef] [PubMed]
70. Li, L.; Choi, J.-Y.; Lee, K.-M.; Sung, H.; Park, S.K.; Oze, I.; Pan, K.-F.; You, W.-C.; Chen, Y.-X.; Fang, J.-Y.; et al. DNA Methylation in Peripheral Blood: A Potential Biomarker for Cancer Molecular Epidemiology. *J. Epidemiol.* **2012**, *22*, 384–394. [CrossRef] [PubMed]
71. Zhang, Z.; Chen, A.C.-Q.; Manev, H. Enzymatic Regional Methylation Assay for Determination of CpG Methylation Density. *Anal. Chem.* **2004**, *76*, 6829–6832. [CrossRef] [PubMed]
72. Galm, O.; Rountree, M.R.; Bachman, K.E.; Jair, K.-W.; Baylin, S.B.; Herman, J.G. Enzymatic Regional Methylation Assay: A Novel Method to Quantify Regional CpG Methylation Density. *Genome Res.* **2001**, *12*, 153–157. [CrossRef] [PubMed]
73. Harrison, A. Parle-McDermott, A. DNA Methylation: A Timeline of Methods and Applications. *Front. Genet.* **2011**, *2*, 74. [CrossRef]
74. Kunde-Ramamoorthy, G.; Coarfa, C.; Laritsky, E.; Kessler, N.; Harris, R.; Xu, M.; Chen, R.; Shen, L.; Milosavljevic, A.; Waterland, R.A. Comparison and quantitative verification of mapping algorithms for whole-genome bisulfite sequencing. *Nucleic Acids Res.* **2014**, *42*, e43. [CrossRef]
75. Smallwood, S.A.; Lee, H.J.; Angermueller, C.; Krueger, F.; Saadeh, H.; Peat, J.; Andrews, S.R.; Stegle, O.; Reik, W.; Kelsey, G. Single-cell genome-wide bisulfite sequencing for assessing epigenetic heterogeneity. *Nat. Methods* **2014**, *11*, 817–820. [CrossRef]
76. Hong, S.R.; Shin, K.-J. Bisulfite-Converted DNA Quantity Evaluation: A Multiplex Quantitative Real-Time PCR System for Evaluation of Bisulfite Conversion. *Front. Genet.* **2021**, *12*, 618955. [CrossRef]
77. Lance, V.; Valenzuela, N.; Hildebrand, P. A hormonal method to determine sex of hatchling giant river turtles, Podocnemis expansa: Application to endangered species. *Am. Zool.* **1992**, *270*, 16A.
78. Gómez-Saldarriaga, C.; Valenzuela, N.; Ceballos, C.P. Effects of Incubation Temperature on Sex Determination in the Endangered Magdalena River Turtle, *Podocnemis lewyana*. *Chelonian Conserv. Biol.* **2016**, *15*, 43–53. [CrossRef]

79. Radhakrishnan, S.; Literman, R.; Neuwald, J.; Severin, A.; Valenzuela, N. Transcriptomic responses to environmental temperature by turtles with temperature-dependent and genotypic sex determination assessed by RNAseq inform the genetic architecture of embryonic gonadal development. *PLoS ONE* **2017**, *12*, e0172044. [CrossRef] [PubMed]
80. Radhakrishnan, S.; Literman, R.; Neuwald, J.L.; Valenzuela, N. Thermal Response of Epigenetic Genes Informs Turtle Sex Determination with and without Sex Chromosomes. *Sex. Dev.* **2018**, *12*, 308–319. [CrossRef] [PubMed]
81. Mizoguchi, B.; Valenzuela, N. Alternative splicing and thermosensitive expression of *Dmrt1* during urogenital development in the painted turtle, *Chrysemys picta*. *PeerJ* **2020**, *8*, e8639. [CrossRef]
82. Valenzuela, N.; Neuwald, J.L.; Literman, R. Transcriptional evolution underlying vertebrate sexual development. *Dev. Dyn.* **2012**, *242*, 307–319. [CrossRef]
83. Valenzuela, N. Relic Thermosensitive Gene Expression In A Turtle With Genotypic Sex Determination. *Evolution* **2007**, *62*, 234–240. [CrossRef]
84. Valenzuela, N.; LeClere, A.; Shikano, T. Comparative gene expression of steroidogenic factor 1 in Chrysemys picta and Apalone mutica turtles with temperature-dependent and genotypic sex determination. *Evol. Dev.* **2006**, *8*, 424–432. [CrossRef]
85. Berry, J.F.; Shine, R. Sexual size dimorphism and sexual selection in turtles (order testudines). *Oecologia* **1980**, *44*, 185–191. [CrossRef]
86. Ernst, C.H. Sexual Cycles And Maturity Of The Turtle, Chrysemys Picta. *Biol. Bull.* **1971**, *140*, 191–200. [CrossRef]
87. Xiao, F.-H.; Chen, X.-Q.; He, Y.; Kong, Q.-P. Accelerated DNA methylation changes in middle-aged men define sexual dimorphism in human lifespans. *Clin. Epigenetics* **2018**, *10*, 133. [CrossRef]
88. Weidner, C.I.; Lin, Q.; Koch, C.M.; Eisele, L.; Beier, F.; Ziegler, P.; Bauerschlag, D.O.; Jöckel, K.-H.; Erbel, R.; Mühleisen, T.W.; et al. Aging of blood can be tracked by DNA methylation changes at just three CpG sites. *Genome Biol.* **2014**, *15*, R24. [CrossRef] [PubMed]
89. Lofton-Day, C.; Model, F.; DeVos, T.; Tetzner, R.; Distler, J.; Schuster, M.; Song, X.; Lesche, R.; Liebenberg, V.; Ebert, M.; et al. DNA Methylation Biomarkers for Blood-Based Colorectal Cancer Screening. *Clin. Chem.* **2008**, *54*, 414–423. [CrossRef] [PubMed]
90. Marsit, C.J.; Koestler, D.C.; Christensen, B.C.; Karagas, M.R.; Houseman, E.A.; Kelsey, K.T. DNA Methylation Array Analysis Identifies Profiles of Blood-Derived DNA Methylation Associated With Bladder Cancer. *J. Clin. Oncol.* **2011**, *29*, 1133–1139. [CrossRef] [PubMed]

Article

Global Terrapin Character-Based DNA Barcodes: Assessment of the Mitochondrial COI Gene and Conservation Status Revealed a Putative Cryptic Species

Mohd Hairul Mohd Salleh [1,2], Yuzine Esa [1,3,*] and Rozihan Mohamed [1]

1 Department of Aquaculture, Faculty of Agriculture, Universiti Putra Malaysia, Serdang 43400, Malaysia
2 Royal Malaysian Customs Department, Persiaran Perdana, Presint 2, Putrajaya 62596, Malaysia
3 International Institute of Aquaculture and Aquatic Sciences, Universiti Putra Malaysia, Lot 960 Jalan Kemang 6, Port Dickson 71050, Malaysia
* Correspondence: yuzine@upm.edu.my; Tel.: +60-397694933

Simple Summary: This study evaluated 26 sequences of terrapins worldwide through COI DNA barcoding and phylogenetic analysis, which included 12 species and three families. Moreover, 16 haplotypes were found; they were either misidentified, or a potential cryptic species was determined between *B. baska* and *B. affinis affinis*. Thus, COI remains an effective barcode marker for the terrapin species.

Abstract: Technological and analytical advances to study evolutionary biology, ecology, and conservation of the Southern River Terrapin (*Batagur affinis* ssp.) are realised through molecular approaches, including DNA barcoding. We evaluated the use of COI DNA barcodes in Malaysia's Southern River Terrapin population to better understand the species' genetic divergence and other genetic characteristics. We evaluated 26 sequences, including four from field specimens of Southern River Terrapins obtained in Bota Kanan, Perak, Malaysia, and Kuala Berang, Terengganu, Malaysia, as well as 22 sequences from global terrapins previously included in the Barcode of Life Database (BOLD) Systems and GenBank. The species are divided into three families: eight Geoemydidae species (18%), three Emydidae species (6%), and one Pelomedusidae species (2%). The IUCN Red List assigned the 12 species of terrapins sampled for this study to the classifications of critically endangered (CR) for 25% of the samples and endangered (EN) for 8% of the samples. With new haplotypes from the world's terrapins, 16 haplotypes were found. The intraspecific distance values between the COI gene sequences were calculated using the K2P model, which indicated a potential cryptic species between the Northern River Terrapin (*Batagur baska*) and Southern River Terrapin (*Batagur affinis affinis*). The Bayesian analysis of the phylogenetic tree also showed both species in the same lineage. The BLASTn search resulted in 100% of the same species of *B. affinis* as *B. baska*. The Jalview alignment visualised almost identical sequences between both species. The Southern River Terrapin (*B. affinis affinis*) from the west coast of Peninsular Malaysia was found to share the same haplotype (Hap_1) as the Northern River Terrapin from India. However, *B. affinis edwardmolli* from the east coast of Peninsular Malaysia formed Hap_16. The COI analysis found new haplotypes and showed that DNA barcodes are an excellent way to measure the diversity of a population.

Keywords: Southern River Terrapin; genetics; haplotype; phylogenetic tree; Peninsular Malaysia; population diversity

Citation: Mohd Salleh, M.H.; Esa, Y.; Mohamed, R. Global Terrapin Character-Based DNA Barcodes: Assessment of the Mitochondrial COI Gene and Conservation Status Revealed a Putative Cryptic Species. *Animals* **2023**, *13*, 1720. https://doi.org/10.3390/ani13111720

Academic Editor: Ettore Olmo

Received: 16 January 2023
Revised: 14 February 2023
Accepted: 17 February 2023
Published: 23 May 2023

1. Introduction

Terrapins inhabit either freshwater or brackish water [1]. There is no clear taxonomic group for terrapins, which may be unrelated. Numerous species belong to the families of Geoemydidae and Emydidae [2]. The only terrapin species not in this group is the *Pelusios seychellensis* from Seychelles [3].

The "Barcode of Life" Consortium is a global effort to conduct a molecular inventory of the planet's biodiversity [4]. After it was demonstrated that the cytochrome c oxidase subunit I (COI) gene of the mitochondrial DNA (mtDNA) could be used to successfully identify North American bird species, such as *Sturnella magna*, *Tringa solitaria*, and *Hirundo rustica* [5], numerous other vertebrate COI barcodes have been developed [6–8]. Ref. [9] also reported that the COI marker was better for barcoding than sequences from the mitochondrial control region.

Traditional taxonomy frequently fails to distinguish between the different terrapin species because they lack essential morphological characteristics. Currently, molecular methods are required to identify certain species [10,11]. A complementing tool to traditional taxonomy and systematics research, DNA barcoding allows for a more accurate understanding of the existing fauna around the world [12]. Especially in species with complicated, accessible anatomy, DNA barcoding is proposed as a method for quickly and readily identifying species using a short DNA sequence [12,13]. DNA barcoding has been used to identify freshwater turtles all over the world, even in Malaysia [14].

Batagur affinis ssp. [15] is among 24 species of turtles found in Peninsular Malaysia [16] and Sumatra, Indonesia, and was initially believed to be conspecific with *B. baska*, a species native to the North (Bangladesh and India) [17]. According to [18], *B. baska* consisted of at least two heritably distinct species: *B. affinis* ssp. populations in the Kedah River systems and *B. affinis affinis* populations in the Perak River systems, both on the west coast of Peninsular Malaysia. In contrast, individuals in the Terengganu River basin were identified as *B. affinis edwardmolli*. According to [19], this species is one of the world's 25 most endangered freshwater turtles and tortoises.

B. affinis ssp. used to live in a large river in Southeastern Asia, including the Tonle Sap in Cambodia and the Mekong delta in Vietnam. However, many of its wild populations have been severely reduced or wiped out [20–24]. *Batagur affinis* ssp. is found only on the west coast of Peninsular Malaysia and is extinct in Sumatra, Indonesia [25,26].

In contrast, the subspecies *B. a. edwardmolli*, located on the east coast of Peninsular Malaysia that once reached from Singapore to Southeast Asia, is now thought to have vanished from Vietnam, Thailand, Singapore, and Indonesia [23,26,27]. Currently, only Peninsular Malaysia and Cambodia are home to this species [18,23,24,28]. Moreover, according to [23], there are still populations of *B. a. edwardmolli* in Cambodia and along the east coast of Peninsular Malaysia. This implies that the Malaysian and Cambodian populations are the only ones whose genes have remained constant across the species' range.

Unfortunately, this study was carried out during a difficult period, namely the COVID-19 pandemic. Due to the Malaysian Movement Control Order (MCO), or lockdown, we were only permitted to gather four specimens of the Southern River Terrapin from Peninsular Malaysia by the Malaysian government authority. The samples are limited due to the conservation status of *B. affinis* ssp., which has been listed as critically endangered on the IUCN Red List since 2000 [16]. This study compares them to the other eleven terrapin species listed by [3,26] and accessed from the public database portal.

In addition, we were the first to upload COI *B. affinis* ssp. sequences to the GenBank database portal. The objectives of this study were to determine if terrapin DNA barcoding could be used all over the world by comparing the unique COI sequences to other COI sequences that were already available from the Barcode of Life Data (BOLD) Systems and GenBank, and to analyse the phylogenetic relationships among terrapins, including the recently collected specimens from Malaysia.

2. Materials and Methods

2.1. Study Sites

Four *Batagur affinis* ssp. individuals from two distinct population locations on the east and west coasts of Peninsular Malaysia were randomly chosen for this study, and the sampling was carried out in 2020 (Figure 1). The captive hatchling population at the Bota Kanan head-starting facility (BK; GPS coordinates: 4.3489° N and 100.8802° E) in

Perak, Malaysia, provided the blood samples of *B. affinis affinis* (N = 1). The facilities were developed beside the Perak River, which is a habitat for the wild Southern River Terrapin population. There was no uncertainty regarding the genetic origin of that sample. In addition, blood samples from three wild *B. affinis edwardmolli* hatchlings (translocated eggs) were taken from a population in Bukit Paloh, Kuala Berang (KB; GPS coordinates: 5.0939° N, 102.7821° E), which is in Terengganu, Malaysia. According to [29], blood was drawn from the species using venipuncture methods through the internal jugular vein and subcarapacial venous plexus (SVP). In a 2 mL microcentrifuge tube, 1.5 mL of blood was preserved with 0.5 mL of EDTA in a 1:3 ratio before being kept at −20 °C. The Department of Wildlife and National Parks, Peninsular Malaysia, issued the study and field permit approval number, which is B-00335-16-20.

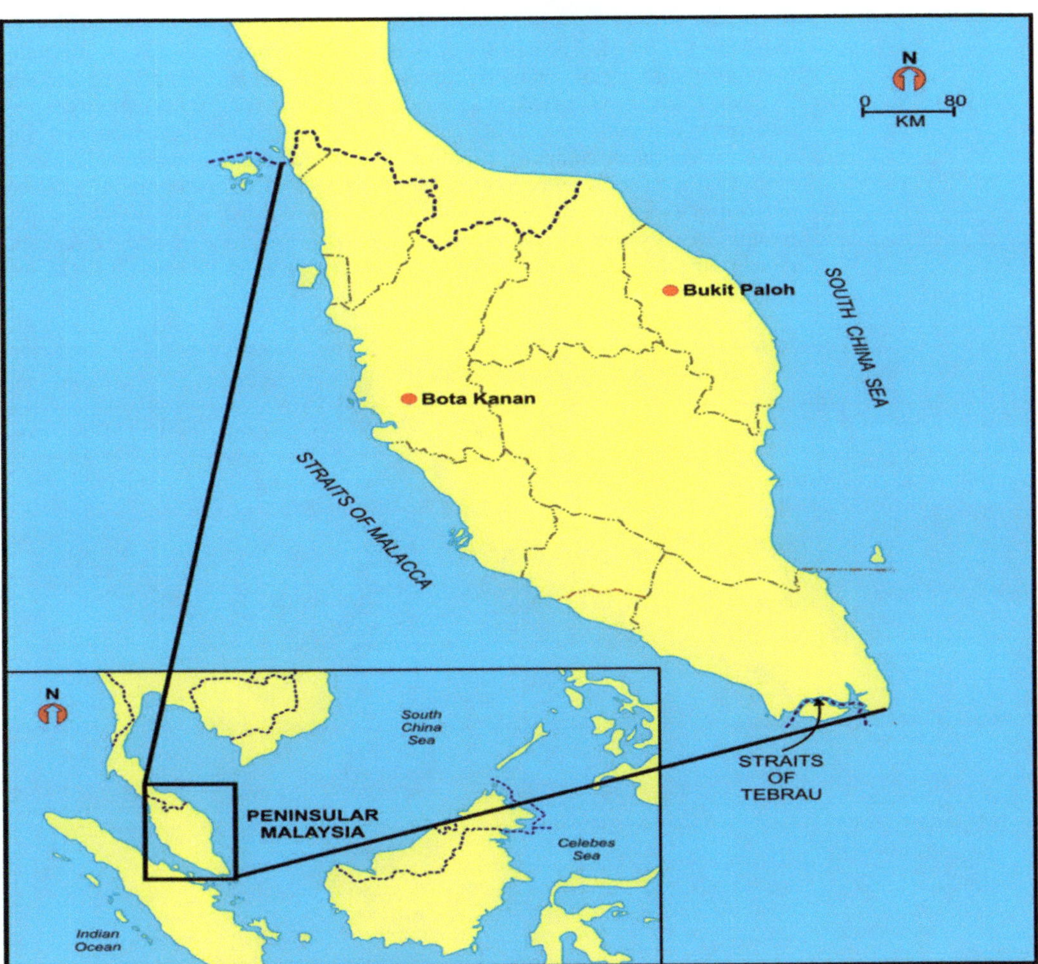

Figure 1. Sampling sites of *Batagur affinis* ssp.

2.2. DNA Isolation, PCR, and Sequencing

For each sample, 200 µL of EDTA whole blood was used to extract the nucleic acids. After cell lysis and protein denaturation, DNA was extracted using the ReliaPrep^{TM} Blood gDNA Miniprep System with binding column technology (Promega, Madison, WI, USA) according to the manufacturer's instructions. The final volume extracted was adjusted to 200 μL based on the input volume of the EDTA whole-blood sample. Using the Thermo Scientific™ NanoDrop 2000 c spectrophotometer model ND-2000, the amount and purity of the extracted DNA samples were evaluated (Thermo Fisher Scientific, Waltham, MA, USA). After quantifying the extracted nucleic acids, the DNA samples were put onto a 1% (*w*/*v*) agarose gel with molecular markers. Electrophoresis was performed to assess the integrity and intactness of the high molecular weight DNA band.

The cross-species primer derived from Painted Terrapin, *Batagur borneoensis*, was utilised for PCR. Ref. [30] made the "Tuntong" primer pair, which targets the COI marker gene. The forward primer (5-CGCGGAATTAAGCCAACCAG-3) and the reverse primer (5-TTGGTACAGGATTGGGTCGC-3) are designed. The COI gene fragment PCR amplification was carried out in a Go Taq Flexi PCR (Promega, Madison, WI, USA) reaction mixture containing 2 μL of DNA template, 0.4 μL of primers, 4 μL of 5× PCR buffer, 1.6 μL of 25 mM $MgCl_2$, 0.4 μL of dNTPs, 0.2 μL of Taq DNA polymerase, and 11 μL of distilled water (ddH_2O). Following an initial denaturation at 94 °C for 4 min, 35 cycles of denaturation at 94 °C for 45 s, annealing at 55 °C for 35 s, and extension at 72 °C for 1 min were performed, followed by a 10 min extension at 72 °C. Finally, the purified PCR products were forwarded to a local laboratory company (First BASE Laboratories Sdn Bhd) for Sanger sequencing of the COI gene of the mitochondrial DNA (mtDNA-COI). In addition, 17 COI sequences of terrapin were extracted from GenBank and downloaded, while five COI sequences of terrapin were extracted from the BOLD Systems. This analysis led to the discovery of four novel sequences (GenBank accession numbers: OL658844–OL658847) for 26 sequences (Table 1).

Table 1. List of terrapin species studied through DNA barcoding with the BOLD IDs of their respective COI sequences and the GenBank accession of each species.

Scientific Name	English Name	GenBank	BOLD ID	Haplotype	BLASTn Result	Locallity	IUCN Red List	References
Batagur baska	Northern River Terrapin	KF894752	GBGCR2852-19	Hap_1	100% with *B. affinis* (OL658844)	India	CR	[31]
Batagur baska	Northern River Terrapin	HQ329671	GBGCR2716-19	Hap_1	99% with *B. affinis* (OL658844)	India	CR	[32]
Batagur borneoensis	Painted Terrapin	HQ329672	GBGCR2717-19	Hap_2	95% with *B. trivittata* (HQ329675)	Indonesia	CR	[32]
Batagur borneoensis	Painted Terrapin	None	BENT109-08	Hap_2	95% with *B. trivittata* (HQ329675)	Indonesia	CR	[33]
Morenia ocellata	Bengal Eyed Terrapin	HQ329690	GBGCR2724-19	Hap_3	90–91% with *M. petersi* (MH157788)	Myanmar	EN	[32]
Morenia ocellata	Bengal Eyed Terrapin	None	BENT264-09	Hap_3	90–91% with *M. petersi* (KF894774)	Myanmar	EN	[33]
Malaclemys terrapin	Diamondback Terrapin	HQ329654	GBGC11262-13	Hap_4	95% with *Graptemys barbouri* (MG728234)	America	VU	[32]
Malaclemys terrapin	Diamondback Terrapin	KX559038	GBGCR2938-19	Hap_5	95% with *Graptemys geographica* (MG728245)	America	VU	[34]
Emys orbicularis	European Pond Terrapin	HQ329643	GBGC11273-13	Hap_6	98% with *E. trinacris* (KX559027)	Unknown	NT	[32]
Emys orbicularis	European Pond Terrapin	KP697925	None	Hap_7	98% with *E. trinacris* (KX559027)	Germany	NT	[35]

Table 1. *Cont.*

Scientific Name	English Name	GenBank	BOLD ID	Haplotype	BLASTn Result	Locallity	IUCN Red List	References
Melanochelys trijuga	Indian Pond Terrapin	KC354725	GBGC11418-13	Hap_8	96% with *M. tricarinata* (KF894770)	India	LC	[31]
Melanochelys trijuga	Indian Pond Terrapin	KC354724	GBGC11419-13	Hap_9	95% with *M. tricarinata* (KF894770)	India	LC	[31]
Rhinoclemmys rubida	Mexican Spotted Terrapin	HQ329701	GBGCR2766-19	Hap_10	91% with *R. annulata* (MH274599)	Mexico	NT	[32]
Trachemys scripta elegans	Red-eared Terrapin	KX559044	GBGCR1038-18	Hap_11	96–100% with *T. s. elegans* (TSU49047)	America	LC	[34]
Trachemys scripta elegans	Red-eared Terrapin	KM216748	GBGCR1008-15	Hap_12	97–100% with *T. s. elegans* (TSU49047)	America	LC	[36]
Pelusios sinuatus	Serrated Hinged Terrapin	None	BENT174-08	Hap_13	100%	Southern Africa	LC	[33]
Pelusios sinuatus	Serrated Hinged Terrapin	HQ329735	GBGC11221-13	Hap_13	100%	Southern Africa	LC	[32]
Siebenrockiella crassicollis	Smiling Terrapin	HQ329704	GBGCR2769-19	Hap_14	100%	Unknown	EN	[32]
Siebenrockiella crassicollis	Smiling Terrapin	None	BENT190-08	Hap_14	100%	Unknown	EN	[33]
Mauremys caspica	Striped-neck Terrapin	AY337348	GBGC0806-06	Hap_15	95% with *Chinemys nigricans* (AF348264)	Iran	LC	[37]
Mauremys caspica	Striped-neck Terrapin	AY337347	GBGC0805-06	Hap_15	95% with *Chinemys nigricans* (AF348264)	Bahrain	LC	[37]
Batagur affinis	Southern River Terrapin	None	MTD042-21	Hap_1	100% with *B. baska* (HQ329671)	Malaysia	CR	[33]
Batagur affinis affinis	Southern River Terrapin	OL658844	HYT001-21	Hap_1	99–100% with *B. baska* (KF894752)	Malaysia	CR	This study
Batagur affinis edwardmolli	Southern River Terrapin	OL658845	HYT002-21	Hap_16	98% with *B. baska* (KF894752)	Malaysia	CR	This study
Batagur affinis edwardmolli	Southern River Terrapin	OL658846	HYT003-21	Hap_16	98% with *B. baska* (KF894752)	Malaysia	CR	This study
Batagur affinis edwardmolli	Southern River Terrapin	OL658847	HYT004-21	Hap_16	98% with *B. baska* (KF894752)	Malaysia	CR	This study

2.3. *DNA Barcode Sequence Quality Control Measures and Analysis*

Chromatograms displaying the nucleotide sequences of both DNA strands for each sample were created—trimmed chromatograms with more than 2% unclear bases and low-quality noisy sequences on both ends. The bidirectional reads were eliminated by benchmarking against a quality value greater than 40. The consensus sequences were obtained by combining the forward and reverse chromatograms in SeqScape, version 2.7 (Applied Biosystems), and comparing them with reference sequences from the NCBI nucleotide (NT) database using BLASTn [38,39]. Additionally, using our COI sequences in a BLASTn search of GenBank, the species that most closely matched our sequences were noted. The sequences' accession codes and BOLD sequence identifiers were confirmed against GenBank and the BOLD Systems (Table 1). Using the BOLD Systems' sequence analysis [40], the Kimura 2 Parameter (K2P) model was used to calculate the pairwise sequencing divergences for the distance analyses. MEGAX was used to find the polymorphic sites (PS) or variable sites [41].

2.4. Analyses of Molecular Phylogenetics and Divergence Times

The best-fitting evolutionary model for each sequence analysed was determined using the Akaike information criterion (AIC) with sample size correction implemented in jModelTest2 on XSEDE (2.1.6) [42]. The phylogenetic studies used models of sequence evolution selected as best with jModelTest2 for coding and non-coding sequences. maximum likelihood (ML) analyses [43] were performed. As a result, the alignments were carried out in MEGAX using ClustalW [41]. All sequences produced multiple alignments with the same length and beginning point. However, Jalview, Ref. [44], was used to accomplish various sequence alignments, functional site analyses, and web postings of alignments between *B. affinis affinis* and *B. baska* [45]. IQ-tree was used for phylogenetic reconstruction by [46] on XSEDE and [47] via the online CIPRES Science Gateway V.3.3 [48]. The trees were visualised in FigTree v1.4.4 [49].

On the other hand, using the BEAST v2.6.6 tool, the phylogenetic tree topology and divergence dates were computed concurrently [50,51]. BEAUti 2 [52] was used to unlink the substitution models of the data partitions and implement the sequence evolution models selected with jModelTest2 as optimal. The "Clock Model" was set to a rigorous clock with uncorrelated rates, while the "Tree Model" was assigned to a Yule speciation process. The sequences were examined using a relaxed molecular clock model, which permits substitution rates to vary among branches based on an uncorrelated lognormal distribution [50]. We established the species tree before the Yule process. Two simultaneous assessments were conducted utilising Bayesian Markov Chain Monte Carlo (MCMC) simulations with a sampling frequency of 5000 for 100,000,000 generations. The nucleotide substitution model for ML was empirically set to TN93. Bootstrap analysis (1000 pseudoreplicates) provided branch support, and all other parameters were left at their default settings.

After that, the phylogenetic trees were plotted using FigTree v1.4.4. To create the phylogenetic trees, the whole mitochondrial COI sequences of *Batagur affinis* (MTD042-21) and the out-group species *Ophiophagus hannah* (MH153655) were chosen from the GenBank online database [33,53]. Then, using the software DnaSP 6.12.03, we analysed the haplotype of each specimen [54–56]. A Median Joining (MJ) network analysis by [57] was performed with NETWORK 10.2.

3. Results

3.1. Taxonomic Range and Red List Coverage

Table 1 contains the details on the taxa used in this study. The final data collection includes 12 species from the Testudines order, two previously unrepresented in the barcode database. One is not available in the BOLD Systems, and five were not sent to GenBank. We initially deposited our novel COI gene of the mitochondrial DNA (mtDNA-COI) samples (*Batagur affinis* ssp.) in the GenBank database portal.

As a result, the IUCN Red List assigned the 12 species of terrapins sampled for this study to the classifications of least concern (LC) for 33% of the samples, critically endangered (CR) for 25% of the samples, vulnerable (VU) for 8% of the samples, and endangered (EN) and near-threatened (NT) for 17% of the samples (Figure 2).

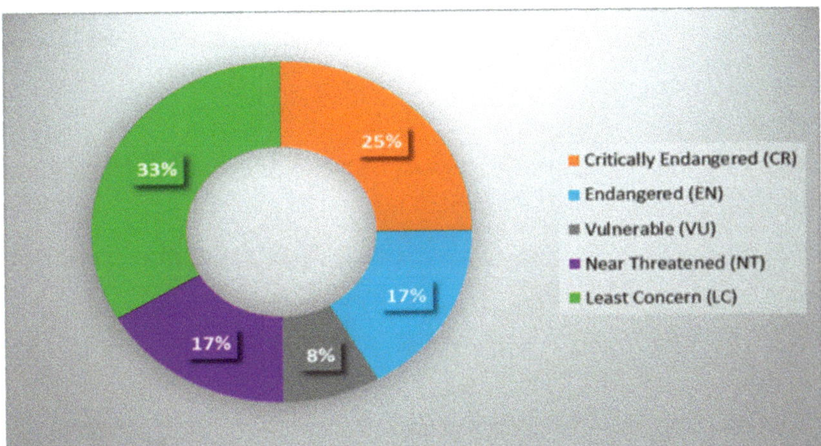

Figure 2. The conservation status of the terrapins is based on the IUCN Red List.

3.2. COI Divergence Assessment

All 26 produced barcodes had sequence lengths of more than 503 bp with no indels or stop codons found. The nucleotide composition was as follows: 16.88% Guanine, 27.21% Cytosine, 27.5% Adenine, and 38.41% Tyrosine. GC Codon position 1 was 52.62% followed by GC Codon position 2 (43.21%) and GC Codon position 3 (36.46%). Almost all species (83.33%, ten species) were represented by dual specimens with a single specimen representing another species and five specimens representing another species (Table S1).

The genetic divergences of the COI sequences within the order Testudines were studied at various taxonomic levels (Table 2). The genetic divergence rose with the taxonomic rank as expected. The hierarchical taxonomic relationship was directly associated with increased K2P genetic divergence. The conspecific K2P levels ranged from 0% to 2.14% with a mean of 0.68% (SE = 0.04). The mean K2P divergence amongst the congeneric species specimens was 5.49% (SE = 0.15; range 0–9.14%). The average K2P divergence between the specimens from various genera in the same family was 17.10% (SE = 0.03; range: 4.98–22.48%). This range, though they overlap, indicates intraspecific (S) and intragenus (G) distances (Figure S1).

Table 2. K2P divergence values from the examined specimens of varying taxonomic levels. SE = standard error.

Category	n	Taxa	Comparisons	Min (%)	Mean (%)	Max (%)	SE (%)
Within Species	25	11	20	0	0.68	2.14	0.04
Within Genus	9	1	24	0	5.49	9.14	0.15
Within Family	24	2	125	4.98	17.10	22.48	0.03

Deep intraspecific K2P divergences were identified in a *Batagur baska* (2.14%) that exceeded the conventional threshold distance of 2% [12,58] (Table 3). A barcode gap analysis revealed that practically all species represented by multiple sequences had a barcode gap (Figure 3). Notably, just one species, *Batagur baska*, had its maximum intraspecific and nearest neighbour distances (0%).

Table 3. The summary statistics include the BIN of each species, their maximum intraspecific K2P distances, and the nearest neighbour K2P distances (i.e., minimum interspecific distance).

Scientific Name	BIN	Nearest Species	Max. Intraspecifc Distance (%)	Nearest Neighbour Distance (%)
Emys orbicularis	BOLD:AAF8183	*Malaclemys terrapin*	0.62	11.97
Malaclemys terrapin	BOLD:AAX3718	*Trachemys scripta*	0.16	4.98
Trachemys scripta	BOLD:AAF5910	*Malaclemys terrapin*	0.14	4.98
Batagur affinis	BOLD:AAW2850 & ADX0374	*Batagur baska*	2.14	0
Batagur baska	BOLD:AAW2850	*Batagur affinis*	0	0
Batagur borneoensis	BOLD:AAW2847	*Batagur affinis*	0	7.67
Mauremys caspica	BOLD:AAJ1604	*Malaclemys terrapin*	0	14.09
Melanochelys trijuga	BOLD:AAX4497	*Mauremys caspica*	0.62	15.07
Morenia ocellata	BOLD:AAX4362	*Batagur borneoensis*	0	13.43
Rhinoclemmys rubida	BOLD:AAY0332	*Malaclemys terrapin*	0	15.62
Siebenrockiella crassicollis	BOLD:AAJ6683	*Batagur borneoensis*	0	17.82
Pelusios sinuatus	BOLD:AAX1329	*Batagur affinis*	0	25.66

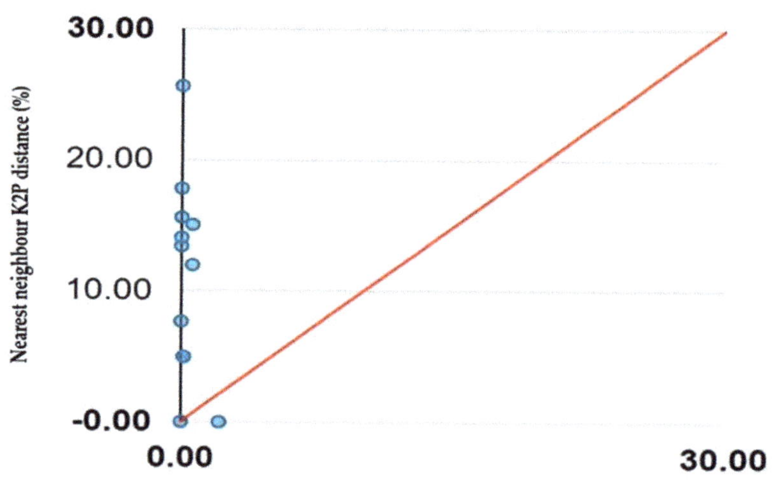

Figure 3. Maximum intraspecific distances plotted against nearest neighbour distances.

3.3. Population Relationships

The nucleotide diversity at 199 nucleotide positions and transitions is approximately 55% saturated (Table S2). When all codon locations are analysed, transitions and transversions are displayed against the pairwise sequence divergence Tajima-Nei Method (TN84) for the terrapins utilising 503bp of the COI DNA barcode (Figure 4). DAMBE [59] uses these substitution models to perform various molecular phylogenetic analyses. DAMBE also includes functions for determining the optimum substitution models for particular sequences.

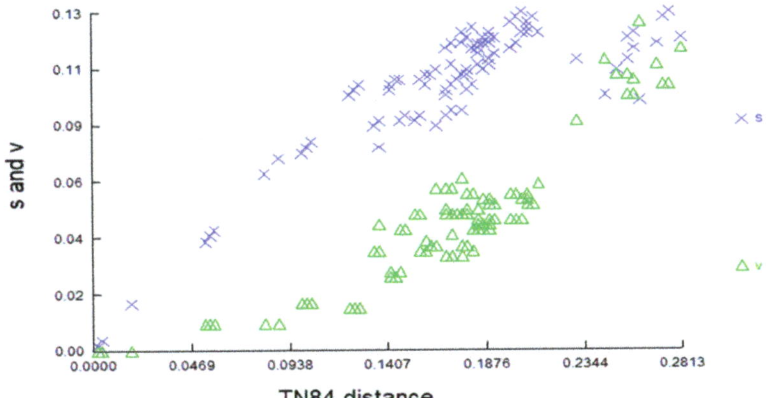

Figure 4. Transitions and transversions are plotted against the pairwise sequence divergence using the Tajima-Nei Method for the terrapins using 503 bp of the COI DNA barcode.

The network had 16 haplotypes (Figure 5), which were confirmed with DNAsp 6.12.03 analysis (Table 1). Different haplotypes were found in *Malaclemys terrapin, Emys orbicularis, Melanochelys trijuga, Trachemys scripta elegans*, and *Batagur affinis* ssp. Furthermore, *Batagur baska* and *Batagur affinis affinis* shared a single haplotype (Hap_1), which was shown to be the most variable haplotype. The remaining haplotype only had two specimens and one species.

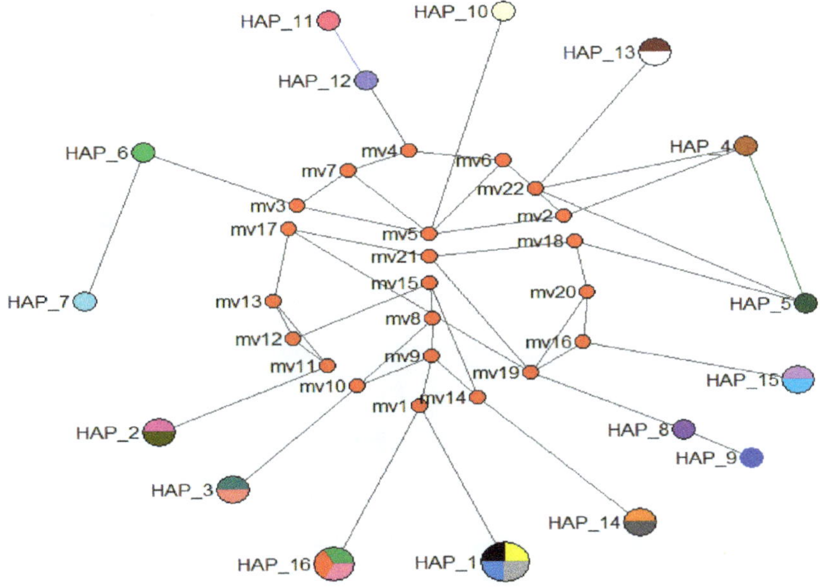

Figure 5. Median-joining network of mtDNA COI haplotypes in the terrapins. The sizes of the circles are proportional to the haplotype frequencies, and the colour-coding corresponds to the locations. The black squares on the lines linking the haplotypes represent the number of mutations.

4. Discussion

This study examined 26 terrapin COI sequences from the order Testudines. The species are divided into three families: eight Geoemydidae species (18%), three Emydidae species (6%), and one Pelomedusidae species (2%) (Figure S2 and Table S3). Based on the IUCN Red List of the 12 species of terrapins, 25% were critically endangered (CR) and 8% were endangered (EN). The terrapins studied all inhabit fresh or brackish water [26]. Furthermore, "terrapin" refers to more or less aquatic, hard-shelled turtles [60]. Notably, refs. [3,26] identified 13 terrapin species worldwide but ignored a previously thought-to-be-extinct Seychelles black terrapin species (*Pelusios seychellensis*). However, a genetic analysis of the lectotype revealed that this terrapin is not extinct and is now known as *Pelusios castaneus*. Before the Zoological Museum Hamburg bought a private collection of specimens in 1901 [26,61], the specimens could have been mislabelled or mixed up.

Therefore, the discovery of species-specific COI sequences allows for the identification of terrapin species using DNA barcodes to supplement taxonomy. This can also be used in the field when identifying lost nests or those caught as bycatch in fishing nets. When no other material is available, terrapin eggs or meat are used in the forensic investigations [4].

Additionally, DNA barcoding holds excellent promise for species identification and other conservational genetic applications in terrapins, which are distinct in the evolutionary tree of terrapins for inhabiting the river realm and are well-known for their lengthy migrations. One of the main objectives of the DNA barcoding initiative, species identification, was accomplished using their COI sequences. Even though these ancient taxa have undergone relatively slow molecular evolution [62,63], diagnostic sites at the COI gene were found for all 12 species of terrapins. Ref. [9] found that the distance-based analysis of COI sequences always put members of the same species together, even though the phenetic methods required a total baseline sample for a correct assignment. Using distinct nucleotide combinations, unique COI barcodes were generated for each of the 12 previously defined terrapin species (Table S2). The diagnoses were reliable with species-specific haplotypes [9] (Table 1; Figure 5).

If a phenetic technique based on a BLAST search was used without a comprehensive baseline sample, such as the one available in GenBank prior to this work, query sequences could be assigned to the wrong species. There were no *Batagur affinis* ssp. COI sequences in GenBank, for example, and a query on a Southern River Terrapin (*B. affinis affinis*) grouped it with a Northern River Terrapin (*B. baska*). The BLASTn search validated it, showing 100% similarity between the *B. affinis*-MTD042-21 COI sequences and *B. baska*-HQ329671 COI sequences (Table 1). So, Jalview's alignment and visualisation (Figure 6) showed that the sequences of *B. baska* (GenBank Accession Number: HQ329671) and *B. affinis* (BOLD ID: MTD042-21) were very similar. Similarly, *Emys orbicularis*, a species with COI sequences in GenBank, may be confused with *Emys trinacris* or a cryptic species due to 98% identical COI BLASTn results (Table 1).

Figure 6. *B. baska* with GenBank Accession Number HQ329671 vs. *B. affinis* BOLD ID MTD042-21 alignment and visualisation with Jalview.

Furthermore, in the BOLD Systems, the identical sequence of the Northern River Terrapin has two different BIN numbers (AAW2850 and ADX0374), which could be misinterpreted as Southern River Terrapin or a cryptic species.

The detection of the so-called "barcode gap," which can be measured by comparing the highest intraspecific distance with the minimum interspecific distance (also known as the nearest neighbour genetic distance), is one of the premises of DNA barcoding [64]. Moreover, DNA barcodes are helpful in the investigation of cryptic species [65], particularly those that appear similar but differ genetically [66]. A morphological species gap is strong evidence for species-level cryptic diversity [67]. On the other hand, the absence of a gap between two morphological species implies that they are different forms within the same species, or that they share ancestral polymorphism and/or hybridisation followed by introgression. In this case, it would be helpful to use a multigene (i.e., genomic) method to figure out the reciprocal taxonomic status of the two morphological species [68].

Table 3 shows that the DNA barcoding method revealed possible hidden variety within a species while failing to discover a meaningful difference between two biological species (*B. baska* and *B. affinis*). Such findings demand additional taxonomic research. In comparison to the mean congeneric divergence (5.49%), the mean conspecific K2P divergence (0.68%) was eight times smaller. Thus, as predicted, there was less genetic diversity between the conspecific individuals than between the congeneric species. It makes sense that there would be a rise in the taxonomic levels and an increase in the genetic divergence [69]. Therefore, both mean genetic estimations are comparable to those that have already been noted. In most fish molecular analyses, the conspecific divergence was found to be 0.25–0.39%, while the congeneric divergence was found to be 4.56–9.93% [70–74].

4.1. Population Relationships

This research began by examining the terrapins' DNA barcodes and mitochondrial COI gene haplotypes worldwide. Some existing terrapins and sea turtles are reported to carry mitochondrial COI gene haplotypes [4,9,26,30]. Nonetheless, our study contributes significantly by discovering new sequences from previously unknown areas in Malaysia and around the world. Previous research employing the COI gene in DNA barcoding of terrapins and sea turtles identified 1–10 haplotypes [4,9,30]. This study revealed 16 haplotypes (Table 1; Figure 5) of terrapins from around the world. The BOLD Systems differ from those previously described in Bota Kanan, Perak, and Kuala Berang, Terengganu. Also, the novel *B. affinis* ssp. COI gene sequences from Malaysia were submitted to GenBank (Table 1). They may serve as a reference for future genetic research of populations. A more comprehensive analysis involving additional sites and samples will be necessary to find common haplotypes. Previous studies by [28,75] described the divergence of *Batagur baska* and *Batagur affinis* ssp. Our research checks the sequences between the Indian and Malaysian populations. Moreover, the sequences from the Malaysian specimens are novel, and we hypothesise that this population is exclusive to this region (Figure 1).

Thus, clustering analyses and haplotype networks indicate that the three families are separated by four significant unique lineages (Figure 7). Figure 5 demonstrates that Hap_1 and Hap_16 are more closely related than other haplotypes. Hap_1 contains two *B. baska* specimens and two *B. affinis affinis* specimens, while Hap_16 contains three *B. affinis edwardmolli* specimens, which are in line with [14] that only found a haplotype in the Kuala Berang, Terengganu population; it has been proven that this is a random sampling, and we are not focusing on a clutch. In this case, it appears to be a cryptic species between *B. baska* and *B. affinis affinis*. We would need a more extensive set of genes and many markers from the nuclear genome [66,76,77] to decide if these groups should be called species or subspecies. Perhaps revision is required following the separation of *B. baska* and *B. affinis* ssp. by [28,75]. Even though it can be challenging to identify the morphological diagnostic features in morphologically cryptic species [78,79], the usefulness of such diagnoses may be in doubt [80]. We now recognise that cryptic species are relatively abundant [81,82] and widespread across most animal phyla [83,84]. Moreover, recent DNA research discovered cryptic species in many aquatic taxa [85], raising the possibility that aquatic biodiversity is higher and speciation possibilities have occurred more frequently than previously thought [86].

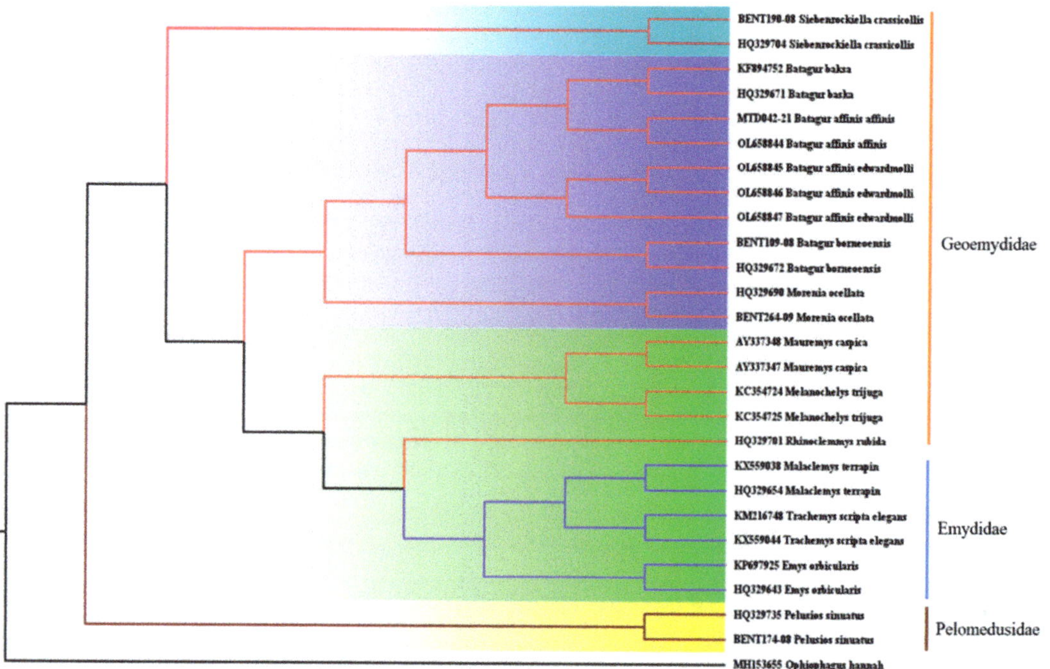

Figure 7. A maximum likelihood tree was constructed with a Bayesian analysis based on the COI sequences belonging to the order Testudines.

In addition, using Bayesian analysis, the maximum likelihood phylogeny of the investigated dataset revealed coherent, monophyletic clustering of all studied species (Figure 7). On the phylogenetic tree, cohesion was also detected between the database reference sequences for the representative species and the created sequences. The species were classified according to their family with Geoemydidae being the most abundant. The evolutionary tree indicates that *B. baska* originated in India and is closely related to *B. affinis affinis* from Malaysia, which is supported as a potential cryptic species. *Melanochelys trijuga* is similar to the Persian Gulf's *Mauremys caspica*, but the *Malaclemys terrapin* in North America is identical to *Trachemys scripta elegans*.

4.2. Conservation Status

The International Union maintains the Red List for biodiversity for the Conservation of Nature (IUCN). The IUCN is essential for guiding and igniting conservation and policy change activities; it is much more than a list of species and their states. The preservation of the natural resources that humans depend on is essential [87,88]. The IUCN Red List Categories and Criteria are designed to offer a clear framework for locating species in danger of going extinct globally. According to [87], species can be "Not Evaluated," "Data Deficient," "Least Concern," "Near Threatened," "Vulnerable," "Endangered," "Critically Endangered," "Extinct in the Wild," or "Extinct".

Nearly every nation with native species has its own conservation effort (Table 4). Three *Batagur* species of terrapin, *B. affinis*, *B. baska*, and *B. borneoensis*, are listed as having Critically Endangered (CR) status in Table 1. Moreover, *B. affinis* ssp. falls under the Extinct in the Wild (EW) category in Southeast Asian nations, including Indonesia, Singapore, Thailand, and Vietnam [23,25]. *B. affinis* ssp. is currently restricted to Malaysia and Cambodia. Ref. [89] also states that *B. baska* may be threatened in Thailand and Myanmar. Additionally, *B. borneoensis* was discovered in Brunei, Malaysia, and Indonesia, although it was virtually extinct in Thailand [90].

Table 4. Conservation centre records for *Batagur* sp. in indigenous species country.

Scientific Name	Common Name	Conservation Centre	Country
Batagur affinis	Southern River Terrapin	Sre Ambel River, Koh Kong Reptile Conservation Centre	Cambodia
Batagur affinis	Southern River Terrapin	Angkor Center for Conservation of Biodiversity	Cambodia
Batagur affinis	Southern River Terrapin	Kedah River, Kepala Batas, Kedah	Malaysia
Batagur affinis	Southern River Terrapin	Perak River, Bota Kanan, Perak	Malaysia
Batagur affinis	Southern River Terrapin	Terengganu River, Kuala Berang, Terengganu	Malaysia
Batagur affinis	Southern River Terrapin	Kemaman River, Kemaman, Terengganu	Malaysia
Batagur affinis	Southern River Terrapin	Setiu Wetlands State Park, Terengganu	Malaysia
Batagur baska	Northern River Terrapin	Vawal National Park	Bangladesh
Batagur baska	Northern River Terrapin	Sajnekhali Forest Station	India
Batagur borneoensis	Painted Terrapin	Langkat, North Sumatera	Indonesia
Batagur borneoensis	Painted Terrapin	Ujung Tamiang, Aceh	Indonesia
Batagur borneoensis	Painted Terrapin	Setiu Wetlands State Park, Terengganu	Malaysia
Batagur borneoensis	Painted Terrapin	Pengkalan Balak, Melaka	Malaysia

5. Conclusions

In conclusion, COI remains an effective barcode marker for terrapin species, contributing vital information that can be utilised to distinguish and identify genera and species. Compatibility with traditional taxonomy could provide a solid and dependable instrument for accurate species identification and biodiversity assessment facilitation. However, more markers and specimens from new sites should be added to the collection to more accurately compare terrapin populations. The detailed results provided fresh insights into the taxonomic classification of terrapins and revealed the existence of potential cryptic species. This investigation found compelling evidence of potential cryptic species between *B. baska* and *B. affinis affinis*. Our research shows that *B. affinis affinis* might be the same species as *B. baska*, but *B. affinis edwardmolli* might be its own species. However, further research is required. Therefore, the genomic and bioinformatics analysis of terrapins described here could serve as a reference for future global studies of this species and permit a more rational attempt to conserve terrapins. The proposed conservation units are based on the fact that phylogeny and phylogeography change over time and space.

Supplementary Materials: The following supporting information can be downloaded at: https://www.mdpi.com/article/10.3390/ani13111720/s1, Figure S1: The family of sampled terrapins; Figure S2: The within-species distribution is normalised to reduce bias in sampling at the species level. This distribution is shown in the table below, and the histogram compares the distribution of normalised divergences between species (blue) and genera (red); Table S1: Terrapin COI sequence composition (from 26 samples); Table S2: DNA barcodes for terrapins based on pure diagnostic characters at selected nucleotide positions. The transition site was highlighted in yellow; Table S3: The number of terrapin sequences, species, genera, and families is analysed in the present study.

Author Contributions: Conceptualisation, M.H.M.S., Y.E. and R.M.; methodology, M.H.M.S. and Y.E.; software, M.H.M.S.; validation, Y.E. and R.M.; formal analysis, M.H.M.S.; investigation, M.H.M.S.; resources, Y.E. and R.M.; data curation, Y.E. and R.M.; writing—original draught preparation, M.H.M.S.; writing—review and editing, M.H.M.S. and Y.E.; visualisation, R.M.; supervision, Y.E.; project administration, Y.E.; funding acquisition, Y.E.; All authors have read and agreed to the published version of the manuscript.

Funding: Article Processing Charge (APC) was funded by the Publication Fund, Research Management Centre (RMC), Universiti Putra Malaysia.

Institutional Review Board Statement: The study was conducted in accordance with the rules and with permission from the Department of Wildlife and National Parks, Peninsular Malaysia (B-00335-16-20).

Informed Consent Statement: Not applicable.

Data Availability Statement: The data presented is authentic and was not improperly selected, manipulated, enhanced, or fabricated. The data can be found in GeneBank (accession numbers OL658844-OL658847) and BOLD Systems (Sequence IDs HYT001-21 to HYT004-21).

Acknowledgments: We would like to express our gratitude to the Breeding Genetics Laboratory, Department of Aquaculture, Faculty of Agriculture, Universiti Putra Malaysia, for the facilities and support of this research project. Furthermore, we thank the Turtle Conservation Society of Malaysia and the Department of Wildlife and National Parks, Peninsular Malaysia, for their collaboration and permitted (B-00335-16-20). Last, but not least, we thank the anonymous reviewers for their insights and suggestions to improve this paper.

Conflicts of Interest: The authors declare no conflict of interest.

References

1. Jualaong, S.; Songnui, A.; Thongprajukaew, K.; Ninwat, S.; Khwanmaung, S.; Hahor, W.; Khunsaeng, P.; Kanghae, H. Optimal salinity for head-starting northern river terrapins (*Batagur baska* Gray, 1831). *Animals* **2019**, *9*, 855. [CrossRef]
2. Sayers, I.; Kubiak, M. Chapter 20: Terrapins. In *Handbook of Exotic Pet Medicine*; Kubiak, M.B., Ed.; John Wiley & Sons Ltd.: Hoboken, NJ, USA, 2020; pp. 387–413. [CrossRef]
3. Turtle Owner. *What Are Terrapins? (General Info, Pictures and Care Guide)*. Available online: https://turtleowner.com/what-are-terrapins-general-info-pictures-and-care-guide/ (accessed on 1 January 2022).
4. Vargas, S.M.; Araújo, F.C.; Santos, F.R. DNA barcoding of Brazilian sea turtles (Testudines). *Genet. Mol. Biol.* **2009**, *32*, 608–612. [CrossRef] [PubMed]
5. Hebert, P.D.N.; Stoeckle, M.Y.; Zemlak, T.S.; Francis, C.M. Identification of Birds through DNA Barcodes. *PLoS Biol.* **2004**, *2*, e312. [CrossRef] [PubMed]
6. Vilaça, S.T.; Lacerda, D.R.; Sari, H.E.R.; Santos, F.R. DNA based identification to Thamnophilidae (Passeriformes) species: The first barcodes of Neotropical birds. *Rev. Bras. Ornitol.* **2006**, *14*, 7–13.
7. Clare, E.L.; Lim, B.K.; Engstrom, M.D.; Eger, J.L.; Hebert, P.D. DNA barcoding of Neotropical bats: Species identification and discovery within Guyana. *Mol. Ecol. Notes* **2007**, *7*, 184–190. [CrossRef]
8. Chaves, A.V.; Clozato, C.L.; Lacerda, D.R.; Sari, E.H.R.; Santos, F.R. Molecular taxonomy of Brazilian tyrant-flycatchers (Passeriformes, Tyrannidae). *Mol. Ecol. Resour.* **2008**, *8*, 1169–1177. [CrossRef]
9. Naro-Maciel, E.; Reid, B.; Fitzsimmons, N.N.; Le, M.; Desalle, R.O.B.; Amato, G. DNA barcodes for globally threatened marine turtles: A registry approach to documenting biodiversity. *Mol. Ecol. Resour.* **2010**, *10*, 252–263. [CrossRef]
10. Alacs, E.A.; Jansen, F.J.; Scribner, K.T. Genetic issues in freshwater turtle and tortoise conservation. *Chelonian Res. Monogr.* **2007**, *4*, 107–123.
11. Kundu, S.; Kumar, V.; Laskar, B.A.; Tyagi, K.; Chandra, K. Pet and turtle: DNA barcoding identified twelve Geoemydid species in northeast India. *Mitochondrial DNA Part B* **2018**, *3*, 513–518. [CrossRef]
12. Hebert, P.D.N.; Cywinska, A.; Ball, S.L.; de Waard, J.R. Biological identifications through DNA barcodes. *Proc. R. Soc. London Ser. B Biol. Sci.* **2003**, *270*, 313–322. [CrossRef]
13. Bertolazzi, P.; Felici, G.; Weitschek, E. Learning to classify species with barcodes. *BMC Bioinform.* **2009**, *10* (Suppl. 14), S7. [CrossRef] [PubMed]

14. Salleh, M.H.M.; Esa, Y. The mtDNA D-loop Marker Identifies the Genetic Variability of Indochina's *Batagur affinis*. In *1st Postgraduate Seminar on Agriculture and Forestry 2021 (PSAF 2021)*; Universiti Putra Malaysia: Bintulu, Malaysia, 2021.
15. Cantor, T. Catalogue of reptiles inhabiting the Malayan peninsula and islands. *J. Asiat. Soc. Bengal* **1847**, *16*, 607–656, 897–952, 1026–1078.
16. Mohd Salleh, M.H.; Esa, Y.; Salleh, S.M.; Mohd Sah, S.A. Turtles in Malaysia: A Review of Conservation Status and a Call for Research. *Animals* **2022**, *12*, 2184. [CrossRef]
17. Buhlmann, K.A.; Dijk, P.P.V.; Iverson, J.B.; Mittermeier, R.A.; Pritchard, P.C.H.; Rhodin, A.G.; Saumure, R.A. Conservation Biology of Freshwater Turtles and Tortoises: A Compilation Project of the IUCN/SSC Tortoise and Freshwater Turtle Specialist Group, IUCN: International Union for Conservation of Nature. Available online: https://iucn-tftsg.org/cbftt/ (accessed on 3 April 2022).
18. Praschag, P.; Hundsdörfer, A.K.; Fritz, U. Phylogeny and taxonomy of endangered South and Southeast Asian freshwater turtles elucidated by mtDNA sequence variation (Testudines: Geoemydidae: *Batagur*, *Callagur*, *Hardella*, *Kachuga*, *Pangshura*). *Zool. Scr.* **2007**, *36*, 429–442. [CrossRef]
19. Turtle Conservation Coalition. *Turtles in Trouble: The World's 25+ Most Endangered Tortoises and Freshwater Turtles—2011*; Rhodin, A.G.J., Walde, A.D., Horne, B.D., van Dijk, P.P., Blanck, T., Hudson, R., Eds.; IUCN/SSC Tortoise and Freshwater Turtle Specialist Group, Turtle Conservation Fund, Turtle Survival Alliance, Turtle Conservancy, Chelonian Research Foundation, Conservation International, Wildlife Conservation Society, and San Diego Zoo Global: Lunenburg, MA, USA, 2011; p. 54.
20. Moll, E.O. Natural history of the River Terrapin, *Batagur baska* (Gray) in Malaysia (Testudines: Emydidae). *Malays. J. Sci.* **1980**, *6*, 23–62.
21. Platt, S.G.; Stuart, B.L.; Sovannara, H.; Kheng, L.; Kimchhay, H. Rediscovery of the critically endangered river terrapin, *Batagur baska*, in Cambodia, with notes on occurrence, reproduction, and conservation status. *Chelonian Conserv. Biol.* **2003**, *4*, 691–694.
22. Kalyar, J.; Thorbjarnarson, K.; Thirakhupt. An overview of the current population and conservation status of the Critically Endangered River Terrapin, *Batagur baska* (Gray, 1831) in Myanmar, Thailand and Malaysia. *Nat. Hist. J. Chulalongkorn Univ.* **2007**, *7*, 51–65.
23. Moll, E.O.; Platt, S.G.; Chan, E.H.; Horne, B.D.; Platt, K.; Praschag, P.; Chen, P.N.; Van Dijk, P.P. *Batagur affinis* (Cantor 1847) Southern river terrapin, tuntong. *Chelonian Res. Monogr.* **2015**, *5*, 090.1–090.17. [CrossRef]
24. Chen, P.N. Conservation of the Southern River Terrapin *Batagur affinis* (Reptilia: Testudines: Geoemydidae) in Malaysia: A case study involving local community participation. *J. Threat. Taxa* **2017**, *9*, 10035–10046. [CrossRef]
25. Mistar, S.A.J.; Singleton, I. *Presence and Distribution of the Southern River Terrapin Batagur Affinis and Painted Terrapin Batagur Borneoensis in Eastern Coast of Sumatra*; Unpublished Report to Auckland Zoo; Auckland Zoo: Auckland, New Zealand, 2012; p. 25.
26. Salleh, M.H.M.; Esa, Y. Conservation Status of Globally Testudines Terrapins Based on COI Mitochondrial Markers. In Proceedings of the MOL2NET'22, Conference on Molecular, Biomedical & Computational Sciences and Engineering, Biscay, Spain, 25–30 December 2022; MDPI: Basel, Switzerland, 2022. [CrossRef]
27. Çilingir, F.G.; Seah, A.; Horne, B.D.; Som, S.; Bickford, D.P.; Rheindt, F.E. Last exit before the brink: Conservation genomics of the Cambodian population of the critically endangered Southern river terrapin. *Ecol. Evol.* **2019**, *9*, 9500–9510. [CrossRef]
28. Praschag, P.; Sommer, R.S.; Mccarthy, C.; Gemel, R. Naming one of the world's rarest chelonians, the southern *Batagur*. *Zootaxa* **2008**, *1758*, 61–68. [CrossRef]
29. Salleh, M.H.M.; Esa, Y. Minimally invasive blood collection techniques as a source of gDNA for genetic studies on turtles and tortoises. *Int. J. Aquat. Biol.* **2022**, *10*, 145–150.
30. Guntoro, J.; Riyanto, A. The very low genetic variability on Aceh Tamiang's (Indonesia) population of Painted Terrapin (*Batagur borneoensis*) inferred by cytochrome oxidase I (CO I) and D-loop (control region). *Biodivers. J. Biol. Divers.* **2020**, *21*, 6. [CrossRef]
31. Kundu, S.; Das, K.C.; Ghosh, S.K. Taxonomic rank of Indian tortoise: Revisit with DNA barcoding perspective. *DNA Barcodes* **2013**, *1*, 39–45. [CrossRef]
32. Reid, B.N.; LE, M.; McCord, W.P.; Iverson, J.B.; Georges, A.; Bergmann, T.; Amato, G.; Desalle, R.; NaroMaciel, E. Comparing and combining distance-based and character-based approaches for barcoding turtles. *Mol. Ecol. Resour.* **2011**, *11*, 956–967. [CrossRef]
33. Ratnasingham, S.; Hebert, P.D. BOLD: The Barcode of Life Data System. *Mol. Ecol. Notes* **2007**, *7*, 355–364. [CrossRef]
34. Spinks, P.Q.; Thomson, R.C.; McCartney-Melstad, E.; Shaffer, H.B. Phylogeny and temporal diversification of the New World pond turtles (Emydidae). *Mol. Phylogenet. Evol.* **2016**, *103*, 85–97. [CrossRef] [PubMed]
35. Hawlitschek, O.; Morinière, J.; Dunz, A.; Franzen, M.; Rödder, D.; Glaw, F.; Haszprunar, G. Comprehensive DNA barcoding of the herpetofauna of Germany. *Mol. Ecol. Resour.* **2016**, *16*, 242–253. [CrossRef]
36. Yu, D.; Fang, X.; Storey, K.B.; Zhang, Y.; Zhang, J. Complete mitochondrial genomes of the yellow-bellied slider turtle *Trachemys scripta scripta* and anoxia tolerant red-eared slider *Trachemys scripta elegans*. *Mitochondrial DNA Part A* **2016**, *27*, 2276–2277. [CrossRef]
37. Feldman, C.R.; Parham, J.F. Molecular systematics of Old World stripe-necked turtles (Testudines: *Mauremys*). *Asiat. Herpetol. Res.* **2004**, *10*, 28–37.
38. Morgulis, A.; Coulouris, G.; Raytselis, Y.; Madden, T.L.; Agarwala, R.; Schäffer, A.A. Database indexing for production MegaBLAST searches. *Bioinformatics* **2008**, *24*, 1757–1764. [CrossRef]
39. Altschul, S.F.; Gish, W.; Miller, W.; Myers, E.W.; Lipman, D.J. Basic local alignment search tool. *J. Mol. Biol.* **1990**, *215*, 403–410. [CrossRef]
40. Kimura, M. A simple method for estimating evolutionary rates of base substitutions through comparative studies of nucleotide sequences. *J. Mol. Evol.* **1980**, *16*, 111–120. [CrossRef]

41. Kumar, S.; Stecher, G.; Li, M.; Knyaz, C.; Tamura, K. MEGA X: Molecular evolutionary genetics analysis across computing platforms. *Mol. Biol. Evol.* **2018**, *35*, 1547. [CrossRef] [PubMed]
42. Darriba, D.; Taboada, G.L.; Doallo, R.; Posada, D. jModelTest 2: More models, new heuristics and parallel computing. *Nat. Methods* **2012**, *9*, 772. [CrossRef] [PubMed]
43. Felsenstein, J. Evolutionary trees from DNA sequences: A maximum likelihood approach. *J. Mol. Evol.* **1981**, *17*, 368–376. [CrossRef]
44. Waterhouse, A.M.; Procter, J.B.; Martin, D.M.A.; Clamp, M.; Barton, G.J. Jalview Version 2—A multiple sequence alignment editor and analysis workbench. *Bioinformatics* **2009**, *25*, 1189–1191. [CrossRef] [PubMed]
45. Procter, J.B.; Carstairs, G.; Soares, B.; Mourão, K.; Ofoegbu, T.C.; Barton, D.; Lui, L.; Menard, A.; Sherstnev, N.; Roldan-Martinez, D.; et al. Alignment of Biological Sequences with Jalview. In *Multiple Sequence Alignment—Methods in Molecular Biology*; Katoh, K., Ed.; Springer: New York, NY, USA, 2021; Volume 2231, pp. 203–224. [CrossRef]
46. Nguyen, L.T.; Schmidt, H.A.; von Haeseler, A.; Minh, B.Q. IQ-TREE: A fast and effffective stochastic algorithm for estimating maximum-likelihood phylogenies. *Mol. Biol. Evol.* **2015**, *32*, 268–274. [CrossRef] [PubMed]
47. Towns, J.; Cockerill, T.; Dahan, M.; Foster, I.; Gaither, K.; Grimshaw, A.; Hazlewood, V.; Lathrop, S.; Lifka, D.; Peterson, G.D.; et al. XSEDE: Accelerating Scientific Discovery. *Comput. Sci. Eng.* **2014**, *16*, 62–74. [CrossRef]
48. Miller, M.A.; Pfeiffer, W.; Schwartz, T. The CIPRES science gateway: A community resource for phylogenetic analyses. In Proceedings of the 2011 TeraGrid Conference: Extreme Digital Discovery, Salt Lake City, UT, USA, 18–21 July 2011. [CrossRef]
49. Rambaut, A. FigTree v1.4.5, a Graphical Viewer of Phylogenetic Trees. Available online: http://tree.bio.ed.ac.uk/software/figtree/ (accessed on 13 August 2022).
50. Drummond, A.J.; Ho, S.Y.W.; Phillips, M.J.; Rambaut, A. Relaxed phylogenetics and dating with confidence. *PLoS Biol.* **2006**, *4*, e88. [CrossRef]
51. Drummond, A.J.; Rambaut, A. BEAST: Bayesian evolutionary analysis by sampling trees. *BMC Evol. Biol.* **2007**, *7*, 214. [CrossRef] [PubMed]
52. Bilderbeek, R.J.; Etienne, R.S. babette: BEAUti 2, BEAST2 and Tracer for R. *Methods Ecol. Evol.* **2018**, *9*, 2034–2040. [CrossRef]
53. Zhou, C.; Gan, S.; Zhang, J.; Fan, Y.; Li, B.; Wan, L.; Nie, J.; Wang, X.; Chen, J. Application of DNA Barcoding for the Identification of Snake Gallbladders as a Traditional Chinese Medicine. *Rev. Bras. Farm.* **2022**, *32*, 663–668. [CrossRef]
54. Tajima, F. Statistical method for testing the neutral mutation hypothesis by DNA polymorphism. *Genetics* **1989**, *123*, 585–595. [CrossRef]
55. Tajima, F. The amount of DNA polymorphism maintained in a finite population when the neutral mutation rate varies among sites. *Genetics* **1996**, *143*, 1457–1465. [CrossRef] [PubMed]
56. Rozas, J.; Ferrer-Mata, A.; Sánchez-DelBarrio, J.C.; Guirao-Rico, S.; Librado, P.; Ramos-Onsins, S.E.; Sánchez-Gracia, A. DnaSP 6: DNA sequence polymorphism analysis of large data sets. *Mol. Biol. Evol.* **2017**, *34*, 3299–3302. [CrossRef]
57. Bandelt, H.J.; Forster, P.; Rohl, A. Median-joining networks for inferring intraspecific phylogenies. *Mol. Biol. Evol.* **1999**, *16*, 37–48. [CrossRef]
58. Hebert, P.D.; Ratnasingham, S.; De Waard, J.R. Barcoding animal life: Cytochrome c oxidase subunit 1 divergences among closely related species. *Proc. R. Soc. London Ser. B Biol. Sci.* **2003**, *270*, S96–S99. [CrossRef]
59. Xia, X.; Xie, Z. DAMBE: Software package for data analysis in molecular biology and evolution. *J. Hered.* **2001**, *92*, 371–373. [CrossRef]
60. Ernst, C.H.; Lovich, J.E. *Turtles of the United States and Canada*; Smithsonian Institution Press: Washington, DC, USA, 1994; 578p.
61. Stuckas, H.; Gemel, R.; Fritz, U. One Extinct Turtle Species Less: *Pelusios seychellensis* Is Not Extinct, It Never Existed. *PLoS ONE* **2013**, *8*, e57116. [CrossRef]
62. Avise, J.C.; Bowen, B.W.; Lamb, T.; Meylan, A.B.; Bermingham, E. Mitochondrial DNA evolution at a turtle¢s pace: Evidence for low genetic variability and reduced microevolutionary rate in the testudines. *Mol. Biol. Evol.* **1992**, *9*, 457–473.
63. FitzSimmons, N.N.; Moritz, C.; Moore, S.S. Conservation and dynamics of microsatellite loci over 300 million years of marine turtle evolution. *Mol. Biol. Evol.* **1995**, *12*, 432–440.
64. Puillandre, N.; Lambert, A.; Brouillet, S.; Achaz, G. ABGD, automatic barcode gap discovery for primary species delimitation. *Mol. Ecol.* **2012**, *21*, 1864–1877. [CrossRef] [PubMed]
65. Trivedi, S.; Aloufifi, A.A.; Sankar, A.A.; Ghosh, S.K. Role of DNA barcoding in marine biodiversity. *Saudi J. Biol. Sci.* **2016**, *23*, 161–171. [CrossRef] [PubMed]
66. Zhang, Z.P.; Wang, X.Y.; Zhang, Z.; Yao, H.; Zhang, X.M.; Zhang, Y.; Zhang, B.G. The impact of genetic diversity on the accuracy of DNA barcoding to identify species: A study on the genus *Phellodendron*. *Ecol. Evol.* **2019**, *9*, 10723–10733. [CrossRef] [PubMed]
67. Meier, R.; Zhang, G.; Ali, F. The use of mean instead of smallest interspecifc distances exaggerates the size of the "barcoding gap" and leads to misidentifcation. *Syst. Biol.* **2008**, *57*, 809–813. [CrossRef]
68. Abidin, D.H.Z.; Nor, S.A.M.; Lavoué, S.; Rahim, M.A.; Jamaludin, N.A.; Akib, N.A.M. DNA-based taxonomy of a mangrove-associated community of fishes in Southeast Asia. *Sci. Rep.* **2021**, *11*, 17800. [CrossRef] [PubMed]
69. Hubert, N.; Meyer, C.P.; Bruggemann, H.J.; Guerin, F.; Komeno, R.J.; Espiau, B.; Causse, R.; Williams, J.T.; Planes, S. Cryptic diversity in Indo-Pacific coral-reef fishes revealed by DNA-barcoding provides new support to the centre-of-overlap hypothesis. *PLoS ONE* **2012**, *7*, e28987. [CrossRef]

70. Ward, R.D.; Zemlak, T.S.; Innes, B.H.; Last, P.R.; Hebert, P.D. DNA barcoding Australia's fish species. *Philos. Trans. R. Soc. B Biol. Sci.* **2005**, *360*, 1847–1857. [CrossRef]
71. Xu, L.; Wang, X.; Van Damme, K.; Huang, D.; Li, Y.; Wang, L.; Ning, J.; Du, F. Assessment of fish diversity in the South China Sea using DNA taxonomy. *Fish. Res.* **2020**, *233*, 105771. [CrossRef]
72. Lakra, W.S.; Verma, M.S.; Goswami, M.; Lal, K.K.; Mohindra, V.; Punia, P.; Gopalakrishnan, A.; Singh, K.V.; Ward, R.D.; Hebert, P. DNA barcoding Indian marine fishes. *Mol. Ecol. Resour.* **2011**, *11*, 60–71. [CrossRef] [PubMed]
73. Landi, M.; Dimech, M.; Arculeo, M.; Biondo, G.; Martins, R.; Carneiro, M.; Carvalho, G.R.; Brutto, S.L.; Costa, F.O. DNA barcoding for species assignment: The case of Mediterranean marine fishes. *PLoS ONE* **2014**, *9*, e106135. [CrossRef]
74. Mecklenburg, C.W.; Møller, P.R.; Steinke, D. Biodiversity of arctic marine fishes: Taxonomy and zoogeography. *Mar. Biodivers.* **2011**, *41*, 109–140. [CrossRef]
75. Praschag, P.; Holloway, R.; Georges, A.; Paeckert, M.; Hundsdoerfer, A.K.; Fritz, U.A. New subspecies of *Batagur affinis* (Cantor, 1847), one of the world's most critically endangered chelonians (Testudines: Geoemydidae). *Zootaxa* **2009**, *2233*, 57–68. [CrossRef]
76. Komoroske, L.M.; Jensen, M.P.; Stewart, K.R.; Shamblin, B.M.; Dutton, P.H. Advances in the application of genetics in marine turtle biology and conservation. *Front. Mar. Sci.* **2017**, *4*, 156. [CrossRef]
77. Jensen, M.P.; FitzSimmons, N.N.; Bourjea, J.; Hamabata, T.; Reece, J.; Dutton, P.H. The evolutionary history and global phylogeography of the green turtle (*Chelonia mydas*). *J. Biogeogr.* **2019**, *46*, 860–870. [CrossRef]
78. Fišer, C.; Zagmajster, M. Cryptic species from cryptic space: The case of *Niphargus fongi* sp. n. (Amphipoda, Niphargidae). *Crustaceana* **2009**, *82*, 593–614. [CrossRef]
79. Schlick-Steiner, B.C.; Steiner, F.M.; Seifert, B.; Stauffer, C.; Christian, E.; Crozier, R.H. Integrative Taxonomy: A Multisource Approach to Exploring Biodiversity. *Annu. Rev. Entomol.* **2010**, *55*, 421–438. [CrossRef]
80. Jugovic, J.; Jalžić, B.; Prevorčnik, S.; Sket, B. Cave shrimps *Troglocaris* s. str. (Dormitzer, 1853), taxonomic revision and description of new taxa after phylogenetic and morphometric studies. *Zootaxa* **2012**, *3421*, 1–31. [CrossRef]
81. Adams, M.; Raadik, T.A.; Burridge, C.P.; Georges, A. Global biodiversity assessment and hyper -cryptic species complexes: More than one species of elephant in the room? *Syst. Biol.* **2014**, *63*, 518–533. [CrossRef]
82. Fišer, C.; Robinson, C.T.; Malard, F. Cryptic species as a window into the paradigm shift of the species concept. *Mol. Ecol.* **2018**, *27*, 613–635. [CrossRef] [PubMed]
83. de León, G.P.P.; Poulin, R. Taxonomic distribution of cryptic diversity among metazoans: Not so homogeneous after all. *Biol. Lett.* **2016**, *12*, 20160371. [CrossRef] [PubMed]
84. Pfenninger, M.; Schwenk, K. Cryptic animal species are homogeneously distributed among taxa and biogeographical regions. *BMC Evol. Biol.* **2007**, *7*, 121. [CrossRef]
85. Knowlton, N. Molecular genetic analyses of species boundaries in the sea. *Hydrobiologia* **2000**, *420*, 73–90. [CrossRef]
86. Dawson, M.N.; Jacobs, D.K. Molecular evidence for cryptic species of *Aurelia aurita* (Cnidaria, Scyphozoa). *Biol. Bull.* **2001**, *200*, 92–96. [CrossRef] [PubMed]
87. IUCN. *IUCN Red List Categories and Criteria: Versian 3.1*; IUCN Species Survival Commission: Gland, Switzerland; IUCN: Cambridge, UK, 2001; p. 30.
88. Vié, J.C.; Hilton-Taylor, C.; Pollock, C.; Ragle, J.; Smart, J.; Stuart, S.N.; Tong, R. The IUCN Red List: A Key Conservation Tool. In *Wildlife in a Changing World—An Analysis of the 2008 IUCN Red List of Threatened, Species*; Vié, J.C., Hilton-Taylor, C., Stuart, S.N., Eds.; IUCN: Gland, Switzerland, 2009; p. 13.
89. Praschag, P.; Singh, S. Batagur Baska. In *The IUCN Red List of Threatened Species*; IUCN: Cambridge, UK, 2019. [CrossRef]
90. Jenkins, M.D. *Tortoises and Freshwater Turtles: The Trade in Southeast Asia*; TRAFFIC International: Cambridge, UK, 1995; p. 48.

Article

Concerted and Independent Evolution of Control Regions 1 and 2 of Water Monitor Lizards (*Varanus salvator macromaculatus*) and Different Phylogenetic Informative Markers

Watcharaporn Thapana [1,2,†], Nattakan Ariyaraphong [1,2,3,†], Parinya Wongtienchai [2,†], Nararat Laopichienpong [1,3], Worapong Singchat [1,3], Thitipong Panthum [1,3], Syed Farhan Ahmad [1,3], Ekaphan Kraichak [4], Narongrit Muangmai [5], Prateep Duengkae [1,3] and Kornsorn Srikulnath [1,2,3,6,*]

1 Animal Genomics and Bioresource Research Center (AGB Research Center), Faculty of Science, Kasetsart University, 50 Ngamwongwan, Bangkok 10900, Thailand; golf_foureye@hotmail.com (W.T.); nattakan.ari@ku.th (N.A.); nararat.l@ku.th (N.L.); fsciwos@ku.ac.th (W.S.); thitipong.pa@ku.th (T.P.); syedfarhan.a@ku.th (S.F.A.); fforptd@ku.ac.th (P.D.)
2 Laboratory of Animal Cytogenetics and Comparative Genomics (ACCG), Department of Genetics, Faculty of Science, Kasetsart University, 50 Ngamwongwan, Bangkok 10900, Thailand; parinya_wongtienchai@hotmail.com
3 Special Research Unit for Wildlife Genomics (SRUWG), Department of Forest Biology, Faculty of Forestry, Kasetsart University, 50 Ngamwongwan, Bangkok 10900, Thailand
4 Department of Botany, Faculty of Science, Kasetsart University, 50 Ngamwongwan, Bangkok 10900, Thailand; ekaphan.k@ku.th
5 Department of Fishery Biology, Faculty of Fisheries, Kasetsart University, 50 Ngamwongwan, Bangkok 10900, Thailand; ffisnrm@ku.ac.th
6 Amphibian Research Center, Hiroshima University, 1-3-1, Kagamiyama, Higashihiroshima 739-8526, Japan
* Correspondence: kornsorn.s@ku.ac.th; Tel.: +662-562-5555; Fax: +662-942-8290
† These authors contributed equally to this work.

Citation: Thapana, W.; Ariyaraphong, N.; Wongtienchai, P.; Laopichienpong, N.; Singchat, W.; Panthum, T.; Ahmad, S.F.; Kraichak, E.; Muangmai, N.; Duengkae, P.; et al. Concerted and Independent Evolution of Control Regions 1 and 2 of Water Monitor Lizards (*Varanus salvator macromaculatus*) and Different Phylogenetic Informative Markers. *Animals* **2022**, *12*, 148. https://doi.org/10.3390/ani12020148

Academic Editor: Ettore Olmo

Received: 13 December 2021
Accepted: 5 January 2022
Published: 8 January 2022

Simple Summary: The evolutionary patterns and phylogenetic utility of duplicate control regions (CRs) in 72 individuals of *Varanus salvator macromaculatus* and other varanids have been observed. Divergence of the two CRs from each individual revealed a pattern of independent evolution in CRs of varanid lineage. This study is a first step towards developing new phylogenetic evolutionary models of the varanid lineage, with accurate evolutionary inferences to provide basic insights into the biology of mitogenomes.

Abstract: Duplicate control regions (CRs) have been observed in the mitochondrial genomes (mitogenomes) of most varanids. Duplicate CRs have evolved in either concerted or independent evolution in vertebrates, but whether an evolutionary pattern exists in varanids remains unknown. Therefore, we conducted this study to analyze the evolutionary patterns and phylogenetic utilities of duplicate CRs in 72 individuals of *Varanus salvator macromaculatus* and other varanids. Sequence analyses and phylogenetic relationships revealed that divergence between orthologous copies from different individuals was lower than in paralogous copies from the same individual, suggesting an independent evolution of the two CRs. Distinct trees and recombination testing derived from CR1 and CR2 suggested that recombination events occurred between CRs during the evolutionary process. A comparison of substitution saturation showed the potential of CR2 as a phylogenetic marker. By contrast, duplicate CRs of the four examined varanids had similar sequences within species, suggesting typical characteristics of concerted evolution. The results provide a better understanding of the molecular evolutionary processes related to the mitogenomes of the varanid lineage.

Keywords: varanid; control region; ortholog; paralog

1. Introduction

The mitochondrial control region (mtCR) is a major noncoding segment of the vertebrate mitochondrial genome (mitogenome). The region includes the displacement loop (D-loop), which comprises the third strand of DNA, thus creating a semi-stable structure [1]. The mtCR plays an important role in transcriptional and translational regulation of protein-coding sequences, or it serves as the origin of DNA replication [2]. The nucleotide CR sequence is the most rapidly evolving region of the mitogenome, and it lacks coding sequences; thus, it is widely used as a molecular marker in population genetics, phylogenetic studies, and phylogeographic studies [3–5]. Vertebrates such as birds, snakes, turtles, and fish exhibit segmental duplications within the CR or an entire duplication of the CR, leading to the formation of repeats or possible homogenization between the duplicated copies of CR [6–8].

GenBank contains 17,489 complete mitogenomes for squamate reptiles (as of April 2021, http://www.ncbi.nlm.nih.gov/genome), with several duplicate CRs observed in varanids and snakes [6,7]. A comparison between two CRs (CR1 and CR2) revealed identical or highly similar nucleotide sequences, similar to the concerted evolution as found in Bothidae and Samaridae [9]. By contrast, orthologous copies of duplicate CRs from different species such as in varanid, gecko lizard, and platysternid lineages, are genetically closer to each other than to paralogous copies of duplicate CRs (CR1 and CR2) within the same species [6]. This might be a result of the independent evolution of the two copies after an ancient duplication event, although the mechanism behind such an event is not clearly understood [8].

Varanids or monitor lizards comprise a single extant genus, *Varanus*, within the family Varanidae. To date, around 80 extant species have been described and distributed in Afro-Arabia, Western to Southeast Asia, the Indonesian Archipelago, Papua New Guinea, and Australia [10]. Mitogenomes of the Komodo dragon (*V. komodoensis*; Ouwens 1912) [11,12] and Nile monitor (*V. niloticus*; Linnaeus 1758) [13,14] have unique gene organization features. Genes between the NADH dehydrogenase subunit 6 (*ND6*) gene and proline tRNA gene are extensively shuffled, and the CR has been duplicated in an ancestral varanid lineage during the Paleocene age or earlier [15]. This is consistent with the Cenozoic over-water dispersal of Southeast Asian varanids, such as the water monitor (*V. salvator macromaculatus*; Deraniyagala 1944) [16] across the Indonesian Archipelago and Komodo dragon (*V. komodoensis*) [15]. The presence of duplicate CRs in varanid mitogenomes is an intriguing structural phenomenon and raises basic questions concerning how the nucleotide sequences of duplicate CRs remained similar over time. Variations in CRs at the population and species level in varanids have not been fully elucidated [5,17]. In light of this scenario, we propose two hypotheses: (1) orthologous copies of duplicate CRs in different individuals are genetically similar due to independent evolution, or (2) two CRs (CR1 and CR2) as paralogous copies exhibit identical or highly similar nucleotide sequences from concerted evolution. To characterize the variations in varanid CRs, we conducted this study to analyze the CR sequences of four varanids, namely *V. salvator* (*V. salvator macromaculatus* and *V. salvator komaini*), *V. exanthematicus*, *V. komodoensis* and *V. niloticus*, and 72 water monitors (*V. salvator macromaculatus*) (Table 1). We also compared and analyzed the sequence variations of mtCRs. These analyses have important implications in the selection of priority mitochondrial regions to assess the evolution and genetic diversity of varanid populations.

2. Materials and Methods

2.1. Specimen Collection and DNA Extraction

Blood specimens of water monitors (*V. salvator macromaculatus*) were collected from the ventral tail vein using a 23-gauge needle attached to a 2 mL disposable syringe containing 10 mM ethylenediaminetetraacetic acid for DNA extraction as previously reported by Wongtienchai et al. [5] (Supplementary Table S1). Samples were collected from 47 individuals at the Bang Kachao Peninsula, Samut Prakan, (13°59′2″ N, 99°59′38″ E) and from 25 individuals at Varanus Farm Kamphaeng Saen, Nakhon Pathom (14°00′59.9″ N, 99°57′46.8″ E). Permission was granted by the Sri Nakhon Khuean Khan Park (Royal Forest Department, Ministry of Natural Resources and Environment) and Kasetsart University (0909.6/15779). All experimental procedures involving animals conformed to the guidelines established by the Animal Care Committee of Kasetsart University, Thailand. Total genomic DNA was extracted according to the standard salting-out protocol, as previously described [18]. DNA quality and concentration were determined using 1% agarose gel electrophoresis and spectrophotometry.

2.2. CR1 Sequencing

The positions of duplicate CRs were determined from the locations of tRNA genes (tRNAPro and tRNAVal for CR1; tRNATyr and tRNASer for CR2) as the flanking regions of CR1 and CR2 in the mitogenome of *V. salvator* (GenBank accession number: EU747731). The CR1 fragments were amplified using the following primers: VSA_CR1 F (5′-ATTAATACCCAATTTTCCTTGCTC-3′) and VSA_CR1 R (5′-GCCCAGTGACCATTAATAT CAACT-3′), which were designed based on five varanid mtDNA sequences, namely *V. salvator macromaculatus* (GenBank accession number: AB980995), *V. salvator komaini* (GenBank accession number: AB980996), *Varanus exanthematicus* (GenBank accession number: AB738957), *Varanus komodoensis* (GenBank accession number: AB080276), and *Varanus niloticus* (GenBank accession number: AB185327). The positions of all primers were located in tRNA genes (tRNAPro and tRNAVal for CR1; tRNATyr and tRNASer for CR2) that are highly conserved along all varanus mitogenomes; therefore, entire sequences of both CRs were collected. Polymerase chain reaction (PCR) amplification was performed using 20 μL of 1× ThermoPol buffer containing 1.5 mM $MgCl_2$, 0.2 dNTPs, 5.0 μM primers, 0.5 U of *Taq* polymerase (Apsalagen Co., Ltd., Bangkok, Thailand), and 25 ng of genomic DNA. The PCR conditions were as follows: initial denaturation at 94 °C for 3 min, followed by 35 cycles at 94 °C for 30 s, 57 °C for 30 s, and 72 °C for 40 s, and a final extension at 72 °C for 5 min [5]. The PCR products were separated via electrophoresis on 1% agarose gels, and they were then cloned using the pGEM®-T Easy vector (Promega Corporation, Madison, WI, USA). Nucleotide sequences of DNA fragments were determined using the DNA sequencing service of First Base Laboratories Sdn Bhd (Seri Kembangan, Selangor, Malaysia). BLASTn programs (http://blast.ncbi.nlm.nih.gov/Blast.cgi) were used to search nucleotide sequences in the National Center for Biotechnology Information database to confirm the identities of amplified DNA fragments. The generated sequences were deposited in the DNA Data Bank of Japan. The mitochondrial CR2 dataset used in our previous study was retrieved from the database [5].

2.3. Positional Annotation

Three functional regions, including the terminal-associated sequence (TAS), central conserved domain (CD), and conserved sequence blocks (CSB), were tentatively investigated in both CRs by recognizing sequences similar to those found in other vertebrates [3,6,12]. Variable number of tandem repeats (VNTRs) have been reported to exist only at the 3′ end of both CRs [19,20]. Tandem repeat sequences, including the motif, length of repeats, and copy number in the CR region, were investigated using the Tandem Repeats Finder 4.09 program [21].

2.4. Comparison of Genetic Variability Based on CR Sequences at the Population Level

Multiple sequence alignment was performed for 72 sequences in both CRs using the default parameters of Molecular Evolutionary Genetics Analysis X (MEGAX) software (Center for Evolutionary Functional Genomics, The Biodesign Institute, Tempe, PA, USA; [22]). Estimates of haplotype (h), nucleotide (π) diversity [23], and number of haplotypes (H) were calculated based on CR1 and CR2 sequences, as implemented in DnaSP version 6 [24]. A statistical parsimony network of consensus sequences was constructed using the Templeton, Crandall, and Sing (TCS) algorithm implemented in PopART version 1.7. to address haplotype grouping [25]. The mitochondrial CR2 dataset employed in our previous study [5] was used for all analyses, similar to the CR1 dataset. The means and standard deviations of h and π diversity of both CRs were used to calculate t-statistics and p-values for two-sample t-test comparisons, following the formula in [26] in R version 4.0.3 [27].

2.5. Phylogenetic Analysis Based on Mitochondrial CR1 and CR2 Sequences at the Species Level

Substitution saturation decreases the amount of phylogenetic signal to the point that sequence similarities could be a result of chance alone rather than homology. Consequently, when saturation is achieved, the phylogenetic signal is lost, and the sequences no longer reveal the underlying evolutionary mechanisms [28]. The saturation of substitutions was evaluated by plotting the number of transitions (s) and transversions (v) against the K80 [29] sequence divergences as well as by comparing the information entropy-based index (I_{ss}) with critical values ($I_{ss.c}$) [30,31], as implemented in DAMBE7 [32]. If I_{ss} is significantly lower than $I_{ss.c}$, the sequences do not experience substitution saturation. Phylogenetic analyses were performed using CR1 and CR2 datasets with the maximum likelihood (ML) reconstructed in IQ-TREE [33] using a model finder with the options TEST and –AICc, a tree search with 1000 bootstrap replicates, and Bayesian inference (BI) with MrBayes version 3.2.6 [34]. The GenBank database of four varanids is shown in Table 1 as of April 2021. The best-fit model of DNA substitution was determined for each CR using Kakusan4 [35]. The Markov chain Monte Carlo process was used to simultaneously run four chains for one million generations. After stabilization of the log-likelihood value, a sampling procedure was performed every 100 generations to obtain 10,000 trees, from which a majority-rule consensus tree with average branch lengths was generated. All sample points were discarded before attaining convergence as burn-in, and the Bayesian posterior probability in the sampled tree population was calculated as a percentage. The genetic distances of p-distance between CR sequences were calculated using the MEGAX program [22].

Table 1. Species used with accession numbers.

Species	GenBank Accession Number	CRs		Reference
		CR1	**CR2**	
Varanus salvator macromaculatus	LC326253-LC326324	CR1	-	This study
Varanus salvator macromaculatus	LC326325-LC326396	-	CR2	Wongtienchai et al. [5]
Varanus salvator	EU747731	CR1	CR2	Castoe et al. [36]
Varanus salvator macromaculatus	AB980995	CR1	CR2	Chaiprasertsri et al. [37]
Varanus salvator komaini	AB980996	CR1	CR2	Chaiprasertsri et al. [37]
Varanus exanthematicus	AB738957	CR1	CR2	-
Varanus komodoensis	AB080276	CR1	CR2	Kumazawa and Endo [11]
Varanus niloticus	AB185327	CR1	CR2	Kumazawa [13]

2.6. Recombination Testing

Discordant evolutionary signals were detected when the phylogenetic trees were separately reconstructed from different regions. These conflicting signals are due to the recombination of duplicate CRs. To further analyze these signals, the following recombination tests were conducted for both the CRs: (1) Recombination Detection Program (RDP) [38], (2) Geneconv [39], (3) Maxchi [40], and (4) Chimaera [41]. These analyses were performed using the Recombination Detection Program, RDP5 [42], with previously described parameters [43]. All analyses were performed for all individuals to check for recombination occurrence in both CRs of water monitors.

3. Results

3.1. Positional Annotation in the Control Regions of *V. salvator macromaculatus*

Three conserved functional sections, including the TAS, CD, and CSB domains, were analyzed in both CR1 and CR2 of all the 72 individuals; however, no CD was observed in either CR. The TAS domain contained 78 bp for CR1 and 80 bp for CR2 in water monitor lizards. The conserved nucleotide sequence of TAS between CR1 and CR2 was 5′-TAGTT-3′. The CSB domain contained three conserved blocks, namely CSB-1 (5′-TTAATGGTCDCNGGRHAT -3′), CSB-2 (5′-DHWDBYMYNYHHDCYYYC -3′), and CSB-3 (5′-GCYHWDYRKTYAHMMAA-3′) for CR1, and CSB-1 (5′-TTCATYWYHAWWWWTTBDN-3′), CSB-2 (5′-WWWWYCMYYWWHYYYY -3′), and CSB-3 (5′-GCYHWWYRKTYAHMAA -3′) for CR2. VNTRs were identified in CR1 (TCGCGCCACCTCCAGGATT), with two copies for one individual only, and CR2, (TTTTTTAAAAAAATTTTTTAT), (AAAAAAATTTTTTTA), (TTAAAAAAAATTTTTT), and (AAAAAAATTTTTTATTTTTTTTAA), ranging from to 2 to 4 copies in all individuals (Figure 1).

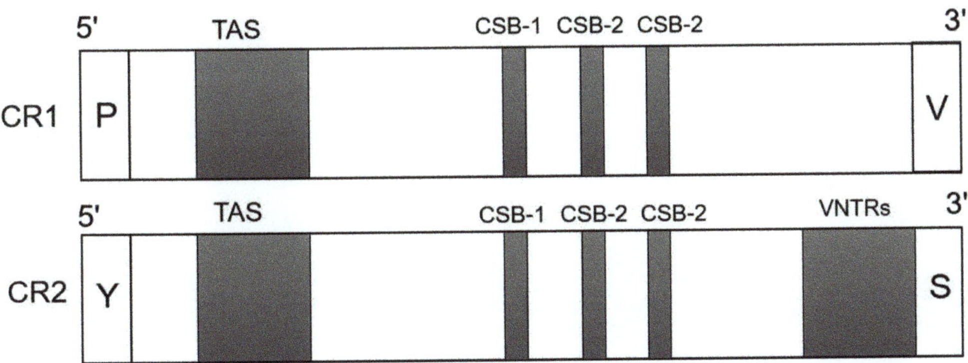

Figure 1. Structures of duplicate control regions (CRs) in all individuals of *Varanus salvator macromaculatus* (Deraniyagala 1944 [16]) in this study. Two functional regions, TAS and CSB, were detected in both the CRs of all individuals. The core sequences of these regions were found to be identical in both CR1 and CR2. Variable numbers of tandem repeats were detected only in CR2.

3.2. Sequence Variation in the CRs of *V. salvator macromaculatus*

The alignment lengths of CR1 and CR2 sequences were 663 and 867 bp, respectively. The number of haplotypes in CR1 was 44, and the number of haplotypes in CR2 was 52. The overall haplotype and nucleotide diversities were 0.935 ± 0.019 and 0.004 ± 0.001 for CR1 and 0.968 ± 0.013 and 0.004 ± 0.001 for CR2. Results of the *t*-test showed that the means of h and π significantly differed between CRs (t = 3.974, df = 141.77, p-value < 0.01 for h value and t = 6.8043, df = 127.83, p-value < 0.01 for π value). Meanwhile, results of the *t*-test showed that the means of h and π significantly differed between CSB (t = 32.678, df = 116.3, p-value < 0.01 for h value and t = 18.607, df = 126, p-value < 0.01 for π value) and TAS (t = −29.507, df = 125.8, p-value < 0.01 for h value and t = −9.779, df = 135.6, p-value < 0.01 for π value). Complex haplotype networks of both CR1 and CR2 were constructed from a large number of polymorphic sites and haplotypes, showing a striking star-shaped topology (Figure 2). The average sequence divergences between CR1 and CR2 (the paralogous CRs) of the same species (p-distance) were $0.39\% \pm 0.11\%$ and $0.37\% \pm 0.10\%$, respectively, whereas those of orthologous CRs in different species were $10.83\% \pm 0.58\%$ for CR1 and $17.08\% \pm 0.45\%$ for CR2 (Table 2).

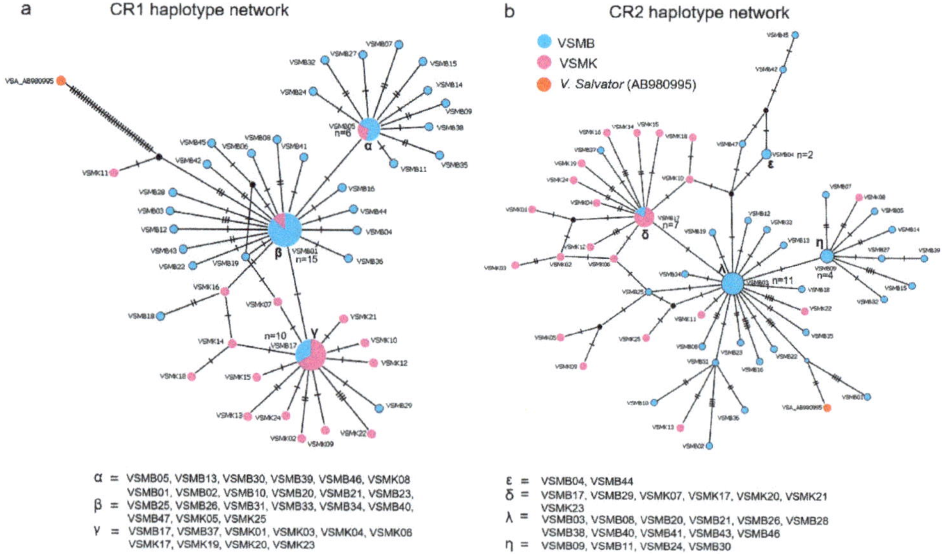

Figure 2. Haplotype network based on mitochondrial control region (mtCR) region sequence data of water monitors from Bang Kachao Peninsula (VSMB) and Varanus Farm Kamphaeng Saen (VSMK) populations, constructed using statistical parsimony with the TCS network. The numbers of individuals possessing haplotypes are indicated by different colors inside the circles. Missing haplotypes are indicated by black circles. (**a**) mtCR1 haplotype network (**b**) mtCR2 haplotype network.

Table 2. Percentage of D-loop sequence diversity for Asian water monitor (*Varanus salvator macromaculatus*; Deraniyagala 1944).

Region	Within Group	Between Group
	p-Distance	*p*-Distance
CR1	0.39 ± 0.11	-
CR2	0.37 ± 0.10	-
CSB_CR1	0.40 ± 0.18	-
CSB_CR2	0.15 ± 0.09	-
TAS_CR1	011 ± 0.11	-
TAS_CR2	0.21 ± 0.08	-
CR1 *	10.83 ± 0.58	25.11 ± 5.50
CR2 *	17.08 ± 0.45	38.88 ± 1.54
CSB_CR1 *	11.23 ± 1.96	34.70 ± 6.40
CSB_CR2 *	14.33 ± 1.88	32.49 ± 6.00
TAS_CR1 *	13.78 ± 0.17	41.10 ± 5.50
TAS_CR2 *	17.08 ± 0.45	38.88 ± 1.55

* Comparison between CRs of *Varanus salvator macromaculatus* with CRs of *V. salvator*, *V. salvator macromaculatus*, *V. salvator komaini*, *V. exanthematicus*, *V. komodoensis*, and *V. niloticus*.

Substitution saturation was estimated for both CR1 and CR2 datasets. No saturation was detected in CR2, as reflected by the linear correlation of the number of transitions and transversions plotted against sequence divergence (Figure 3) as well as from a significantly lower value of I_{SS} as compared to $I_{SS.C}$ (Table 3). By contrast, in CR1, the number of transitions was higher than that of transversions, and substitution saturation occurred when the frequency of transitions exceeded the frequency of transversions (Figure 3).

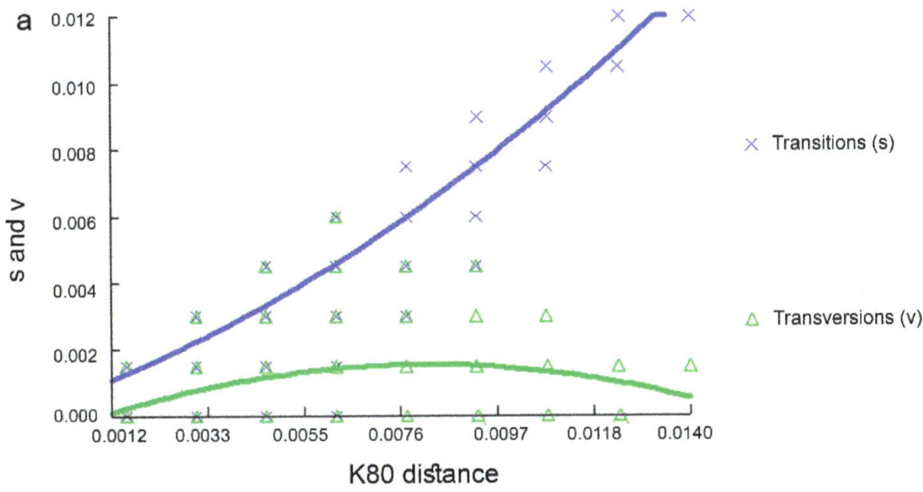

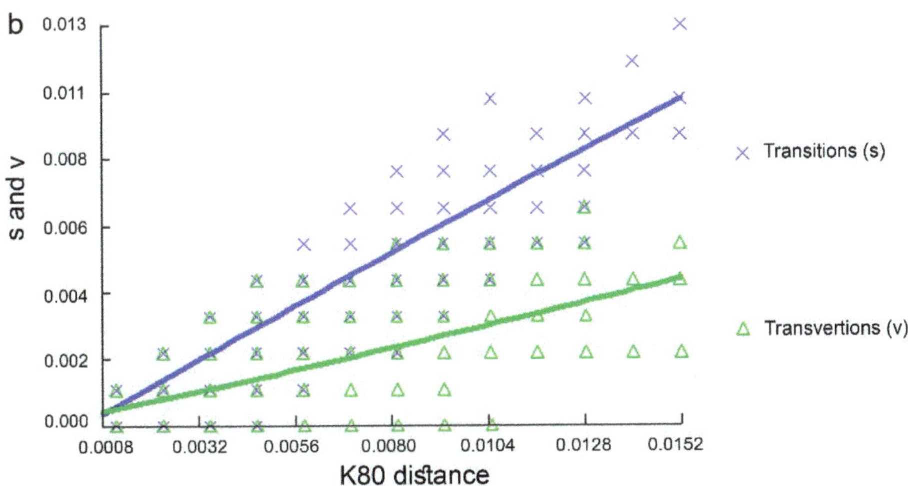

Figure 3. DAMBE7 substitution saturation plots for (**a**) CR1 and (**b**) CR2. Numbers of transitions (s) and transversions (v) are plotted against the K80 distance; lines indicate mean values (thick lines) and standard deviations (fine lines) of s and v.

3.3. Phylogenetic Relationship Based on the Control Regions of V. salvator macromaculatus

Phylogenetic analyses based on CR1 and CR2 were supported with high posterior probabilities and bootstrap values. Although the phylogenetic trees shared similar topologies, general differences existed (Supplementary Figures S1–S4). When combining both CR sequence datasets, paralogous CRs of an individual in the same dataset did not group together, while orthologous CRs of different individuals always clustered. When reconstructed separately using the separated regions (Supplementary Figures S5–S10), phylogenetic trees of TAS and CSB sequences shared the same topologies, indicating that orthologous copies from different individuals always clustered together rather than with paralogous copies from the same individuals. However, CR1 and CR2 from the same species are always clustered together (Supplementary Figures S1–S4).

Table 3. Substitution saturation analysis of CRs based on the index of substitution saturation as implemented in DAMBE7 [32].

Region	Number of OUT [a]	I_{ss} [b]	$I_{ss.cSym}$ [c]	Df [d]	*p*-Value [e]	$I_{ss.cAsym}$ [f]	df	*p*-Value
CR1	4	0.028	0.805	662	<0.00001	0.774	662	0.0000
	8	0.033	0.766	662	<0.00001	0.656	662	0.0000
	16	0.039	0.745	662	<0.00001	0.535	662	0.0000
	32	0.046	0.719	662	<0.00001	0.393	662	0.0000
CR2	4	0.007	0.815	866	<0.00001	0.784	866	0.0000
	8	0.008	0.780	866	<0.00001	0.673	866	0.0000
	16	0.008	0.763	866	<0.00001	0.560	866	0.0000
	32	0.009	0.738	866	<0.00001	0.424	866	0.0000

[a] number of sequences used in random resampling; OTP: operational taxonomic unit, [b] index of substitution saturation, [c] critical value for a symmetrical tree topology, [d] degrees of freedom, [e] probability that I_{ss} is significantly different from the critical value ($I_{ss.cSym}$ / $I_{ss.cAsym}$), [f] critical value for an asymmetrical tree topology.

3.4. Recombination Events in CRs

To explain the presence of discordant signals between phylogenetic trees derived from CR1 and CR2 constructs, multiple recombination points were investigated using RDP software (Figure 4). Non-significant recombination events in CRs were observed at positions 138 bp and 672 bp in CR2 (MaxChi, $p = 0.137$; RDP, Geneconv, and Chimaera, no evidence of recombination was found) whereas no events were found in CR1. However, when we combined CR1 and CR2 of all individuals, a recombination event was observed at position 555 (Geneconv, $p < 0.05$) while no evidence of recombination was found in RDP, MaxChi, and Chimaera.

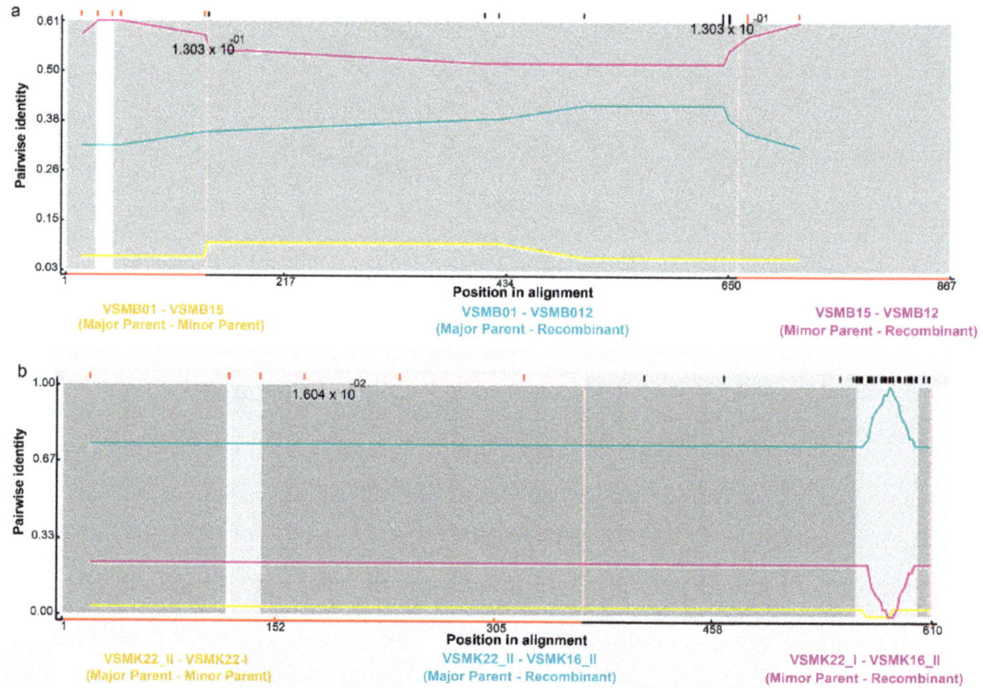

Figure 4. Assessment of recombination in duplicate CRs of all individuals: (**a**) CR2; (**b**) CRs detected using RDP software.

4. Discussion

Vertebrate mitogenome sequences are important systems that are predominantly utilized for molecular evolutionary studies, phylogenetics, and systematic taxonomy [3,5,7,44–50]. A special phenomenon has been observed in mitogenomes involving several vertebrates with duplicate CRs [9]. Comparisons of the four mitogenomes of varanids have revealed that all species possess duplicate CRs that tally with the process of mitogenomic rearrangement, and they can be reshuffled by investigating the use of PCR-based DNA marker analysis across 11 varanids [15]. The CR structures in vertebrate mitogenomes predominantly contain conserved sequences known as TASs and CSBs, which were observed in both CR1 and CR2. These conserved sequences are known to play important roles in the replication and expression of mitogenomic genes [6,51]. However, no CD was observed for either CR1 or CR2. This result suggests the plasticity of mitogenomes with the CD motif across vertebrates.

Entire sequences of CRs within species were highly similar, and the paralogs of CR1 and CR2 from each species showed a closer resemblance than those of their orthologs from other lineages. This suggests that CRs concertedly evolved in each species [9]. Duplicated CRs might remain conserved during reproduction and thus maintained in the mitogenome during speciation of varanids [6]. Alternatively, duplicate CRs probably play different roles in the replication of mitogenomes under evolutionary selective pressure and may have evolved independently within a particular species [52]. We observed this appearance in sequence divergences between CR1 and CR2 in 72 *V. salvator macromaculatus* individuals from the two populations [5]. The observed values were 5.91% and 17.4% between the two CRs. Homologs of CRs from different individuals were genetically more similar than paralogous CRs from the same individual within a species (Supplementary Figure S3), thereby agreeing with phylogenetic analyses, although both populations studied exhibited a high degree of population-level genetic diversity [5]. This suggests the independent evolution of evolutionary patterns for duplicate CRs within *V. salvator macromaculatus*. The CRs would have evolved independently within each varanid species after the divergence of species as a result of different mutations between CR1 and CR2 during evolution.

However, with evolutionary variations in characteristics, we found evidence of substantial variation in the inferred usage of CRs in *V. salvator macromaculatus*. Saturation analyses also showed that plots of transitions and transversions related linearly, with sequence divergence indicating no saturation in the CR2 data set [53]. The non-linear pattern in the CR1 data set suggested substitution saturation (Figure 3); therefore, CR2 may be a more informative phylogenetic marker at the population level in varanids. Different mitochondrial genome coding regions with diverse mutation rates and evolutionary trajectories are required to further elucidate the varanid phylogenetic lineage. Further analysis on their differential expression should be performed across varanid species to clearly understand the role of duplicate CRs. Although phylogenetic trees of CR1 and CR2 shared generally similar topologies, only minor differences existed in the placement of specimens (Supplementary Figures S1–S10). Additionally, ML and BI trees were constructed based on the TAS and CSB datasets derived from all individuals with three other variants. Phylogenetic trees of TAS and CSB sequences were largely congruent with each other, and homologs within an individual always formed closer clusters than their paralogs from other individuals. The phylogenetic analysis detected certain discordant signals among different CRs, suggesting that recombination might have reshaped the evolution of duplicate CRs in the mitogenomes of water monitors. The results of this study provide several recombination points for CR2. The breakpoints of recombination tend to occur at 138 bp and 672 bp in the TAS motif. It is essential to further investigate the recombination sites of CRs to better understand the evolutionary conflict and accurately detect the phylogenetic patterns [6,54,55]. This can be achieved by analyzing heterologous sequences that contribute to mitogenomic recombination [6]. In the mtDNA of water monitors, heterogeneous regions (A and T arrays) were detected in the VNTR motif [56]. There are four types of compound CT arrays. Different studies have identified the mitogenomic recombination sites associated with heterologous in various vertebrates [6,56,57]. Specifically, VNTRs in CR2 of different species or individuals were heterologous with respect to their sequence size and motifs. This suggests that VNTRs have recombination roles. Further analyses with more varanid species are required to investigate more basic questions such as how mitochondrial genomes with duplicate CRs evolved and how the nucleotide sequences of duplicate CRs remained identical or highly similar over evolution under the concept of concerted and independent evolution.

5. Conclusions

This paper reports foundational knowledge on the dynamics of duplicated CRs in varanids mitogenomes. Our data suggest that these sequences might follow independent evolution within the same species. CRs seem to have acquired concerted evolution across different species. This hypothesis provides a baseline to study mitogenomic evolutionary events such as recombination, gene rearrangement, and concerted evolution between duplicates. A thorough understanding of nucleotide substitution in varanid CRs is important for advances in evolutionary model construction, with accurate evolutionary inferences to provide basic insights into the biology of mitogenomes.

Supplementary Materials: The following are available online at https://www.mdpi.com/article/10.3390/ani12020148/s1: Figure S1: Phylogenetic relationship between all individual water monitors (*Varanus salvator macromaculatus;* Deraniyagala 1944) and one GenBank accession: AB167711 constructed with the help of Bayesian inference (BI) analysis using CR1 sequence. Support values at each node are bootstrap values of Bayesian posterior probability. Figure S2: Phylogenetic relationship between all individual water monitors (*Varanus salvator macromaculatus;* Deraniyagala 1944) and one GenBank accession: AB167711 constructed with the help of Bayesian inference (BI) analysis using CR2 sequence. Support values at each node are bootstrap values of Bayesian posterior probability. Figure S3: Phylogenetic relationship between all individual water monitors (*Varanus salvator macromaculatus;* Deraniyagala 1944) and one GenBank accession: AB167711 constructed with the help of Bayesian inference (BI) analysis using CRs sequence. Support values at each node are bootstrap values of Bayesian posterior probability. Figure S4: Phylogenetic relationship between all individual water monitors (*Varanus salvator macromaculatus;* Deraniyagala 1944) and one GenBank accession: AB167711 constructed with the help of maximum likelihood (ML) analysis using of CRs sequence. Support values at each node are bootstrap values of maximum likelihood posterior probability. Figure S5: Phylogenetic relationship between all individual water monitors (*Varanus salvator macromaculatus;* Deraniyagala 1944) and one GenBank accession: AP018114 constructed with the help of Bayesian inference (BI) analysis using CSB of CR1 sequence. Support values at each node are bootstrap values of Bayesian posterior probability. Figure S6: CSB CR2: Phylogenetic relationship between all individual water monitors (*Varanus salvator macromaculatus;* Deraniyagala 1944) and one GenBank accession: AP018114 constructed with the help of Bayesian inference (BI) analysis using CSB of CR2 sequence. Support values at each node are bootstrap values of Bayesian posterior probability. Figure S7: CSB CRs: Phylogenetic relationship between all individual water monitors (*Varanus salvator macromaculatus;* Deraniyagala 1944) and one GenBank accession: AP018114 constructed with the help of Bayesian inference (BI) analysis using CSB of CRs sequence. Support values at each node are bootstrap values of Bayesian posterior probability. Figure S8: TAS CR1: Phylogenetic relationship between all individual water monitors (*Varanus salvator macromaculatus;* Deraniyagala 1944) and one GenBank accession: AP018114 constructed with the help of Bayesian inference (BI) analysis using TAS of CR1 sequence. Support values at each node are bootstrap values of Bayesian posterior probability. Figure S9: TAS CR2: Phylogenetic relationship between all individual water monitors (*Varanus salvator macromaculatus;* Deraniyagala 1944) and one GenBank accession: AP018114 constructed with the help of Bayesian inference (BI) analysis using TAS of CR2 sequence. Support values at each node are bootstrap values of Bayesian posterior probability. Figure S10: TAS CRs: Phylogenetic relationship between all individual water monitors (*Varanus salvator macromaculatus;* Deraniyagala 1944) and one GenBank accession: AP018114 constructed with the help of Bayesian inference (BI) analysis using TAS of CRs sequence. Support values at each node are bootstrap values of Bayesian posterior probability. Table S1: Summary of water monitor lizard (*Varanus salvator macromaculatus*) specimens.

Author Contributions: Conceptualization, K.S.; methodology, W.T., N.A., P.W. and K.S.; software, N.A. and E.K.; validation, W.T., N.A., P.W. and K.S.; formal analysis, N.A., T.P. and E.K.; investigation, K.S.; data curation, W.T., N.A. and K.S.; writing—original draft preparation, N.A. and K.S.; writing—review and editing, W.T., N.A., P.W., N.L., W.S., T.P., S.F.A., E.K., N.M., P.D. and K.S.; visualization, N.A., T.P. and K.S.; supervision, K.S.; project administration, K.S.; funding acquisition, K.S. All authors have read and agreed to the published version of the manuscript.

Funding: This research was funded by the National Research Council of Thailand (NRCT; grant number 2560096003012) awarded to K.S., Domestic Graduate Degree of Civil Servants Bangkok scholarship (Type 1), Department of Education, BMA, and Science Achievement Scholarship of Thailand (SAST) from the Office of the Higher Education Commission (no. 5717400071) awarded to W.T., Thailand Research Fund (TRF; nos. RSA6180075) awarded to K.S., the Center for Advanced Studies in Tropical Natural Resources, National Research University-Kasetsart University (CASTNAR, NRU-KU, Thailand) awarded to P.D. and K.S., the Office of the Ministry of Higher Education, Science, Research and Innovation; and the Thailand Science Research and Innovation through The Kasetsart University Reinventing University Program 2021 awarded to N.A., N.L., T.P. and K.S. No funding source was involved in the study design, collection, analysis, and interpretation of the data, writing the report and the decision to submit the article for publication.

Institutional Review Board Statement: Permission was granted by the Sri Nakhon Khuean Khan Park (Royal Forest Department, Ministry of Natural Resources and Environment) and Kasetsart University (0909.6/15779). All experimental procedures involving animals conformed to the guidelines established by the Animal Care Committee of Kasetsart University, Thailand.

Data Availability Statement: DNA sequences: GenBank accessions LC315243–LC315386. Microsatellite data submitted: Dryad (https://doi.org/10.5061/dryad.v6wwpzgt4).

Acknowledgments: The authors are also grateful to Rujira Mahaprom (Bureau of Conservation and Research, Zoological Park Organization under the Royal Patronage of His Majesty the King, Bangkok, Thailand) for providing water monitor samples, and the Center of Excellence on Agricultural Biotechnology, Science and Technology Postgraduate Education and Research Development Office, Office of the Permanent Secretary, Ministry of Higher Education, Science, Research and Innovation. (AG-BIO/MHESI).

Conflicts of Interest: The authors declare no conflict of interest.

References

1. Taanman, J.W. The mitochondrial genome: Structure, transcription, translation and replication. *Biochim. Biophys. Acta Bioenerg.* **1999**, *1410*, 103–123. [CrossRef]
2. Gadaleta, M.C.; Noguchi, E. Regulation of DNA replication through natural impediments in the eukaryotic genome. *Genes* **2017**, *8*, 98. [CrossRef]
3. Areesirisuk, P.; Muangmai, N.; Kunya, K.; Singchat, W.; Sillapaprayoon, S.; Lapbenjakul, S.; Thapana, W.; Kantachumpoo, A.; Baicharoen, S.; Rerkamnuaychoke, B.; et al. Characterization of five complete *Cyrtodactylus* mitogenome structures reveals low structural diversity and conservation of repeated sequences in the lineage. *PeerJ* **2018**, *6*, e6121. [CrossRef] [PubMed]
4. Vinson, C.C.; Mangaravite, E.; Sebbenn, A.M.; Lander, T.A. Using molecular markers to investigate genetic diversity, mating system and gene flow of Neotropical trees. *Rev. Bras. Bot.* **2018**, *41*, 481–496. [CrossRef]
5. Wongtienchai, P.; Lapbenjakul, S.; Jangtarwan, K.; Areesirisuk, P.; Mahaprom, R.; Subpayakom, N.; Singchat, S.; Sillapaprayoon, S.; Muangmai, N.; Songchan, R.; et al. Genetic management of a water monitor lizard (*Varanus salvator macromaculatus*) population at Bang Kachao Peninsula as a consequence of urbanization with Varanus Farm Kamphaeng Saen as the first captive research establishment. *J. Zool. Syst. Evol. Res.* **2021**, *59*, 484–497. [CrossRef]
6. Zheng, C.; Nie, L.; Wang, J.; Zhou, H.; Hou, H.; Wang, H.; Liu, J. Recombination and evolution of duplicate control regions in the mitochondrial genome of the Asian big-headed turtle, *Platysternon megacephalum*. *PLoS ONE* **2013**, *8*, e82854. [CrossRef] [PubMed]
7. Singchat, W.; Areesirisuk, P.; Sillapaprayoon, S.; Muangmai, N.; Baicharoen, S.; Suntrarachun, S.; Chanhome, L.; Peyachoknagul, S.; Srikulnath, K. Complete mitochondrial genome of Siamese cobra (*Naja kaouthia*) determined using next-generation sequencing. *Mitochondrial DNA Part B Resour.* **2019**, *4*, 577–578. [CrossRef]
8. Lallemand, T.; Leduc, M.; Landès, C.; Rizzon, C.; Lerat, E. An overview of duplicated gene detection methods: Why the duplication mechanism has to be accounted for in their choice. *Genes* **2020**, *11*, 1046. [CrossRef]
9. Li, D.H.; Shi, W.; Munroe, T.A.; Gong, L.; Kong, X.Y. Concerted evolution of duplicate control regions in the mitochondria of species of the flatfish family Bothidae (Teleostei: Pleuronectiformes). *PLoS ONE* **2015**, *10*, e0134580. [CrossRef] [PubMed]
10. Uetz, P.; Freed, P.; Aguilar, R.; Hošek, J. The Reptile Database. 2021. Available online: http://www.reptile-database.org (accessed on 23 May 2006).

11. Ouwens, P.A. On a large *Varanus* species from the island of Komodo. *Bull. Jard. Bot. Buitenzorg.* **1912**, *6*, 1–3.
12. Kumazawa, Y.; Endo, H. Mitochondrial genome of the Komodo dragon: Efficient sequencing method with reptile-oriented primers and novel gene rearrangements. *DNA Res.* **2004**, *11*, 115–125. [CrossRef]
13. Fitzinger, L. *Neue Classification der Reptilien nach ihren natürlichen Verwandtschaften nebst einer Verwandschafts-Tafel und einem Verzeichnisse der Reptilien-Sammlung des K. K. Zoologischen Museums zu Wien*; J.G., Heubner: Wien, Austria, 1826; p. 66.
14. Kumazawa, Y. Mitochondrial genomes from major lizard families suggest their phylogenetic relationships and ancient radiations. *Gene* **2007**, *388*, 19–26. [CrossRef] [PubMed]
15. Amer, S.A.; Kumazawa, Y. Timing of a mtDNA gene rearrangement and intercontinental dispersal of varanid lizards. *Genes Genet. Syst.* **2008**, *83*, 275–280. [CrossRef]
16. Deraniyagala, P.E.P. Four New Races of the "Kabaragoya" Lizard *Varanus salvator*. *Spolia Zeylan.* **1944**, *24*, 59–62.
17. Hague, M.T.J.; Routman, E.J. Does population size affect genetic diversity? A test with sympatric lizard species. *J. Hered.* **2016**, *116*, 92–98. [CrossRef]
18. Supikamolseni, A.; Ngaoburanawit, N.; Sumontha, M.; Chanhome, L.; Suntrarachun, S.; Peyachoknagul, S.; Srikulnath, K. Molecular barcoding of venomous snakes and species-specific multiplex PCR assay to identify snake groups for which antivenom is available in Thailand. *Genet. Mol. Res.* **2015**, *14*, 13981–13997. [CrossRef] [PubMed]
19. Parham, J.F.; Feldman, C.R.; Boore, J.L. The complete mitochondrial genome of the enigmatic big-headed turtle (Platysternon): Description of unusual genomic features and the reconciliation of phylogenetic hypotheses based on mitochondrial and nuclear DNA. *BMC Evol. Biol.* **2006**, *6*, 11. [CrossRef]
20. Peng, Q.L.; Nie, L.W.; Pu, Y.G. Complete mitochondrial genome of Chinese big-headed turtle, *Platysternon megacephalum*, with a novel gene organization in vertebrate mtDNA. *Gene* **2006**, *380*, 14–20. [CrossRef] [PubMed]
21. Easterling, K.A.; Pitra, N.J.; Morcol, T.B.; Aquino, J.R.; Lopes, L.G.; Bussey, K.C.; Matthews, P.D.; Bass, H.W. Identification of tandem repeat families from long-read sequences of *Humulus lupulus*. *PLoS ONE* **2020**, *15*, e0233971. [CrossRef]
22. Kumar, S.; Stecher, G.; Li, M.; Knyaz, C.; Tamura, K. MEGA X: Molecular evolutionary genetics analysis across computing platforms. *Mol. Biol. Evol.* **2018**, *35*, 1547–1549. [CrossRef] [PubMed]
23. Nei, M. *Molecular Evolutionary Genetics*; Columbia University Press: New York, NY, USA, 1987.
24. Rozas, J.; Ferrer-Mata, A.; Sánchez-DelBarrio, J.C.; Guirao-Rico, S.; Librado, P.; Ramos-Onsins, S.E.; Sánchez-Gracia, A. DnaSP 6: DNA sequence polymorphism analysis of large data sets. *Mol. Biol. Evol.* **2017**, *34*, 3299–3302. [CrossRef]
25. Clement, M.; Snell, Q.; Walker, P.; Posada, D.; Crandall, K. TCS: Estimating gene genealogies. 16th International Parallel and Distributed Processing Symposium. *Int. Proc.* **2002**, *2*, 184.
26. Dalgaard, P. Power and the computation of sample size. In *Introductory Statistics with R*; Springer: New York, NY, USA, 2008; pp. 155–162.
27. R Core Team. *R: A Language and Environment for Statistical Computing*; R Foundation for Statistical Computing: Vienna, Austria, 2020.
28. Salemi, M. Nucleotide substitution models. PRACTICE: The PHYLIP and TREEPUZZLE software packages. In *The Phylogenetic Handbook a Practical Approach to Phylogenetic Analysis and Hypothesis Testing*; Salemi, M., Vandamme, A.M., Eds.; Cambridge University Press: Cambridge, UK, 2003; pp. 88–97.
29. Felsenstein, J. Distance methods for inferring phylogenies: A justification. *Evolution* **1984**, *38*, 16–24. [CrossRef]
30. Xia, X.; Xie, Z.; Salemi, M.; Chen, L.; Wang, Y. An index of substitution saturation and its application. *Mol. Phylogenet. Evol.* **2003**, *26*, 1–7. [CrossRef]
31. Xia, X.; Lemey, P. *Assessing Substitution Saturation with DAMBE. The Phylogenetic Handbook: A Practical Approach to DNA and Protein Phylogeny*; Cambridge University Press: Cambridge, UK, 2009; pp. 615–630.
32. Xia, X. DAMBE7: New and improved tools for data analysis in molecular biology and evolution. *Mol. Biol. Evol.* **2018**, *35*, 1550–1552. [CrossRef]
33. Nguyen, L.T.; Schmidt, H.A.; Von Haeseler, A.; Minh, B.Q. IQ-TREE: A fast and effective stochastic algorithm for estimating maximum-likelihood phylogenies. *Mol. Biol. Evol.* **2015**, *32*, 268–274. [CrossRef] [PubMed]
34. Rambaut, A.; Drummond, A.J.; Xie, D.; Baele, G.; Suchard, M.A. Posterior summarization in Bayesian phylogenetics using Tracer 1.7. *Sys. Biol.* **2018**, *67*, 901. [CrossRef] [PubMed]
35. Tanabe, A.S. Kakusan4 and Aminosan: Two programs for comparing nonpartitioned, proportional and separate models for combined molecular phylogenetic analyses of multilocus sequence data. *Mol. Ecol. Resour.* **2011**, *11*, 914–921. [CrossRef] [PubMed]
36. Castoe, T.A.; Jiang, Z.J.; Gu, W.; Wang, Z.O.; Pollock, D.D. Adaptive evolution and functional redesign of core metabolic proteins in snakes. *PLoS ONE* **2008**, *3*, e2201. [CrossRef]
37. Chaiprasertsri, N.; Uno, Y.; Peyachoknagul, S.; Prakhongcheep, O.; Baicharoen, S.; Charernsuk, S.; Nishida, C.; Mutsuda, Y.; Koga, A.; Srikulnath, K. Highly species-specific centromeric repetitive DNA sequences in lizards: Molecular cytogenetic characterization of a novel family of satellite DNA sequences isolated from the water monitor lizard (*Varanus salvator macromaculatus*, Platynota). *J. Hered.* **2013**, *104*, 798–806. [CrossRef] [PubMed]
38. Martin, D.; Rybicki, E. RDP: Detection of recombination amongst aligned sequences. *Bioinformatics* **2000**, *16*, 562–563. [CrossRef]
39. Smith, J.M. Analysing the mosaic structure of genes. *J. Mol. Evol.* **1992**, *35*, 126–129.
40. Sawyer, S. Statistical tests for detecting gene conversion. *Mol. Biol. Evol.* **1989**, *6*, 526–538. [PubMed]

41. Posada, D.; Crandall, K.A. Evaluation of methods for detecting recombination from DNA sequences: Computer simulations. *Proc. Natl. Acad. Sci. USA* **2001**, *98*, 13757–13762. [CrossRef]
42. Martin, D.P.; Varsani, A.; Roumagnac, P.; Botha, G.; Maslamoney, S.; Schwab, T.; Kelz, Z.; Kumar, V.; Murrell, B. RDP5: A computer program for analyzing recombination in, and removing signals of recombination from, nucleotide sequence datasets. *Virus Evol.* **2021**, *7*, veaa087. [CrossRef] [PubMed]
43. Ujvari, B.; Dowton, M.; Madsen, T. Mitochondrial DNA recombination in a free-ranging Australian lizard. *Biol. Lett.* **2007**, *3*, 189–192. [CrossRef]
44. Srikulnath, K.; Thongpan, A.; Suputtitada, S.; Apisitwanich, S. New haplotype of the complete mitochondrial genome of *Crocodylus siamensis* and its species-specific DNA markers: Distinguishing *C. siamensis* from *C. porosus* in Thailand. *Mol. Biol. Rep.* **2012**, *39*, 4709–4717. [CrossRef] [PubMed]
45. Laopichienpong, N.; Muangmai, N.; Supikamolseni, A.; Twilprawat, P.; Chanhome, L.; Suntrarachun, S.; Peyachoknagul, S.; Srikulnath, K. Assessment of snake DNA barcodes based on mitochondrial *COI* and *Cytb* genes revealed multiple putative cryptic species in Thailand. *Gene* **2016**, *594*, 238–247. [CrossRef] [PubMed]
46. Laopichienpong, N.; Ahmad, S.F.; Singchat, W.; Suntronpong, A.; Pongsanarm, T.; Jangtarwan, K.; Bulan, J.; Pansrikaew, T.; Panthum, T.; Ariyaraphong, N.; et al. Complete mitochondrial genome of Mekong fighting fish, *Betta smaragdina* (Teleostei: Osphronemidae). *Mitochondrial DNA Part B Resour.* **2021**, *6*, 776–778. [CrossRef] [PubMed]
47. Lapbenjakul, S.; Thapana, W.; Twilprawat, P.; Muangmai, N.; Kanchanaketu, T.; Temsiripong, Y.; Unajak, S.; Peyachoknagul, S.; Srikulnath, K. High genetic diversity and demographic history of captive Siamese and Saltwater crocodiles suggest the first step toward the establishment of a breeding and reintroduction program in Thailand. *PLoS ONE* **2017**, *12*, e0184526. [CrossRef] [PubMed]
48. Ahmad, S.F.; Laopichienpong, N.; Singchat, W.; Suntronpong, A.; Pongsanarm, T.; Panthum, T.; Ariyaraphong, N.; Bulan, J.; Pansrikaew, T.; Jangtarwan, K.; et al. Next-generation sequencing yields complete mitochondrial genome assembly of peaceful betta fish, *Betta imbellis* (Teleostei: Osphronemidae). *Mitochondrial DNA Part B Resour.* **2020**, *5*, 3856–3858. [CrossRef] [PubMed]
49. Ariyaraphong, N.; Laopichienpong, N.; Singchat, W.; Panthum, T.; Ahmad, S.F.; Jattawa, D.; Duengkae, P.; Muangmai, N.; Suwanasopee, T.; Koonawootrittriron, S.; et al. High-level gene flow restricts genetic differentiation in dairy cattle populations in Thailand: Insights from large-scale mt D-Loop sequencing. *Animals* **2021**, *11*, 1680. [CrossRef]
50. Ariyaraphong, N.; Pansrikaew, T.; Jangtarwan, K.; Thintip, J.; Singchat, W.; Laopichienpong, N.; Pongsanarm, T.; Panthum, T.; Suntronpong, A.; Ahmad, S.F.; et al. Introduction of wild Chinese gorals into a captive population requires careful genetic breeding plan monitoring for successful long-term conservation. *Glob. Ecol. Conserv.* **2021**, *28*, e01675. [CrossRef]
51. Fonseca, M.M.; Harris, D.J.; Posada, D. The inversion of the control region in three mitogenomes provides further evidence for an asymmetric model of vertebrate mtDNA replication. *PLoS ONE* **2014**, *9*, e106654.
52. Eberhard, J.R.; Wright, T.F.; Bermingham, E. Duplication and concerted evolution of the mitochondrial control region in the parrot genus Amazona. *Mol. Biol. Evol.* **2001**, *18*, 1330–1342. [CrossRef]
53. Bronstein, O.; Kroh, A.; Haring, E. Mind the gap! The mitochondrial control region and its power as a phylogenetic marker in echinoids. *BMC Evol. Biol.* **2018**, *18*, 80. [CrossRef]
54. Abbott, C.L.; Double, M.C.; Trueman, J.W.; Robinson, A.; Cockburn, A. An unusual source of apparent mitochondrial heteroplasmy: Duplicate mitochondrial control regions in *Thalassarche albatrosses*. *Mol. Ecol.* **2005**, *14*, 3605–3613. [CrossRef]
55. Morris-Pocock, J.A.; Taylor, S.A.; Birt, T.P.; Friesen, V.L. Concerted evolution of duplicated mitochondrial control regions in three related seabird species. *BMC Evol. Biol.* **2010**, *10*, 14. [CrossRef]
56. Hoarau, G.; Holla, S.; Lescasse, R.; Stam, W.T.; Olsen, J.L. Heteroplasmy and evidence for recombination in the mitochondrial control region of the flatfish *Platichthys flesus*. *Mol. Biol. Evol.* **2002**, *19*, 2261–2264. [CrossRef]
57. Tsaousis, A.D.; Martin, D.P.; Ladoukakis, E.D.; Posada, D.; Zouros, E. Widespread recombination in published animal mtDNA sequences. *Mol. Biol. Evol.* **2005**, *22*, 925–933. [CrossRef] [PubMed]

 *animals*

Review

Past, Present, and Future of Naturally Occurring Antimicrobials Related to Snake Venoms

Nancy Oguiura [1,*], Leonardo Sanches [1], Priscila V. Duarte [1], Marcos A. Sulca-López [2] and Maria Terêsa Machini [3,*]

1 Laboratory of Ecology and Evolution, Instituto Butantan, Av. Dr. Vital Brasil 1500, São Paulo 05503-900, Brazil
2 Aquatic Microbiology and Technological Applications, Department of Microbiology and Parasitology, Faculty of Biological Sciences, Universidad Nacional Mayor de San Marcos, Av. Venezuela Cdra. 34 s/n, Ciudad Universitaria, Lima 15081, Peru
3 Laboratory of Peptide Chemistry, Department of Biochemistry, Institute of Chemistry, University of São Paulo, Av. Prof. Lineu Prestes, 748, São Paulo 05508-000, Brazil
* Correspondence: nancy.oguiura@butantan.gov.br (N.O.); mtmachini@iq.usp.br (M.T.M.)

Simple Summary: A critical global health problem is microbial resistance to antibiotics. In order to further discuss this issue and search for practical means to overcome such problems, we reviewed the bibliography related to snake venoms, their proteins, and peptides with antimicrobial activity because many of them have the potential to become alternative antimicrobial agents or serve as lead compounds for the development of new ones. Among the proteins classified according to their structures are lectins, metalloproteinases, L-amino acid oxidases, phospholipases type A_2, cysteine-rich secretory proteins, and serine proteinases. Among the oligopeptides are waprins, cardiotoxins, cathelicidins, and β-defensins. The list includes natural and synthetic small peptides, many derived from the proteins and the oligopeptides cited above. In vitro, all these snake-venom components are active against bacteria, fungi, parasites, and/or viruses pathogenic to humans. Some have also been tested in laboratory animals. In addition to organizing and discussing such an expressive amount of information, we propose here a multidisciplinary approach that includes sequence phylogeny as a way to better understand the relationship between amino-acid sequence and antimicrobial activity.

Abstract: This review focuses on proteins and peptides with antimicrobial activity because these biopolymers can be useful in the fight against infectious diseases and to overcome the critical problem of microbial resistance to antibiotics. In fact, snakes show the highest diversification among reptiles, surviving in various environments; their innate immunity is similar to mammals and the response of their plasma to bacteria and fungi has been explored mainly in ecological studies. Snake venoms are a rich source of components that have a variety of biological functions. Among them are proteins like lectins, metalloproteinases, serine proteinases, L-amino acid oxidases, phospholipases type A_2, cysteine-rich secretory proteins, as well as many oligopeptides, such as waprins, cardiotoxins, cathelicidins, and β-defensins. In vitro, these biomolecules were shown to be active against bacteria, fungi, parasites, and viruses that are pathogenic to humans. Not only cathelicidins, but all other proteins and oligopeptides from snake venom have been proteolyzed to provide short antimicrobial peptides, or for use as templates for developing a variety of short unnatural sequences based on their structures. In addition to organizing and discussing an expressive amount of information, this review also describes new β-defensin sequences of *Sistrurus miliarius* that can lead to novel peptide-based antimicrobial agents, using a multidisciplinary approach that includes sequence phylogeny.

Keywords: snake venoms; antimicrobial activity; snake toxins; snake immunity; rattlesnakes; cathelicidins; defensins; genes; peptides

Citation: Oguiura, N.; Sanches, L.; Duarte, P.V.; Sulca-López, M.A.; Machini, M.T. Past, Present, and Future of Naturally Occurring Antimicrobials Related to Snake Venoms. *Animals* **2023**, *13*, 744. https://doi.org/10.3390/ani13040744

Academic Editor: Ettore Olmo

Received: 15 December 2022
Revised: 10 February 2023
Accepted: 11 February 2023
Published: 19 February 2023

1. Introduction

Animals and plants possess an arsenal of potent macromolecules to protect themselves against infections. Such an arsenal is chemically heterogeneous and includes proteins and peptides with antimicrobial activity [1,2].

In the animal kingdom, reptiles are organisms of great adaptability, a feature that allows them to survive in several environments or ecological niches. Therefore, reptiles have undergone significant diversification and have been considered intermediates between ectothermic anamniotes (fish and amphibians) and endothermic amniotic animals (birds and mammals) [3]. Hence, snakes are widely distributed throughout the world [4].

Snake venoms are mixtures of a variety of pharmacologically active chemicals, under study mainly for scientific and medical interest. Many of the published studies focusing these natural sources aim at disclosing the biological activities of toxins or developing new molecules with high therapeutic indexes [5]. Furthermore, expanding the knowledge of snake immunity can be quite useful in the battle against pathogenic microorganisms that are resistant to antibiotics [6]. Indeed, bacterial antimicrobial resistance (AMR) has emerged as one of the leading public health threats of the 21st century, so every year the World Health Organization (WHO) organizes the global campaign, World Antimicrobial Awareness Week (WAAW), aiming to improve awareness and understanding of AMRs as well as to encourage good practices for treating bacterial infections. The theme of WAAW 2022 was "Preventing Antimicrobial Resistance Together".

In view of such relevant information and aiming to contribute to the elucidation of snakes' abilities to survive in different ecological niches, we concluded that it would be particularly interesting to shed light on topics related to snakes' defense against microorganisms. Thus, this review organizes and discusses part of the existing knowledge of snake immunity, snake-venom toxins, and antimicrobial proteins and peptides (AMPs), or host defense peptides (HDPs) found in snake venoms. It is worth stressing here that last June, a Brazilian research group tracked and published the scientific production of our country related to peptides from snake venoms [7], confirming that Brazilian research in this field is strong. Indeed, our pioneering studies mostly focused on accidents and treatments, then on biological activities of toxins and, in the 21st century on new functions, such as anti-inflammatory, antitumor, analgesic, and antimicrobial activities [7].

In comparison with conventional antibiotics, AMPs inhibit the growth of, and/or rapidly kill, pathogenic microorganisms with higher efficiency, because they mainly target bacterial and fungal cell membranes [8,9]. In addition, the most significant advantage of these biopolymers over antibiotics is the fact that they do not induce the generation of resistant mutant microorganisms after sequential exposure at concentrations close to their minimum inhibitory concentrations (MICs) [10,11]. Although all AMPs known so far are catalogued in APD3 (https://aps.unmc.edu/, accessed on 7 January 2023), a database that also includes AMPs related to snake venoms or components of this natural source, it is difficult to order them in terms of potency, because the MICs reported were determined using different experimental approaches (like radial diffusion or standard disc diffusion assay, Bactec TB-460 radiometric method [12], determination of MICs in liquid media using optical density or colony-forming units) or tests with a fixed concentration of AMP. Even so, it is feasible to trace a path to use these biomolecules as candidates for therapeutic drugs, or as lead compounds for the development of novel antimicrobial agents.

2. The Immunity of Snakes

Reptiles are ectothermic animals, since they are not able to control their internal temperature, requiring strong seasonal shifts in behavior to maintain the body temperature [13]. Like mammals, reptile immunity is complex and comprises innate and adaptive immune systems, including cell-mediated and humoral responses [13]. So, this is an interesting group to be studied regarding host defense, since the innate immune system of reptiles—which includes nonspecific leukocytes, antimicrobial peptides, and the complement system—responds vigorously and quickly, allowing these animals to combat a wide range of pathogens and thrive in numerous environments. Such broad feedback is typically followed by a moderate adaptive immune response [14]. Since relatively little is known about it, and even less in snakes, this revision will focus on naturally occurring antimicrobial proteins, oligopeptides, and short peptides (AMPs) found in snake venoms.

Like lizards and amphisbaenians, snakes belong to the order Squamata. These reptiles are distributed throughout almost every environment of the globe, except for the polar caps. There are aquatic and terrestrial snakes. Thus in our planet's environments, these animals occupy fossorial, terrestrial, and arboreal niches; they live in forests, savannas, or deserts; while some are venomous, others are not [4].

According to Grego et al., 2006 [15] the cells commonly found in snake blood are erythrocytes, thrombocytes, and leukocytes. Among the last are lymphocytes, azurophils, heterophils, and basophils. Eosinophils are found in chelonians and lizards; however, their presence in snakes is not sufficiently studied. Snake lymphocytes are mononuclear cells and smaller than erythrocytes; the nucleus has a low standard of dense chromatin; the cytoplasm is basophilic; the number increases in circulation during inflammatory processes, wound healing, parasitemia, and viral diseases. The azurophils, the second most common leukocyte found in the blood of snakes, have a vacuolated cytoplasm and a central or eccentric nucleus; a number increase suggests the occurrence of infectious diseases. Heterophiles are large and eosinophilic and have eccentric nuclei and cytoplasmic granules that can be found intact or degranulated; a number increase is usually associated with an inflammatory response linked to inflammation, microbial and parasitic diseases, stress, and neoplasms. Basophils are small and spherical, with many granules in the cytoplasm. The function of snake basophils is probably the same as in mammals because such reptiles release immunoglobulins and histamine during degranulation [15].

Carvalho et al., 2017 [16] examined the leucocytes of *Boa constrictor*, *Bothrops jararaca*, and *Crotalus durissus* snakes. Cytochemistry and flow cytometry revealed small lymphocytes, large lymphocytes, azurophils, and heterophils. The authors did not detect any difference in the cell populations, but observed heterophils, lymphocytes, and azurophils with phagocytic activity [16]. Farag and El Ridi, 1986 [17] used spleen cells of the *Psammophis sibilans* adult snake to demonstrate that such lymphocytes can be stimulated by concanavalin A. Three years later, Saad, 1989 [18] used concanavalin A, phytohemagglutinin, and *Escherichia coli* lipopolysaccharide as a mitogen to show that mitogenic responsiveness of such snake lymphocytes varies according to the animal's sex.

There are reports of hemolysis tests indicating that the complement system of the *Naja kaouthia* snake's innate immunity (actual species name of *Naja naja kaouthia*, Reptile Database [19]) is similar to that of mammals [20]. Such a complement cascade seems to act in two ways: (1) direct adherence to microbial cell membranes without any involvement with the adaptive immune system; or (2) direct pathogen lysis via the formation of a membrane attack complex that perforates pathogen cell membranes [21].

On the other hand, AMPs are also part of innate immunity. Among them, the best known are cathelicidins and defensins, which belong to the large group of cationic peptides with amphipathic properties. Such a group corresponds to the main part of the host defense in many vertebrates [22], and includes peptide chains of low molecular weights (MW) or short AMPs with antibiotic activity. All these types of AMPs will be further discussed below.

Most published studies on innate immunity in snakes used samples of their plasma for tests on vertebrate erythrocytes aiming to verify the complement activity [23] and lysis of the Gram-negative (G−) bacteria *Escherichia coli*, the Gram-positive (G+) *Staphylococcus aureus*, and the fungus *Candida albicans* [24]. This approach has been widely explored in ecological studies involving snakes, with the results indicating the immunity of reptiles is closely dependent on several intrinsic factors related to the snake or the environment [25]. This type of result and the mitogenic responsiveness of lymphocytes has helped to evaluate the immune capacity of snakes (Table 1). Indeed, studying several mesic snake communities, Brusch et al., 2020 [26] found a correlation between dehydration and the presence of hemoparasites with cellular and humoral immunity.

Table 1. Plasma innate immunity and association to environmental and physiological conditions.

Snake	Microorganisms	Assay	Factors	Reference
Agkistrodon piscivorus	*E. coli* (G−)	Antimicrobial assay using plasma from cottonmouth pregnant relative to non-pregnant females.	Pregnancy	[23]
A. piscivorus	*E. coli* (G−)	Antimicrobial action of the complement system from plasma samples obtained in different seasons.	Temperature	[21]
Antaresia childreni	G−: *E. coli Salmonella enterica*.	Effect of the immune system against the growth of pathogenic G− bacteria in *A. childreni* eggs under normal environmental conditions of incubation and dehydration.	Dehydration	[27]
Boa constrictor	*E. coli* (G−)	Antibacterial activity of blood plasma from *Boa constrictor* fasting and fed with mice.	Feeding	[25]
Crotalus atrox	G−: *E. coli, S. enterica*	Bacterial killing assays of plasma to inhibit the growth of G− microorganism.	Dehydration	[28]
C. durissus	*E. coli* (G−)	Effects of temperature (25 °C to 35 °C) on antibacterial activity and variation of corticosterone levels in plasma.	Temperature	[29]
C. viridis	G−: *E. coli, Klebsiella oxytoca, S. typhimurium* #, *Citrobacter freundii, Shigella flexneri, Enterobacter cloacae;* G+: *Streptococcus pyogenes, Staphylococcus aureus.*	The antimicrobial effect of *C. viridis*'s plasma against the growth of G+ and G− bacteria.	Temperature	[30]
Liasis fuscus	*E. coli* (G−)	Effect of dehydration conditions in adults, measuring osmolality of plasma and response of it against bacteria.	Dehydration	[31]
Natrix piscator	*Saccharomyces cerevisae*	Snake lymphocyte proliferation and phagocytosis against yeasts by snake macrophage.	Testosterone	[32]
N. piscator	*S. cerevisae*	Snake splenocyte proliferation and phagocytosis against yeasts by snake macrophage.	Daily and seasonal rhythms	[33]
N. piscator	*S. cerevisae*	Snake lymphocyte proliferation and phagocytosis against yeasts by snake macrophage.	Photoperiod	[34]
Panterophis guttatus	Not described	Hemagglutination on whole sheep blood.	Feeding	[35]
Sistrurus miliarius	Lipopolysaccharides (LPS) extracted from *E. coli* (G−)	Quantification of the metabolic cost in the immune response of pregnant and non-pregnant snakes using LPS.	Pregnancy	[36]

Table 1. *Cont.*

Snake	Microorganisms	Assay	Factors	Reference
S. miliarius	*E. coli* (G−)	Relationship between environmental factors or the energetic/physiological state of snakes against infections.	Climate	[37]
Thamnophis elegans	*E. coli* (G−)	Comparison of the innate efficiency of the immune system in fast-living and slow-living snakes.	Size and age	[38]
T. elegans	*E. coli* (G−)	Hemolysis and hemagglutination on sheep red blood cells.	Ecotype and age	[39]
T. elegans	*E. coli* (G−), *S. aureus* (G+), *Candida albicans*	Realization of a new method for the microbicidal analysis of blood plasma by spectrophotometry.	Inter- and intraspecific variation	[24]
T. elegans	*E. coli* (G−)	The effect of fasting and stress on altered energy use and immune function.	Stress, food restriction	[40]
T. elegans	*E. coli* (G−)	Snake lymphocyte proliferation.	Pregnancy	[41]
T. elegans	*E. coli* (G−)	Snake lymphocyte proliferation.	Ecotype and parasitosis	[42]
T. elegans	*E. coli* (G−)	Interaction of immunological and endocrinological efficiency in inhibiting the growth of G− microorganisms.	Corticosterone, temperature, climate	[43]
T. elegans	*E. coli* (G−)	Effects of annual climate variation on immunity and bactericidal competence.	Climate	[44]
T. sirtalis	*E. coli* (G−)	Hemolysis on sheep red blood cell.	Physiology	[45]
T. sirtalis	*E. coli* (G−)	Effects of annual climate variation on immunity and bactericidal competence.	Corticosterone, temperature, climate	[43]
T. sirtalis	*E. coli* (G−)	Bacterial killing assays of plasma and hemolysis and hemagglutination on sheep red blood cells.	Climate	[44]
Vipera ammodytes ammodytes	Not described	Snake leukocyte proliferation, quantity of immune complexes and immunoglobulins.	Climate, shedding, hibernation, activity	[46]
Vipera berus berus	Not described	Snake leukocyte proliferation, quantity of immune complexes and immunoglobulins.	Climate, shedding, hibernation, activity	[46]

Microorganisms, microorganisms sensitive to snake plasma; assay, response of snake immunity cells to induction; factors, factors that influence innate immunity; G−, Gram-negative bacteria; G+, Gram-positive bacteria; # *Salmonella enterica* serovar typhi.

3. Antimicrobials Related to Snake Venoms

In 1991, Stiles [47] published a systematic work showing that venoms of 30 Elapidae and Viperidae snakes were active against G− (*Aeromonas hydrophila, Pseudomonas aeruginosa, Escherichia coli*) and G+ (*Staphylococcus aureus, Bacillus subtilis*) bacteria. In addition, the authors observed that L-amino acid oxidase (LAAO) was the main toxin of *Pseudechis australis* venom with antibacterial activity [47]. Nonetheless, the first purified toxin tested against bacteria was an LAAO found in *Crotalus adamanteus* venom by Skarnes, 1970 [48]. Since then, antimicrobial activities have been detected on crude snake venoms, fractions of it, or in purified components [49].

3.1. Toxins—Proteins and Enzymes

In general, the macromolecules produced by living organisms as part of their innate immunity that are capable of inhibiting the growth of, or even killing, microorganisms pathogenic to them, acting as broad-spectrum anti-infectives, belong to the following families of proteins: lectins, metalloproteinases, LAAO, serine proteinases, and phospholipase type A_2 (PLA_2) [50]. See below a brief discussion of the members of each family.

3.1.1. Lectins

Lectins from snake venoms are divided in two classes: C-type, or calcium-dependent, lectins that bind carbohydrate groups (true CTLs) and C-type lectin-like proteins (CLPs) not able to bind sugars [51]. Convulxin (CVX) is a heterodimeric toxin CLP isolated from the venom of South American rattlesnake *Crotalus durissus terrificus*, whose subunits α (CVXα, 13.9 kDa) and β (CVXβ, 12.6 kDa) are joined by inter- and intrachain disulfide bonds arranged in a tetrameric $\alpha4\beta4$ conformation; CVXs activate platelets [52].

Historically, crotacetin (CTC), which is a CVX-like purified from the venom of *C. d. terrificus* [53], was the first of its family described as having antibacterial activity. At 150 µg/mL, both CVX and CTC can inhibit the cellular growth of the G− bacteria *Xanthomonas axonopodis* pv. passiflorae and *Clavibacter michiganensis michiganensis* by 87.8% and 96.4%, respectively. Interestingly, the monomeric subunits of these antimicrobial proteins do not display any antibacterial activity [53].

The homodimer of 33.6 kDa BpLec was isolated from *Bothrops pauloensis* and reported as an efficient inhibitor of *S. aureus* (G+) growth at an MIC of 31.25 µg/mL, although it was not able to affect *E. coli* (G−) growth even after 22 h of incubation [54].

In 2011, Nunes et al. [55] described BlL, a CLP isolated from *B. leucurus* snake venom that has molecular mass of 30 kDa, is composed of two subunits of 15 kDa, and showed activity against the human pathogenic G+ bacteria *S. aureus, Enterococcus faecalis*, and *B. subtilis* (with MICs of 31.25, 62.25, and 125 µg/mL, respectively), but not against the G− bacteria *E. coli* and *Klebsiella pneumoniae*. These data suggested that although lectins can interact with the peptidoglycan present in the cell wall of G+ bacteria, they cannot cross the outer membrane of G− bacteria to reach the periplasmic space. Since BlL showed no antimicrobial activity in the presence of 200 mM galactose, this result indicated that its antibacterial effect involves the carbohydrate-binding property of lectin.

Six years later, Sulca et al. purified another CLP (14/18 kDa) from *Bothriopsis oligolepis*, active against *S. aureus* (G+) ATCC 25923 with an MIC of 100 µg/mL [49], so the authors digested it by incubation with highly purified bovine pancreatic trypsin to search for new AMPs among the resulting peptide fragments.

It was also reported that a CLP from *B. jararacussu* venom did not affect bacterial growth, but was able to inhibit the formation of biofilms of *E. coli* (G−) and *Streptococcus agalactiae* (G+) and disrupt pre-formed staphylococcal biofilms of the G+ bacteria: *S. chromogenes, S. hyicus*, and *S. aureus* [56].

3.1.2. Metalloproteinases

Zn^{2+}-dependent snake-venom metalloproteinases (SVMPs) are specific hemorrhagic toxins derived from the disintegrin A and metalloproteinase (ADAM) cellular family. These enzymes are secreted, single-pass transmembrane proteins [57,58].

SVMPs of the PIII group are the closest homologs of cellular ADAMs because they are large multidomain toxins (60–100 kDa) containing an N-terminal metalloproteinase, a C-terminal disintegrin-like, and cysteine-rich domains. The members of the PII group (30–60 kDa) contain a disintegrin domain at the carboxyl terminus of the metalloproteinase domain. However, PI-metalloproteinases (20–30 kDa) are single-domain proteins. As members of a broad family of proteins formed by 40–100 amino acid (AA) residues, the disintegrins are cysteine-rich polypeptides isolated from the venoms of vipers and rattlesnakes. These proteins can be released in viper venoms by the proteolytic processing of PII SVMP precursors or biosynthesized from short-coding mRNAs [58].

Samy et al. [59] described a viper metalloproteinase (AHM) of *Gloydius halys* (actual name of the species *Agkistrodon halys* Pallas [19]) venom with antimicrobial activity. Once purified, this AMP was characterized as a single-chain polypeptide with a MW of 23.1 kDa, highly similar to other SVMPs present in Viperidae venoms, with antibacterial activity against *S. aureus* (G+, MIC >7.5 μM), *Burkholderia pseudomallei* (also known as *Pseudomonas pseudomallei*, G−, 30 μM), *Proteus vulgaris* (G−, 15 μM), *E. coli* (G−, 60 μM), *P. aeruginosa* (G−, 60 μM), and *Enterobacter aerogenes* (G−, 60 μM). Data obtained in scanning electron microscopy studies indicated that the protein interacts with the peripheral cell wall, causing an explosion-like disruption of the plasma membrane in G+ bacteria [59].

No activity against G+ bacteria has been reported for SVMPs up to 2017, when the research group of Institute of Chemistry-USP isolated and purified a PIII-SVMP (73/60 kDa) from *B. oligolepis*, with an MIC of 20 μg protein/mL against *S. aureus* ATCC 25923 [49]. Sulca-López et al. also found out that one of its tryptic peptide fragments could be modified to produce very effective AMPs active against a few species of *Candida* [49].

It should be mentioned that proteolysis of a *Cerastes cerastes* SVMP generated a disintegrin (1 mg) that can significantly inhibit (84.7%) the growth of the parasite *Leishmania infantum*, a flagellate protozoan and an etiologic agent of visceral leishmaniasis [60].

3.1.3. Serine Proteinases

Snake-venom serine proteinases (SVSPs) are among the best characterized. These enzymes have molecular weights varying from 26 kDa to 67 kDa and various levels of glycosylation [61]. Because SVSPs act on various components of the vertebrate coagulation cascade on the fibrinolytic and kallikrein-kinin systems, they were further denominated as snake venom thrombin-like enzymes (SVTLEs). As to structure, the 30 members of this group share the active site sequence motif. A good example is the serine proteinase found in many snake venoms that resembles, at least in part, thrombin [62].

So far, SVSPs have not been associated with antimicrobial activity. Nevertheless, in 2017, Sulca et al. purified one (27 kDa) from *B. oligolepis* venom with an MIC of 80 μg/mL against *S. aureus* ATCC 25923 (G+) [49].

3.1.4. L-Amino Acid Oxidases (LAAO)

These enzymes are classical flavonoid-containing proteins that catalyze the oxidative deamination of L-amino acids to convert them into keto acids, ammonia, and hydrogen peroxide (H_2O_2) [63]. The content of LAAO in snake venoms varies from 1% to 30% of all proteins [63–65].

As presented in Table 2, svLAAO exhibit antimicrobial activity, as they can inhibit the growth of both G− and G+ bacteria at different concentrations or amounts. It is highly accepted that this biological action is a consequence of H_2O_2 production during the aerobic oxidation of appropriate substrates, an explanation reinforced by the observation that catalase inhibits the antimicrobial activity of LAAO [66].

Table 2. Antimicrobial activity of snake-venom L-amino oxidase (LAAO).

Snake	Microorganisms	Properties	Reference
Agkistrodon blomhoffii	*Staphylococcus aureus* (G+)	>3.75 µg (standard disc-diffusion)	[67]
A. halys [#]	*Escherichia coli* (G−), *Bacillus subtilis* (G+)	8.5 µg/mL (IC_{50})	[68]
Bothriechis schlegelii	*S. aureus* (G+), *Acinetobacter baumanni* (G−)	2–4 µg/mL (MIC); 4–8 µg/mL (MBC)	[69]
Bothrops alternatus	*E. coli* (G−), *S. aureus* (G+)	48 µg (kill 80–90% of bacteria)	[70]
B. jararaca	G+: *Eubacterium lentum*, *Propionibacterium acnes*, *S. aureus*, *S. epidermidis* *Peptostreptococcus anaerobius*; G−: *Porphyromonas gingivalis*, *Prevotella intermédia*, *Pseudomonas aeruginosa*, *S. typhimurium*	40 µg of venom (standard disc-diffusion)	[71]
B. marajoensis	*S. aureus* (G+), *Pseudomonas aeruginosa* (G−), *Candida albicans*	>50 µg/mL (MIC) 2.55–2.86 µg/mL (IC_{50})	[72]
B. mattogrosensis (actually *Bothropoides mattogrosensis*) *	G+: *Bacillus subtilis*, *Enterococcus faecalis*, *S. aureus*, *Streptococcus pyogenes*; G−: *E. coli*, *Klebsiella pneumoniae*, *Proteus mirabilis*, *P. aeruginosa*, *S. typhimurium*	64–256 µg/mL (MIC)	[73]
B. moojeni	G+: *S. aureus*; G−: *P. aeruginosa*, *S. typhimurium*, *E. coli.*	50 µg/mL (inhibit <90% of cfu)	[74]
B. paoloensis	*E. coli* (G−), *S. aureus* (G+)	25 µg/mL (inhibit <90% of bacterial growth)	[75]
B. pirajai	G−: *E. coli*, *P. aeruginosa*	4 µg/mL (inhibit <90%)	[76]
Calloselasma rhodostoma	*E. coli* (G−), *S. aureus* (G+)	0.2–125 µg/mL (MIC)	[77]
Crotalus adamanteus	G+: *Micrococcus aureus*; G−: *S. typhimurium*, *P. vulgaris*, *P. aeruginosa*, *Serratia marcenses*, *E. coli*	1–50 µg (LD_{50}, <50% inhibition of bacterial growth)	[48]
C. durissus cascavella	*Xanthomonas axonopodis pv passiflorae* (G−); *Streptococcus mutans* (G+)	12.3–35 µg/mL (50% of inhibition)	[78]
C. d. cumamensis	*S. aureus* (G+), *A. baumannii* (G−)	8–16 µg/mL (MIC)	[79]
Daboia russellii	G+: *S. aureus*; G−: *P. aeruginosa*, *E. coli*	4.5–36 µg/mL (MIC); 9–72 µg/mL (MBC)	[80]
Naja oxiana (actual of *Naja naja oxiana*) *	*B. subtilis* (G+); *E. coli* (G−)	0.036–0.094 µM (IC_{50})	[81]
Ophiophagus hannah	G−: *K. pneumoniae*, *P. aeruginosa*, *E. coli*; G+: *S. aureus*, *S. epidermidis*	0.78–50 µg/mL (MIC)	[82]
O. hannah	G+: *S. aureus*, *MRSA*, *S. epidermidis*, *B. subtilis*, *B. cereus*; G−: *S. enteritidis*, *P. aeruginosa*, *S. marcescens*, *K. pneumoniae*, *E. coli*, *Enterobacter cloacae*	7.5–110.0 µg/mL (MIC)	[83]
Pseudechis australis	G+: *B. subtilis*, *S. aureus*; G−: *Aeromonas hydrophila*, *A. sobria*	18 µg (standard disc-diffusion)	[47]
Trimeresurus stejnegeri	*HIV*	1.5 nM (EC_{50})	[84]
Vipera lebetina	*E. coli* (G−), *B. subtibilis* (G+)	20 µg/mL (MIC)	[85]

Microorganisms, microorganisms sensitive to antimicrobial activity; MIC, minimum inhibitory concentration; MBC, minimal bactericidal concentration; G−, Gram-negative bacteria; G+, Gram-positive bacteria; MRSA, methicillin-resistant *S. aureus*; * the actual species name was consulted in the Reptile Database [19]; # actual name *Gloydius halys* [19].

3.1.5. Phospholipases A_2 (PLA_2)

Snake-venom PLA_2s can be found in Elapidae and Viperidae snakes and are grouped according to the amino acid sequence (primary structure) and the pattern of disulfide bonds (tertiary structure), as Group I and Group II, respectively [86,87]. They can present as neurotoxic, myotoxic, or both [88]. Group II of PLA_2s presents mainly in Viperidae venoms, shows myotoxic activity, and can be divided into Asp49- or Lys49-PLA_2, the last being enzymatically inactive [86]. Most PLA_2s from snake venoms have a basic character [87] in Viperidae snakes, and correspond to 40–50% of the dry weight of *Crotalus durissus terrificus* venom; it is the main responsible of crotalic venom toxicity [87]. Despite this low toxicity [88], an acidic PLA_2 purified from the venom of *Porthidium nasutum* showed an antibacterial activity against *S. aureus* but not against *E. coli* [88], exposing the importance of the net charge to the antibacterial spectrum. It has been proposed that these phospholipases can inhibit bacterial growth by damaging the cell membrane's lipid bilayer [89]. Unfortunately, the Asp49-PLA_2 myotoxin cited above also causes myonecrosis and kidney failure in mammals, so this enzyme has not been considered a potential antibacterial agent [90]. A table listing other PLA_2 from snake venoms with antimicrobial activity is shown below (Table 3).

Crotoxin, a *C. d. terrificus* PLA_2, shows in vitro activity against yellow fever virus (EC_{50} of 0.04 ng/μL), dengue virus 2 (EC_{50} of 0.05 ng/μL) [91,92]. *B. asper* PLA_2 was shown to be active against dengue virus at 1.7 ng/mL (IC_{90}) as well as Rocio, Mayaro, and Oroupouche viruses (0.0021–0.0078 ng/mL, EC_{50}) [93], prevented the release of HIV-1 strains (ID_{50} of 1 nM) [94], and inhibited the replication of the hepatitis virus C at 6.08 μg/mL (IC_{50}) [95]. In addition to the antimicrobial action, the PLA_2 of *B. jararacussu* displayed antitumoral activity [96].

Table 3. Antimicrobial activity of snake-venom phospholipases A_2 (PLA_2).

Snake	Microorganisms	Properties	Reference
Atropoides nummifer	*Salmonella typhimurium* (G−)	50 µg/mL (>80% inhibition), Lys49-PLA_2	[97]
Bothriechis schlegelii	*S. typhimurium* (G−)	50 µg/mL (>50% inhibition), Lys49-PLA_2	[97]
Bothrops asper	*S. typhimurium* (G−)	100 µg/mL (>50% inhibition), mt-I, II, III, and IV, Lys49-PLA_2	[97]
B. brazili	*Escherichia coli* (G−)	120 µg/mL of Asp49-PLA_2 and Lys49-PLA_2 (80% of bacterial inhibition)	[89]
B. jararacussu	*E. coli* (G−)	5 µg/mL (>50% inhibition), Lys49-PLA_2	[98]
B. jararacussu	*Xanthomonas axonopodis pv passiflorae* (G−)	125 µg/mL (<86% inhibition) BthTx-I, BthTx-II (Lys49-PLA_2)	[99]
B. marajoensis	*Staphylococcus aureus* (G+)	50 µg/mL (MIC), Lys49-PLA_2	[72]
B. neuwiedi (actual of *B. neuwiedi urutu*) *	*Pseudomonas aeruginosa* (G−)	100 µg/mL (60% inhibition), Lys49-PLA_2	[100]
Bungarus fasciatus	*E. coli* (G−), *S. aureus* (G+)	0.4 and 0.1 µM (MIC), Group I PLA_2	[101]
Bungarus multicinctus	*E. coli* (G−)	50 µM (80% inhibition), Group I PLA_2	[102]
Cerrophidion godmani	*S. typhimurium* (G−)	100 µg/mL (>50% inhibition), mt-I and mt-II, Asp49-PLA_2	[97]
Crotalus adamanteus	G+: *S. aureus*, G−: *Burkholderia pseudomallei*, *Enterobacter aerogenes*	7.8–15.6 µg/mL (MIC) wound healing in vivo by topical application	[103]
C. durissus collilineatus	*X. axonopodis pv. Passiflorae* (G−), *Clavibacter m. michiganensis* (G+)	250 µg/mL (>90% inhibition), Lys49-PLA_2 with high enzymatic activity	[104]
C. d. ruruima	*X. axonopodis pv. passiflorae* (G−)	75 µg inhibit about 96% of the bacterial growth, Lys49-PLA_2	[105]
C. d. terrificus	*B. pseudomallei* (G−)	0.5 mg/mL (radial diffusion), Asp49-PLA_2	[106]
C. d. terrificus	G+: *S. aureus*; G−: *E. aerogenes*, *P. aeruginosa*, *E. coli*	100 µg/mL (radial diffusion), Asp49-PLA_2	[107]
C. oreganus abyssus	G+: MRSA; G−: *P. aeruginosa*, *E. coli*	At 125 µg/mL inhibits 25–60% the bacterial growth, Lys49-PLA_2	[108]
Daboia russellii (actual of *D. russellii pulchella*) *	G+: *S. aureus*, *Bacillus subtilis*; G−: *E. coli*, *S. typhimurium*, *Vibrio cholerae*, *Klebsiella pneumoniae*, *S. paratyphi*	12–15 µg/mL (MIC), VRV_PL_V, basic PLA_2	[109]
Daboia russellii (actual of *D. russellii pulchella*) *	G+: *S. aureus*, *B. subtilis*; G−: *E. coli*, *S. typhimurium*, *V. cholerae*, *K. pneumoniae*, *S. paratyphi*	11–19 µg/mL (MIC), VRV-PL-VIIIa, basic PLA_2	[110]
D. russelli (actual of *D. russelli russelli*) *	G−: *E. coli*, *E. aerogenes*, *Proteus vulgaris*, *P. mirabilis*, *P. aeruginosa*, *B. pseudomallei*; G+: *S. aureus*	6.25–100 µg/mL (MBC), VipTx-I and VipTx-II, Asp49-PLA_2	[111]
D. siamensis (actual of *D. russellii siamensis*) *	*B. pseudomallei* (G−)	0.5 mg/mL (radial diffusion), basic PLA_2	[106]
D. siamensis (actual of *D. russellii siamensis*) *	*S. aureus* (G+)	100 µg/mL (radial diffusion), basic PLA_2	[107]

Table 3. *Cont.*

Snake	Microorganisms	Properties	Reference
Echis carinatus	G−: *E. coli, E. aerogenes, P. vulgaris, P. mirabilis, P. aeruginosa, B. pseudomallei*; G+: *S. aureus*	15–60 μg/mL (MIC), Asp49-PLA_2	[112]
Lachesis muta muta	G+: MRSA; G−: *P. aeruginosa, K. pneumoniae*	12.5 μg/mL of Lys49-PLA2 named LmutTX, inhibits about 60% of G+ and ~30–50% of G− bacteria	[113]
Montivipera bornmuelleri	G−: *E. coli, P. aeruginosa*; G+: *S. aureus*	100 μL of no informed concentration (radial diffusion), the type of PLA_2 was not informed	[114]
Naja naja	G+: *S. aureus, B. subtilis*; G−: *E. coli, S. typhi, V. cholerae, K. pneumoniae, S. paratyphi, P. aeruginosa*; *C. albicans, Trichophyton tonsurans*	19–23 μg/mL (MIC), NN-XIa-PLA_2, acidic PLA_2	[115]
Naja naja	G+: *S. aureus, B. subtilis*; G−: *E. coli, S. typhi, V. cholerae, K. pneumoniae, S. paratyphi*	17–120 ug/mL (MIC), NN-XIb-PLA_2, acidic PLA_2	[116]
Porthidium nasutum	*S. aureus* (G+)	32 μg/mL (MIC), acidic PLA_2	[88]
Pseudechis australis	*B. pseudomallei* (G−)	0.5 mg/mL (radial diffusion), Group I PLA_2	[106]
P. australis	*E. aerogenes* (G−)	100 μg/mL (radial diffusion), Group I PLA_2	[107]

Microorganisms, microorganisms sensitive to antimicrobial activity; MIC, minimum inhibitory concentration; MBC, minimal bactericidal concentration; MRSA, methicillin-resistant *S, aureus*; G−, Gram-negative bacteria; G+, Gram-positive bacteria; * the actual species name was consulted in the Reptile Database [19].

3.1.6. Cysteine-Rich Secretory Protein (CRISP)

The protein crovirin with 24.9 kDa was purified from *C. viridis viridis* venom. It was active on different forms of *Trypanosoma cruzi*, *T. brucei rhodesiense*, and *L. amazonensis* with IC_{50} ranging from 1.10 μg/mL to 2.38 μg/mL [117].

Finally, Table 4 presents other snake-venom protein toxins active on fungi and parasites not presented in the previous tables.

Table 4. Snake toxins activity against fungi and parasites.

Snake	Microorganisms	Properties and Toxin	Reference
Bothrops asper	*Plasmodium falciparum*	1.42–22.89 µg/mL (IC_{50}), Asp49-PLA_2	[118]
B. atrox	*Trypanosoma cruzi*	11.3 µM (epimastigote), 0.44 µM (trypomastigote) (IC_{50}, 24 h), batroxicidin (cathelicidin)	[119]
B. brazili	*Candida albicans*, *Leishmania braziliensis* and *L. amazonensis*	120 µg/mL of Asp49-PLA_2 and Lys49-PLA_2 (50% of fungal inhibition, and 80–90% *Leishmania* inhibition)	[89]
B. marajoensis	*Leishmania amazonensis*, *L. chagasi promastigotes*	2.55–2.86 µg/mL (IC_{50}), LAAO	[72]
B. moojeni	*L. amazonensis*, *L. chagasi*, *l. panamensisi*	1.08–1.44 µg/mL (EC_{50}), LAAO	[120]
B. moojeni	*Trypanossoma cruzi*	8 µg/mL (inhibit 60% of *T. cruzi* growth), LAAO	[74]
B. paoloensis	*L. amazonensis*, *L. donovani*, *L. braziliensis*, *L. major*	1.03–1.59 µg/mL (IC_{50}), LAAO	[75]
B. pirajai	*L. donovani*, *L. braziliensis*, *L. amazonensis*, *L. major*	5 µg/mL (inhibit <90% of *Leishmania* growth), LAAO	[76]
Bungarus fasciatus	*Aspergillus terreus*, *A. niculans*, *Chaetomium globosun*, *Candida albicans*, *Pichia pastoris*	0.3–18.7 µg/mL (MIC); Cath-BF (cathelicidin)	[121]
Calloselasma rhodostoma	*C. albicans*; *L. braziliensis*; *T. cruzi*	100 µg/mL (76% of *C. albicans* inhibition); 24.47 µg/mL (IC_{50} against *L. braziliensis*); 32 µg/mL (47% of *T. cruzi* killing), LAAO	[77]
Crotalus durissus cascavella	*Leishmania donovani*	2.39 µg/mL (IC_{50}), LAAO	[78]
C. d. cumanensis	*Plasmodium falciparum*	0.6 µg/mL (IC_{50}), Lys49-PLA_2	[122]
C. d. terrificus	*Candida* spp, *Trichosporon* spp, and *Cryptococcus neoformans*	12.5-125 µg/mL (MIC)	[123]
C. d. terrificus	*Plasmodium falciparum*	1.87 µM (IC_{50}), crotamine	[124]
C. d. terrificus	*Trypanosoma cruzi*	0.22–6.21 µM (EC_{50}, against epimastigote and trypomastigote), Crotalicidin (cathelicidin); 9.5–33.1 (EC_{50}, against trypomastigote), Crotalicidin fragments	[125]
Hydrophis cyanocinctus	*C. albicans*, *C. glabrata*, *Arcyria cinerea*	2.34–9.38 µg/mL (MIC), Hc-CATH(cathelicidin)	[126]
Naja mossambica	*Plasmodium falciparum*	2.3 pM (IC_{50}), Group I PLA_2	[127]
N. scutatus	*Plasmodium falciparum*	2.6 pM (IC_{50}), Group I PLA_2	[127]
Sinonatrix annularis	*C. albicans*, *C. glabrata*	18.75–37.5 µg/mL (MIC), SA-CATH	[128]
Vipera ammodytes	*Plasmodium falciparum*	2.8 pM (IC_{50}), Lys49-PLA_2	[127]

Microorganisms, microorganisms sensitive to antimicrobial activity; MIC, minimum inhibitory concentration; MBC, minimal bactericidal concentration; G−, Gram-negative bacteria; G+, Gram-positive bacteria.

3.2. Oligopeptides with ≥60 Amino Acid Residues

3.2.1. Waprins

These oligopeptides or small proteins show structural similarity to whey acidic proteins (WAPs). Omwaprin, whose structure contains 50 AA residues and four disulfide bridges, was purified from *Oxiuranus microlepidus* venom [129]. Recombinant omwaprin has been produced and tested in a radial diffusion assay; the results revealed activity against the G+ bacteria *B. megaterium* (560.2 μg/mL) and *S. warneri* (1.7 mg/mL), but not against G+ strains of *B. thuringiensis*, *S. aureus*, and *Streptomyces clavuligerus*, or G− strains of *E. coli* (BL21) and *Agrobacterium tumefaciens* (even at the dose of 5.6 mg/mL). This AMP is also reported as relatively salt tolerant (as it was active on bacteria even at 250 mM NaCl), not hemolytic up to 1 mM, and not toxic to Swiss albino male mice at concentrations up to 10 mg/kg. It specifically targets bacterial membranes.

As nawaprin is a very similar structure isolated from the venom of *Naja nigricolis* [130], it also belongs to the waprins family and was expected to display antibacterial activity but, so far, no results have confirmed such ability.

3.2.2. Cardiotoxins

Three-finger toxins are members of a family of highly basic small proteins (MW of approximately 6.5 kDa) commonly found in elapid venoms. Among them are the cardiotoxin produced by *Naja atra* (actual species name of *N. naja atra* [19]) [131] and *Naja nigricolis* gamma toxin [132] that, beyond the cardiotoxicity, are active against *E. coli* (G−) and *S. aureus* (G+). The fusogenic effect on phosphatidylethanolamine (PE)/phosphatidylglycerol (PG) and PG/cardiolipin vesicles of both toxins has been used to explain their antibacterial activity [133].

3.2.3. Peptide VGF-1

Isolated from *Naja atra* venom, this toxin formed by 60 AA residues inhibits the growth of drug-resistant clinical strains of *Mycobacterium tuberculosis* (G+) at the concentration of 8.5 mg/L [12].

3.3. Peptides Containing 2-58 Amino Acid Residues

Most naturally occurring AMPs contain 2-50 AA residues; they are cationic compounds owing to the presence of one or some arginine and lysine residues and, consequently, they have net charges varying from +2 to +6 at a neutral pH. The majority are composed of amphiphilic sequences, meaning that in solution, these AMPs can acquire secondary structures, especially amphipathic α-helices typically characterized by a hydrophobic face exhibiting non-polar AA residues and a hydrophilic face displaying polar or positively charged amino acids [2,8,134].

As already cited, such AMPs inhibit bacterial and fungal growth, and many also kill these microorganisms at low minimum concentrations by different molecular mechanisms of action. Most of these antimicrobials are cell-membrane active, meaning that they act through the disruption or permeabilization of such cellular targets. It has been proposed that this phenomenon occurs by three non-exclusive types of events: detergent action or micellization (carpet model), barrel stave pore formation, and toroidal pore formation. The other possible events are disordered toroidal pore formation, membrane thinning/thickening, charged lipid clustering, formation of non-bilayer intermediate, oxidized lipid targeting, involvement of an anion carrier, non-lytic membrane depolarization, and electroporation. It follows comments on naturally occurring AMPs of low MW [135].

3.3.1. Pep5Bj

Pep5Bj is present in *B. jararaca* venom, with 1370 Da, is active against the phytopathogenic fungi *Fusarium oxysporum*, *Colletotrichum lindemuthianum*, and against the yeasts *Candida albicans* and *Saccharomyces cerevisiae* [136].

3.3.2. β-Defensins

The first β-defensin found in snakes was crotamine, a small basic myotoxin from the venom of the rattlesnake *C. d. terrificus*. It contains 42 AA residues and presents a net charge of +7 at a neutral pH and a motif of six cysteines, characteristic of this AMP family [137]. Crystallography followed by X-ray diffraction indicated that such an AMP structure is organized in a β-sheet-rich fold with a three-dimensional (3D) structure similar to β-defensins, as confirmed by Coronado et al. [138].

Crotamine is a myotoxin that acts on negatively charged plasma membranes, causing bursts to giant unilamellar vesicles (GUVs) [139] and in *E. coli* (G−), *Citrobacter freundii* (G−), *B. subtilis* (G+), and *Micrococcus luteus* (G+) cells [140–142]. It also inhibits the growth of *Candida* spp, *Trichosporon* spp, and *Cryptococcus neoformans* [123], as well as displays antiplasmodial activity, here exemplified by the IC_{50} of 1.87 μM found for *Plasmodium falciparum* [124].

Genomics-based approaches have been used to discover genes of innate immunity related to this group of AMPs [143,144]. Although mature β-defensins have a high variation in the AA sequence, it is known that the untranslated regions and signal peptides are highly conserved. Depending on the snake family, the propeptides are codified in two exons (Boidae, Elapidae, and Colubridae snakes) [145] or three exons (Viperidae snakes) [146]. So, due to the small size of β-defensin genes, the PCR approach was shown to be the most suitable for phylogenetic analysis of β-defensin-like genes in pit vipers [146] and colubrid snakes [145]. Crotamine-like genes identified in Brazilian pit vipers were used to deduce the amino acid sequences codified in the exons, and design and synthesize linear peptides with approximately 4 kDa. They were capable of inhibiting the bacterial growth of *E. coli* (G−), *C. freundii* (G−), *M. luteus* (G+), and *S. aureus* (G+) with MICs ranging from 1.6 μM to 28.4 μM [142].

Our research group working at Instituto Butantan analyzed crotamine-like sequences of *Sistrurus catenatus* and *S. miliarius*, rattlesnakes from the USA [147], using an approach very similar to that developed by Corrêa and Oguiura [146]. The DNA of North American rattlesnakes was used as a template in PCR, sequences were concatenated using Geneious software [148] as described in the Supplementary Materials. Although it was impossible to amplify crotamine-like sequences of *S. catenatus*, the authors analyzed eight sequences from *S. miliarius* derived from two specimens of Florida (accession number MT021631-024638 on GenBank) and found that the propeptides are encoded in two exons and can be grouped into two sets, one with a short intron with approximately 400 bp and the other with a long intron with about 1100 bp. The introns are phase 1 (inserted after the first nucleotide of codon), as are those of other snake β-defensins. The sequences with a short intron (MT024631-02633) codified only one β-defensin sequence. Such gene organization (Figure 1) is similar to the β-defensin genes of the Colubridae, Boidae, and Elapidae snake families [145], but not of the pit vipers [146].

The alignment of the AA sequences (Figure 2) shows a conserved signal peptide, the motif GNA, and the cysteine residues that determine the 3D β-defensin structure as well as the glycine residue at position 31. Interestingly, mature *S. miliarius* β-defensins have glutamine as first amino acid, as have the other snake β-defensins described, except for MT024631, which begins with an arginine.

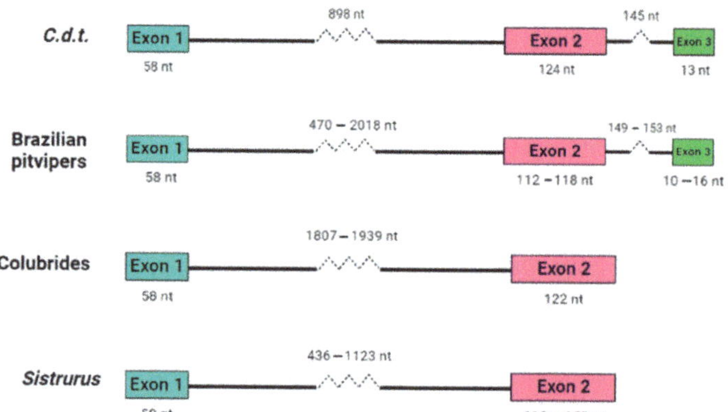

Figure 1. Structural organization of snake β-defensin genes. Crotamine sequence (*C.d.t.*, GenBank AF223947 [149]), crotamine-like sequences of Brazilian pit vipers [146], β-defensin-like sequences of Colubrides (*Phalotris mertensi*, *Thamnodynastes hypoconia*, and *T. strigatus* [145], and crotamine-like sequences of *S. miliarius* (GenBank MT024631-024638). Only exons and introns are represented.

ID	Snake	Sequence (positions 1–65)
defbPm	*Phalotris mertensis*	MKILYLLFALLFLAFLSQPGNAQPRCHSRGGRCYSPR--CPINTKCHGKMDCPRGSICCVP----
defbTs	*Thamnodyastes strigatus*	MKILYLLFALLFLAFLSEPGNAQDLCHNLGGRCFRNR--CSWSLRNHGGQDCPWGSVCCKP----
Crotamine	*Crotalis durissus terrificus*	MKILYLLFAFLFLAFLSEPGNAYKQCHKKGGHCFPKEKICLPPSSDFGKMDCRWRWKCCKKGSGK
Crotasin	*Crotalis durissus terrificus*	MKILYLLSAFLFLAFLSESGNAQPQCRWLDGFCHSSP--CPSGTTSIGQQDCLWYESCCIPRYEK
MT024635	*Sistrurus miliarius*	MKILYLLSAFLFLAFLSEPGNAQPRCHGVSGFCRSWP--CPPGTTDIQQQDCPWGQSCCVK----
MT024634	*Sistrurus miliarius*	MKILYLLSAFLFLAFLSEPGNAQPRCHGVSGFCRPDH--CPPGTTDIRQQDCPWGQSCCVK----
MT024638	*Sistrurus miliarius*	MKILYLLSAFLFLAFLSKPGNAQPRCHGVSGFCRPDH--CPPGTTDIQQQDCPWGQSCCV-----
debIM01	*Lachesis muta*	MKILYLLFPFLFLAFLSEPGNAQEWCRGLGGFCSFYQ--CRPGH-DLGPQDCWPERRCCR--WGK
MT024631	*Sistrurus miliarius*	MKILYLLFAFLFLAFLSEPGNARGQCRRLEGRCFRRR--CPLRHVDLKLQDCGPRRRCCRR----
defbLm02	*Lachesis muta*	MKILYLLFPFLFLAFLSEPGNAQGQCHQQRGRCFLHQ--CPLSHYFLGRLDCGPGRRCCRRRK--
defbBj	*Bothrops jararaca*	MKILYLLFTFLFLAFLSEPGNAQEECLQQGGFCRLIR--CPFGYDSPEQQDCRKGQRCCIRKPRK
defbBd03	*Bothrops diporus*	MKILYLLFTFLFLAFLSEPGNAQEPCLRQGGMCRPRL--CP--YVSLGQLDCQNGHVCCRKKPRK
defbBd02	*Bothrops neuwiedi*	MKILYLLFTFLFLAFLSEPGNAQPECCQEGGECHSKQ--CPLGYSSLGRLDCQLGQRCCIRIFGK
defbBju	*Bothrops jaracussu*	MKILYLLFTFLFLAFLSEPGNAQRRCHQKGGMCLPGP--CPPGYDSLGQQDCRRGQKCCIKRFGK
defbBm02	*Bothrops matogrossensis*	MKILYLLFTFLFLAFLSEPGNAQRRCRQRRGICRPRP--CPPENFSLGRLDCQMGRMCCRRRFGK
defbBm03	*Bothrops matogrossensis*	MKILYLLFTFLFLAFLSEPGNAQRRCRQRRGICRPRP--CPPENFSLGRLDCQMGQMCCRRRFGK

Figure 2. Amino acid sequences of snake β-defensins. Alignment used MUSCLE [150]), and the figure edition employed BioEdit [151] and the BioRender was used to create the art. Non-polar amino acid residues are in green, positively charged amino acid residues in blue, and the polar amino acid residues, including cysteines, glycines, and prolines, in brown.

Figure 3 shows a phylogenetic tree of snake-venom β-defensins built after analyses using maximum likelihood. The sequences were grouped into three main branches: (1) crotamine-like, (2) crotasin-like, and (3) *Bothrops*. (1) The crotamine-like group constitutes sequences of crotamine and *Lachesis* β-defensins that are active against *E. coli* (G−), *M. luteus* (G+), *C. freundii* (G−), and *S. aureus* (G+). Crotasin is a paralogous gene of crotamine found in South American rattlesnakes [152] with no antibacterial activity [142], so (2) crotasin-like group encompasses crotasin, *Sistrurus* sequences closely related to crotasin, and colubrid sequences with antibacterial activity against only *M. luteus* (G+) [142]. (3) The *Bothrops* group shows three subgroups: the DefbBju with no antibacterial activity, the *B. mattogrossensis* sequences with the highest antibacterial activity and active against *E. coli* (G−), *M. luteus* (G+), *C. freundii* (G−), and *S. aureus* (G+). In the remaining subgroup, while DefbBd03_B. diporus and DefbBj_B. jararaca show activity only against *M. luteus* (G+), DefbBn_B. neuwiedi has no antibacterial activity [142]. Of the four translated sequences of *Sistrurus*, only one (MT024631) was grouped with crotamine and the others were grouped with crotasin. Interestingly, in the crotasin group, while the *Sistrurus* sequences (MT024634, MT024635, MT024638) have net charges at pH 7 of +1; in the crotamine group, MT024631 has +11. The MT024631 position in the phylogenetic tree and its high basicity makes this sequence a strong candidate for exhibiting high antimicrobial activity.

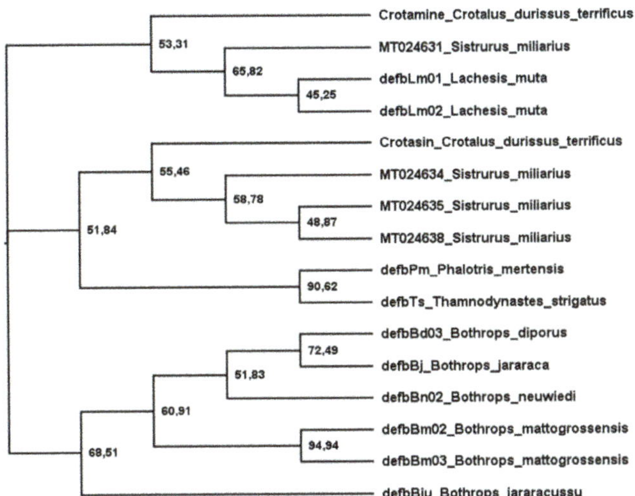

Figure 3. Phylogenetic tree of snake β-defensins. The tree was estimated using translated sequences and maximum likelihood [153]. The Edge LR-ELW support is shown in each node [154]. Details are described in the Supplementary Materials.

3.3.3. Cathelicidins (CATH)

These peptides are multifunctional biomolecules resulting from the propeptide proteolytic cleavage [155]. The first ones discovered were isolated from venoms produced by Asian elapid species [156], including *Bungarus fasciatus* [121] and *Ophiophagus hannah* [156]. These bioactive peptides are members of a group of AMPs that present variations in their amino acid sequences, chemical structures, and sizes. On the other hand, they all have in common two functional domains: one of them has high homology to the cathelin domain from which the name cathelicidins originated, a well-known inhibitor of cathepsin L; the other domain is the antimicrobial one, located at the C-terminus of the structure, also presents wide functional diversity [22,157]. The antimicrobial domains of some cathelicidins have α-helical conformations, others have β-hairpin structures and might contain high content of proline and arginine. Even though the mature peptides contain 12 to 80 or more AA residues, some discussed here contain 30–34 [158].

All CATH are encoded by genes that are made up of four exons [158]. The first exon consists of the sequence encoding the signal peptide (pre-peptide) of 29–30 AA residues, while exons 2 and 3 encode the cathelin domain (pro-peptide) of 99–114 AA residues. Exon 4 encodes the mature peptide, with the antimicrobial domain [158]. Cathelicidin genes have not been described in snakes, but Dalla Valle et al. [159] demonstrated that the genes of the lizard *Anolis carolinensis* have structural organization similar to that of mammals, which is up to four exons with three introns of different sizes. Mature cathelicidins generally exhibit antimicrobial activity against a wide range of Gram+ and Gram- bacterial species [160]. These antimicrobial peptides and proteins were found in transcripts of venom glands and others in genomes.

As experiments using NA-CATH and liposomes have shown, the main event of the general mechanism of action proposed for cathelicidins is the disruption of the bacterial cell membrane [161,162]. However, elapid venom cathelicidins can also inhibit *E. coli* ATP synthase [163]. Their low MICs for Gram+ and Gram- bacteria, resistance to salt and serum, and in vivo activity make these macromolecules promising candidates for new antimicrobial drugs. Further information on 13 different CATH antimicrobials is summarized in Table 5 and in the review published by Barros et al. [164].

Table 5. Activity of snake cathelicidins on bacteria.

Snake	Microorganisms	Properties and Cathelicidin	Reference
Bothrops atrox	G−: *Escherichia coli, Pseudomonas aeruginosa, Klebsiella pneumoniae;* G+: *Enterococcus faecalis, Staphylococcus aureus, Acinetobacter baumanni, Streptococcus pyogenes*	0.25–128 µg/mL (MIC), Batroxicidin	[165]
B. atrox	G−: *K. pneumoniae, P. aeruginosa, E. coli*	8–16 µg/mL (MIC), 16–64 µg/mL (MBC), Batroxicidin	[166]
Bungarus fasciatus	G+: *B. subtilis, B. pumilus, Bacillus cereus, S. aureus, A. calcoaceticus;* G−: *E. coli, P. aeruginosa, Sphingobacterium siyangense, Sacharibacillus kuerlensis, Serratia marcescens, P. luteola, S. typhi, K. pneumoniae; Aspergillus terreus, A. niculans, Chaetomium globosum, Pichia pastoris*	0.3–100 µg/mL (MIC), resistant to 150 mM Na+, Cath-BF	[121]
B. fasciatus	G+: *Propionibacterium acnes, S. epidermidis*	1.2–4.7 µg/mL (MIC), anti-inflammatory activity in vivo, Cath-BF	[167]
B. fasciatus	G−: *E. coli, P. aeruginosa;* G+: *S. aureus*	2–32 µg/mL (MIC) protection against *P. aeruginosa* in infected burns, no resistance until the 8th subculture, Cath-BF	[11]
B. fasciatus	G−: *E. coli*, NDM-1-carrying *E. coli;* G+: *S. aureus*	4–36 µg/mL (MIC), 8–64 µg/mL (MBC), Cath-BF, Cbf-K16, Cbf-A7A13	[168]
B. fasciatus	*E. coli* (G−), *S. aureus* (G+)	4–16 µg/mL (MIC), local treatment in vivo is effective on vaginitis, Cath-BF	[169]
B. fasciatus	*E. coli* (G−)	BF-30 microspheres; antibacterial activity maintained during >15 days of release	[170]
B. fasciatus	*S. typhimurium* (G−)	attenuated the clinical symptoms in mice, Cath-BF	[171]
B. fasciatus	*P. aeruginosa* (G−)	innate immunity activation, pretreatment ameliorate pneumonia in vivo, Cath-BF	[172]
B. fasciatus	G−: *P. aeruginosa, E. coli;* G+: *A. baumannii*	8–64 µg/mL (MIC), Cath-BF recombinant and synthetic, active against *P. aeruginosa* and *A. baumannii* biofilms	[173]
B. fasciatus	G−: *P. aeruginosa, E. coli;* G+: *A. baumannii, MRSA*	8–128 µg/mL (MIC), active against *MRSA* biofilms, recombinant and synthetic Cath-BF	[126]
Crotalus durissus terrificus	G+: *E. faecalis, S. aureus, S. pyogenes, A. baumannii;* G−: *P. aeruginosa, K. pneumoniae, E. coli*	0.25–128 µg/mL (MIC), Crotalicidin	[165]
C. d. terrificus	G−: *P. aeruginosa, E. coli, K. pneumoniae*	2–16 µg/mL (MIC), 8–64 µg/mL (MBC), Crotalicidin	[166]
Hydrophis cyanocinctus	G−: *E. coli, Shigella dysenteriae, K. pneumoniae, K. oxytoca, Proteus mirabilis, Stenotrophomonas maltophilia, P. aeruginosa, S. paratyphi, Aeromonas sobria, A. hydrophila, A. veronni, Vibrio vulnificus, V. harveyi, V. fluvialis, V. alginolyticus, Edwardsiella tarda;* G+: *S. aureus, Bacillus cereus, B. subtilis, E. faecium, Nocardia asteroides*	2–75 µg/mL (MIC), Hc-CATH, Increased survival of mice to *P. aeruginosa* infection	[174]

Table 5. *Cont.*

Snake	Microorganisms	Properties and Cathelicidin	Reference
Naja atra	*Burkholderia thailandesis* (G−)	3.66 µg/mL (EC50, 3 µg/mL inhibit biofilm formation), NA-CATH	[175]
N. atra	*Bacillus anthracis* (G+)	0.29 µg/mL (EC_{50}), inactive against spores, 0.5 µg protect *Galleria mellonella* in vivo, NA-CATH	[176]
N. atra	*Francisella novicida* (G−)	1.54 µg/mL (EC_{50}), NA-CATH	[177]
N. atra	G−: *E. coli*, *Aggregatibacter actinomycetemcomitas*	1.7 µg/mL (EC_{50}), NA-CATH	[178]
Ophiophagus hannah	G−: *E. coli*, *P. aeruginosa*, *Enterobacter aerogenes*, *E. cloacae*	1–20 µg/mL (MIC), resistant to 1% NaCl, OH-CATH	[156]
O. hannah	G−: *E. coli*, *E. cloacae*, *E. aerogenes*, *P. aeruginosa*, *Haemophilus influenzae*, *K. pneumoniae*; G+: *S. aureus*, *E. faecalis*	1.56–25 µg/mL (MIC), resistant to serum, efficacious against *E. coli* bacteremia, OH-CATH	[179]
O. hannah	G−: *E. coli*, *P. aeruginosa*, *K. pneumoniae*, *MRSA*; G+: *E. faecalis*, *S. aureus*, *A. baumannii*, *S. pyogenes*	0.25–128 µg/mL (MIC), Oh-CRAMP	[165]
O. hannah	G+: *Acitenobacter sp.*, *Enterobacter sp.*, *Streptococcus pneumoniae*; G−: *Citrobacter sp.*, *Escherichia sp.*, *Klebsiella sp.*, *Proteus sp.*, *Pseudomonas sp.*, *Salmonella sp.*, *Serratia sp.*, *Stenotrophomonas maltophilia*, *Yersinia sp.*	8–64 µg/mL (MIC_{90}), OH-CATH and D-OH-CATH	[180]
O. hannah	*P. aeruginosa* (G−)	3.25 µM (MIC), 5.07 µM (MIC after exposition to lung proteases), SnE1=OH-CATH	[181]
Pseudonaja textilis	G−: *E. coli*, *P. aeruginosa*, *K. pneumoniae*; G+: *E. faecalis*, *S. aureus*, *A. baumannii*, *S. pyogenes*	2–64 µg/mL (MIC), Pt_CRAMP	[165]
Python bivittatus	G−: *E. coli*, *P. aeruginosa*, *S. tiphimurium*; G+: *B. cereus*	1.5–46 µg/mL (MIC), Pb-CATH	[182]
P. bivittatus	G−: *Dysentery bacillus*, *E. coli*, *K. oxytoca*, *K. pneumoniae*, *S. paratyphi*, *P. aeruginosa*; G+: *Nocardia asteroids*, *S. aureus*, *B. cereus*, *E. faecalis*, *E. faecium*	1.17–75 µg/mL (MIC), 37.47–75.31% biofilm eradication using 2–5 MICs, CATHPb1, protection against MRSA and VRSA in vivo	[183]
Trimerodytes annularis (actual of *Sinonatrix annularis*) *	G−: *E. coli*, *K. pneumoniae*, *Shigela dysenteriae*, *P. aeruginosa*; G+: *S. aureus*, *B. cereus*, *B. subtillis*, *E. faecium*, *N. asteroids*	4.69–75 µg/mL (MIC), inhibit *E. coli* biofilm formation from 3.87% to 40.33% (1.25–40 µg/mL), SA-CATH	[127]

Microorganisms, microorganisms sensitive to antimicrobial activity; MIC, minimum inhibitory concentration; MBC, minimal bactericidal concentration; G−, Gram-negative bacteria; G+, Gram-positive bacteria; * the actual species name was consulted in Reptile Database [19]; NDM-1, New Delhi metallo-beta-lactamase-1; MRSA, methicillin-resistant *S. aureus*; VRSA, vancomycin-resistant *S. aureus*.

Cathelicidins also display anti-inflammatory activity that helps the recovery of organisms with pneumonia [172], other inflammatory diseases [184,185], and pathogen-induced intestinal injury [186]. In vivo, Cath-BF was found to help treat burn and wound infections in rats [11], and protect mice against sepsis caused by *E. coli* (G−), *P. aeruginosa* (G−), and *S. aureus* (G+) [187]. In addition, Cath-BF inhibited intestinal inflammation and enhanced the phagocytosis of immune cells in weanling piglets [186].

Phylogenetic analysis was used to understand the relationship between snake cathelicidins (Figure 4). The cathelicidin sequence tree did not group in species or family snakes. This disconnection between the species tree and the sequence tree is due to the duplications and extinctions that the genes of multigenic families undergo [188]. The tree is grouped into three main branches. The most basic group (1) presents an exception, KAG8148195, with a net charge of +4, and all cathelicidins tested in this group showed antibacterial activity. The second group (2) encompasses an extensive range of net charges and *Python bivittatus* cathelicidins with and without antibacterial activity. The last group (3), with a wide range of net charges, did not have any member tested. Group 1 shows three subgroups, two with Elapidae and Colubridae snake sequences and one with Viperidae. The association of Elapidae and Colubridae sequences was observed in snake β-defensins [145]. Group 2 is also organized into three subgroups, but there is no Viperidae branch (*Crotalus* CATHs are present in all subgroups). Moreover, all *P. bivittatus* sequences are associated in one subgroup independently of antibacterial activity. The last group assembled was not tested for cathelicidins of any family snake.

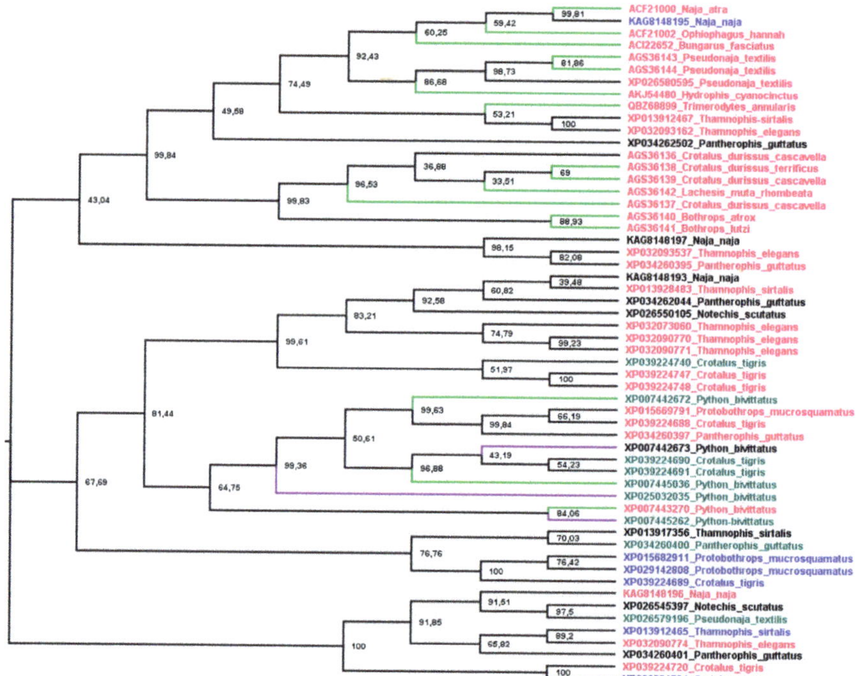

Figure 4. Phylogenetic tree of snake cathelicidins. The tree was estimated based on maximum likelihood [153], and the Edge Support LR-ELW is shown in each node [154]. Details are described in the Supplementary Materials. Branches in green indicated sequences with antimicrobial activity, purple branches indicate no activity, and black ones were not tested. Sequence names in purple indicate net charge < 5, in green 5 < 10, red > 10, and black, not determined, except XP007442673, which shows −4 as net charge at pH 7.

3.3.4. Peptides Derived from Larger AMPs from Snake Venoms (Proteins and Oligopeptides)

Short and medium-sized peptides with pharmacological functions have been widely studied, owing to their potential to become therapeutic drugs or serve as lead compounds for developing new ones. Indeed, such short biopolymers can be much more specific to cellular targets than other non-peptide drugs. On the other hand, in vivo they are prone to enzymatic degradation, can be sensitive to high salt concentrations, be cytotoxic, or interfere with the host immunity. These disadvantages have been extensively studied in order to overcome these problems: mutations and/or modifications of their reactive chemical groups have been tested [189].

Table 6 lists several short AMPs found in snake venoms that correspond to peptide fragments of snake toxins with the ability to inhibit the growth and even kill a variety of pathogenic microorganisms. As the table shows, these peptides represent specific portions of proteins, enzymes, or oligopeptides with the antimicrobial activity described above, such as cathelicidins, myotoxins, PLA_2, and defensins, unmodified or modified. A comparative analysis of their amino acid sequences reveals that practically all are cationic at a neutral pH and, as do most of the short cationic AMPs already described, have amphiphilic structures.

Table 6. Fragments derived from toxins and AMP and their antimicrobial action.

Snake	Microorganisms	Fragment or Peptide Name and Properties	Reference
Agkistrodon contortrix laticinctus	*Leishmania amazonensis, L. infantum chagasi*	Derived from Lys49-PLA$_2$: **pACl**, 50.98–220.32 µM (EC$_{50}$); **pAClR7**, 27.19–70.71 µM (EC$_{50}$); derived from Lys49-PLA$_2$	[190]
A. piscivorus piscivorus, A. c. laticinctus	G−: *P. aeruginosa;* G+: *S. aureus;* cancer cell lines: RAMOS, K562, NB4, and CEM cells	Derived from Lys49-PLA$_2$ toxins (**p-AppK** and **p-Acl**)	[191]
Bothrops asper	G−: *Escherichia coli, Pasteurella multocida, Salmonella montevideo, S. typhi, Shigella sonnei;* G+: *Listeria monocytogenes, Staphylococcus aureus, Streptococcus pyogenes, Vibrio cholerae*	Derived from Lys49-PLA$_2$: **p115-129**, 20–90 µg (MIC)	[192]
B. asper	*E. coli* (G−)	Derived from Lys49-PLA$_2$: **p115-129**, 0.1–1 µM (MBC)	[97]
B. asper	G−: *S. aureus, S. thyphimurium*	Derived from Lys49-PLA$_2$: **pEM1-10**, at 50 µg/mL (10–100% growth inhibition)	[97]
B. asper	G−: *Pseudomonas aeruginosa, Vibrio cholerae, Shigella sonnei, E. coli, S. typhimurium, Klebsiella pneumoniae, Brucella abortus;* G+: *Enterococcus faecalis, S. aureus*	Derived from myotoxin II, Lys49-PLA$_2$: **pEM-2**, 1–250 µg/mL (MBC), at 100 µg/mL (25% protection of mice against *E. coli* and *S. enterica* peritoniti)	[97]
B. brazili	*E. coli* (G−); *Candida albicans*	**pep115-129 MTX-I** (derived from Asp 49 PLA$_2$) and **pep115-129 MTX-II** (derived from Lys49-PLA$_2$), at 120 µg/mL (<60% growth inhibition	[89]
B. jararacussu	*E. coli* (G−), *S. aureus* (G+)	Derived from Bothropstoxin I, Lys49-PLA$_2$: **p-BthTX-I and (p-BthTX-I)$_2$**, 4–128 µM (MIC)	[193]
B. jararacussu	G−: *E. coli,* G+: *S. aureus, S. epidermidis, E. faecium*	Derived from Bothropstoxin I, Lys49-PLA$_2$: **p-BthTX-I and (p-BthTX-I)$_2$**, 4–512 µM (MIC), eradication of *S. epidermidis* biofilm	[194]
B. mattogrossensis	*M. luteus* (G+)	**PS2** (derived from DefbBm02) and **PS4** (derived from DefbBm03), 26.1–26.6 µM (MIC)	[142]
B. mattogrossensis	G+: *Bacillus subtilis, S. aureus, Streptococcus pyogenes;* G−: *E. coli, K. pneumoniae, P. aeruginosa, S. typhimurium*	**BmLAO-f1, BmLAO-f2, BmLAO-f3** (derived from BmLAO, LAAO), 32–256 µg/mL (MIC)	[73]
B. moojeni	G+: S. aureus	Derived from Asp49-PLA$_2$, **pBmTxJ**, 37.5 µM (MIC)	[195]
Bungarus fasciatus	G+: *B. subtilis, S. aureus;* G−: *E. coli, P. aeruginosa, Sacharibacillus kuerlensis, S. typhi, K. pneumoniae; C. albicans, Pichia pastoris*	**BF-15** (derived from cathelicidin Cath-BF), 1.2–75 µg/mL (MIC)	[121]

Table 6. *Cont.*

Snake	Microorganisms	Fragment or Peptide Name and Properties	Reference
B. fasciatus	G−: *E. coli, S. aureus, P. aeruginosa, S. typhi;* G+: *B. subtilis, Enterobacter cloacae; C. albicans*	Fragments of BF-30 (Cath-BF), 1–128 µg/mL (MIC),	[196]
B. fasciatus	G+: *MRSA, S. aureus, S. epidermidis;* G−: *E. coli*	**Cbf-K16** (derived from Cath-BF), 4–64 µg/mL (MIC), synergism with ceftazidime/ampicilin in vivo	[197]
B. fasciatus	G−: *E. coli;* G+: *S. aureus, B. subtilis; C. albicans*	**ZY13** (derived from Cath-BF), 0.59–75 µg/mL (MIC), resistant to 150 mM NaCl	[198]
B. fasciatus	G−: *E. coli, K. pneumoniae, P. aeruginosa;* G+: *S. aureus, S. epidermidis*	**Cbf-14**, derived from Cath BF, 8–64 µg/mL (MIC), 16–128 µg/mL (MBC), anti-inflammatory activity, protection in vivo	[199]
B. fasciatus	G−: *E. coli, K. pneumoniae;* G+: *MRSA*	**Cath-A** and **Cath-B**, derived from Cath-BF	[200]
Cerrophidion. godmani *	G−: *E. coli; Leishmania braziliensis, L. amazonensis promastigotas*	Derived from Asp49-PLA_2, **pCergo**, 75 µM (MIC); 93.69–110.40 µM (EC_{50})	[195]
Crotalus durissus terrificus	G+: *E. faecalis, S. aureus;* G−: *S. pyogenes, P. aeruginosa, K. pneumoniae, E. coli, A. baumannii*	**Ctn(1–14)**, **Ctn(15–34),** (derived from Crotalicidin, cathelicidin), 0.25–128 µg/mL (MIC)	[201]
C. d. terrificus	*Candida krusei, C. glabrata, parapsilosis, C. tropicalis, C. guilliermondii, Trichosporon spp.*	**C1, C2**, derived from crotamine, 2.5–40 µM (MIC), substituição de C por S diminui a atividade	[202]
C. d. terrificus	*Candida parapsilosis, C. krusei, C. tropicalis, C. albicans, Cryptococcus laurenti, Microsporum canis*	**Ctn(15–34),** derived from Crotalicidin, 5–20 µM (MIC),	[203]
C. d. terrificus	G−: *E. coli, S. typhi, Xanthomonas oryzae, X. axonopodis, Vibrio cholerae;* G+: *B. cereus, MRSA; C. albicans, Fusarium solani*	**CyLoP-1**, derived from crotamine, 5–40 µM (MIC)	[204]
C. d. terrificus	*C. albicans*	**Ctn(15–34)** Crotalicidin, synergism with amphotericin B	[205]
C. d. terrificus	G−: *E. coli, P. aeruginosa*	**Ctn(15–34)** Crotalicidin, 3.13–12.5 µM (MIC), 6.25–50 µM (MBC)	[203]
C. d. terrificus	*P. aeruginosa* (G−)	**SnV1**, derived from Crotalicidin, 0.877–5.42 µM (MIC) before and after treatment with lung proteases	[181]
C. d. terrificus	G−: *E. coli, Citrobacter freundii;* G+: *S. aureus, M. luteus.*	**PS1** and **PS6**, derived from crotamine, 28.4–56.5 µM (MIC)	[142]
C. d. terrificus	G+: *Mycobacterium smegmatis, M. fortuitum, M. wolinskyi*	**CyLoP-1**, derived from crotamine, 10–20 µM (MIC), 10–20 µM (MBC)	[206]

Table 6. *Cont.*

Snake	Microorganisms	Fragment or Peptide Name and Properties	Reference
C. oreganus abyssus	G−: *E. coli*, *P. aeruginosa*; G+: *S. aureus*	**pC-CoaTxII** (C-terminal fragment 115–129 from Lys49-PLA$_2$), inhibited <90% of bacterial growth at 5.95 µM.	[207]
Hydrophis cyanocinctus	*P. aeruginosa* (G−)	**Sn1, Sn1A, Sn1b**, derived from Hc-CATH, 0.66–45.6 µM (MIC) before and after treatment with lung proteases	[181]
Lachesis muta	G−: *E. coli*, *K. pneumoniae*, *C. freundii*; G+: *S. aureus*, *M. luteus*.	**PS3** and **PS5**, derived from DefbLm02, 13.9–110 µM (MIC)	[142]
Naja atra	G−: *E. coli*, *Aggregatibacter actinomycetemcomitans*	**ATRA-1** and **ATRA-1A**, fragments of CATH-ATRA, 0.88–160 µg/mL (EC$_{50}$)	[178]
N. atra	*Francisella novicida* (G−)	**ATRA-1** and **ATRA-1A**, fragments of CATH-ATRA, 8.95–147.9 µg/mL (EC$_{50}$)	[177]
N. atra	*S. aureus* (G+)	**ATRA-1** and **ATRA-1A**, fragments of CATH-ATRA, 0.52–18 µg/mL (EC$_{50}$), inhibit biofilm formation	[208]
N. atra	G−: *E. coli*, *P. aeruginosa*; G+: *B. cereus*, *S. aureus*.	**ATRA-1** and **ATRA-1A**, derived from NA-CATH, 1.9–72.9 µg/mL (EC$_{50}$)	[209]
N. atra	*Burkholderia thailandensis* (G−)	**ATRA-1A**, derived from NA-CATH, 7–14 µg/mL (EC$_{50}$)	[175]
N. atra	G−: *E. coli*, *P. aeruginosa*, *K. pneumoniae*, *E. cloacae*, *B. cepacia*, *P. mirabilis*, *Moraxella catarrhalis*; G+: *A. baumannii*, *S. aureus*, *E. hirae*, *S. agalactiae*; *Candida albicans*, *C. glabrata*, *Malassezia pachydermatis*, *Mycobacterium smegmatis*, *M. fortuitum*	**NP-0, NP-2, NCP-3, NCP-3a, NCP-3b,** derived from cardiotoxin, 1.6–50 µg/mL (MBC), active at 250 mM NaCl	[210]
Ophiophagus hannah	G−: *E. coli*, *P. aeruginosa*, *Enterobacter aerogenes*, *E. cloacae*; G+: *S. aureus*,	**Fragments (3–34), (5–34), (1–24) and (3–17)**, derived from OH-CATH, 4–24 µg/mL (MIC)]	[211]
O. hannah	G−: *E. coli*, *E. cloacae*, *E. aerogenes*, *P. aeruginosa*, *H. influenzae*, *K. pneumoniae*; G+: *S. aureus*, *E. faecalis*	**OH-CM6,** derived from OH-CATH-30, 1.56–25 µg/mL (MIC), active against *E. coli* bacteremia in vivo, stable in serum	[187]
O. hannah	*P. aeruginosa* (G−)	**SnE1**, derived from OH-CATH, 3.25–5.07 µM (MIC before and after treatment with lung proteases)	[181]

Microorganisms, microorganisms sensitive to antimicrobial activity; MIC, minimum inhibitory concentration; MBC, minimal bactericidal concentration; G−, Gram-negative bacteria; G+, Gram-positive bacteria; * the actual species name was consulted in the Reptile Database [19].

Among the many examples given is Ctn(15–34), a fragment of 20 AA residues from the 34-mer Crotalicidin, able to kill Gram- and Gram+ bacteria [201,203]. Clinical isolates of fungi were tested associated with fluconazol and presented additive activity [212], as well as damaged tumor cells [213]. Ctn(15–34) also has remarkable stability in human serum, is regarded as a promising anti-infective lead compound, and its mode of action seems to comprise the three stages needed for membrane-active AMPs: (1) initial peptide recruitment; (2) peptide accumulation on the phospholipidic bilayer of the plasma membrane; and (3) cell death caused by disruption of the plasma membrane.

The M.T. Machini research group (Institute of Chemistry-USP) has been developing new short AMPs active against *Candida* species derived from fragments of a metalloprotease and a PLA_2 found in the venom of the Peruvian snake *B. oligolepis*, still very little studied [49].

4. Discussion

This review shows that the innate immunity of snakes is similar to that of mammalian vertebrates in terms of cell-mediated and humoral responses. The blood of these animals contains erythrocytes, thrombocytes, and leukocytes [15], and the lymphocytes have phagocytic activity [16]. Snake immunity can be influenced by hormones, daily and seasonal rhythms, temperature, and dehydration, as shown in Table 1. These factors have been widely studied with an ecological focus using plasma samples. Since their influence on innate immunity does not interfere with snakes' adaptive capacity, these reptiles have spread to different ecosystems and microhabitats.

The ability of snakes to live in different environments, to resist different pathogenic microbes, and to eat different prey makes their venom a rich source of biomolecules that can be explored as a biological tool for science or potential anti-inflammatory, analgesic, antitumor, or antimicrobial agents. The venom has a potent antimicrobial activity, so snakes can keep their prey uncontaminated when digestion takes days.

One of the major problems facing public health is the growing resistance of microbes to antibiotics, so multiple scientific approaches have been employed to find new antimicrobials with high therapeutic indexes. Natural secretions, including snake venoms, have been considered excellent sources of bioactive compounds, with mechanisms of biological and physiological actions alternative to those of the conventional antibiotics. Thus, these proteins, oligopeptides, and short peptides can be seen as potential bactericides and fungicides, or valuable leading molecules [214]. In addition, larger AMPs can be proteolyzed to generate short antimicrobial fragments. The information given here fully agrees with a previous report that also discusses this important matter [215].

In the last century, snake-venom toxins were extensively studied for their antimicrobial activity and other properties, most likely because they are an abundant natural source [216]. As emphasized here, the AMPs studied more recently are cathelicidins (Tables 4–6) and defensins. Indeed, with a few exceptions, these macromolecules can be expressed on demand in low or large amounts, and they fit the pattern described above. Transcriptome and genome databases can help to overcome any difficulty concerning obtaining biomolecules that have a low expression or that are not easily purified.

In this report, we also describe new sequences obtained from the genome of the rattlesnake *S. miliarius* using PCR. Eight were shown to codify four β-defensins, but only one peptide has antimicrobial potential as predicted by the phylogenetic analysis (Figure 3) and calculation of theoretical net charge. This peptide was encoded by MT024631, MT024632, and MT024633 sequences.

The association of phylogenetic analysis and biological activity can provide us with indications to choose the best organism for searching for the molecules that have the necessary biological activity or sequences and help select the best minimal structure to develop [217]. Such an approach was used for cathelicidins. Phylogenetic relationships were established, and the antimicrobial activities and net charges were associated with sequences. In this context, the phylogenetic tree of Figure 4 showing cathelicidin groups with antibacterial activity (1) with and without activities (2), and not tested (3) indicates that the unknown sequences with a larger chance of having antimicrobial activity could be those related to group 1. In order to confirm this hypothesis, more antimicrobial tests need to be done with the molecules of this branch.

Finally, this article reinforced that the peptides of snake venoms are valued biopolymers that could be used in vivo as antimicrobial drugs for activating the cellular and immune response of superior animals, and improving the immune response to infection. An interesting proposal is to employ mixtures of AMPs combined with conventional antibiotics, aiming to potentiate their actions on pathogenic microorganisms and circumvent drug resistance [197,205]. Snake-venom proteins, oligopeptides, and short peptides can also be used for wound healing, preventing infection, and increasing cell regeneration.

Much remains to be done in this field of research after finding a new bioactive molecule, such as maintaining or increasing bioactivity under physiological conditions, decrease cytotoxicity, and increase chemical stability in vivo. The protection of peptides by carboxyamidation can increase the chemical stability and improve antimicrobial activity [9,205].

5. Conclusions

In conclusion, snakes and their secretions are important sources of antimicrobials. Molecular evolution and phylogeny approaches, in addition to traditional techniques such as proteomics, transcriptomics, peptide chemistry, and in silico studies, can increase the success of searching for new molecules with therapeutical potential or peptide-based lead compounds.

Supplementary Materials: The following supporting information can be downloaded at: https://www.mdpi.com/article/10.3390/ani13040744/s1, Table S1: Characteristics of snake cathelicidins. The sequences were obtained from the literature or NCBI databases. The net charge was calculated using the Henderson-Hasselbalch equation and the Lehninger pKa Scale; Table S2: Snake cathelicidins genome position. The contigs containing cathelicidin genes were identified through recursive BLAST searches of the WGS NCBI database, and the approximate positions of the genes were recorded. The sequences represented in this table are all partial and have not been previously reported in the literature. In silico approach has been used to search the sequences, realize the alignments, and generate the Phylogenetic tree [153,218–226].

Author Contributions: Conceptualization, writing and review, N.O. and M.T.M.; β-defensin sequences of *S. miliarius*, L.S. and P.V.D.; bibliography search, L.S., P.V.D. and M.A.S.-L.; cathelicidin genes search, L.S. All authors have read and agreed to the published version of the manuscript.

Funding: This research was funded by São Paulo Research Foundation—FAPESP (www.fapesp.br) for grants 2015/00003-5 (NO) and 2022/01825-2 (MTM).

Institutional Review Board Statement: This work was conducted according to Brazilian laws.

Informed Consent Statement: Not applicable.

Data Availability Statement: β-defensin sequences of *Sistrurus miliarius* can be retrieved at GenBank, accession numbers MT024631 to MT024638.

Acknowledgments: We thank H. Lisle Gibbs (Department of Evolution, Ecology and Organismal Biology—The Ohio State University, Columbus, OH, USA) for providing the *Sistrurus* DNA. We thank the Laboratory of Bacteriology—Instituto Butantan for the sequencing of β-defensin clones.

Conflicts of Interest: The authors declare no conflict of interest. The funders had no role in the design of the study; in the collection, analyses, or interpretation of data; in the writing of the manuscript; or in the decision to publish the results.

References

1. Zasloff, M. Antimicrobial peptides of multicellular organisms. *Nature* **2002**, *415*, 389–395. [CrossRef]
2. Nunes, L.G.P.; Reichert, T.; Machini, M.T. His-Rich Peptides, Gly- and His-Rich Peptides: Functionally Versatile Compounds with Potential Multi-Purpose Applications. *Int. J. Pept. Res. Ther.* **2021**, *27*, 2945–2963. [CrossRef]
3. Santana, F.L.; Estrada, K.; Ortiz, E.; Corzo, G. Reptilian β-defensins: Expanding the repertoire of known crocodylian peptides. *Peptides* **2021**, *136*, 170473. [CrossRef]
4. Pyron, R.A.; Burbrink, F.T.; Wiens, J.J. A phylogeny and revised classification of Squamata, including 4161 species of lizards and snakes. *BMC Evol. Biol.* **2013**, *13*, 93. [CrossRef]
5. Reeks, T.A.; Fry, B.G.; Alewood, P.F. Privileged frameworks from snake venom. *Cell. Mol. Life Sci.* **2015**, *72*, 1939–1958. [CrossRef]
6. Samy, R.P.; Gopalakrishnakone, P.; Satyanarayanajois, S.D.; Stiles, B.G.; Chow, V.T.K. Snake Venom Proteins and Peptides as Novel Antibiotics against Microbial Infections. *Curr. Proteom.* **2013**, *10*, 10–28. [CrossRef]
7. Assis, R.A.; Bittar, B.B.; Amorim, N.P.L.; Carrasco, G.H.; Silveira, E.D.R.; Benvindo-Souza, M.; Santos, L.R.S. Studies about Snake Peptides: A Review about Brazilian Contribution. *Braz. Arch. Biol. Technol.* **2022**, *65*, e22210421. [CrossRef]
8. Hancock, R.E.W. Peptide antibiotics. *Lancet* **1997**, *349*, 418–422. [CrossRef]
9. Carvalho, L.A.C.; Remuzgo, C.; Perez, K.R.; Machini, M.T. Hb40-61a: Novel analogues help expanding the knowledge on chemistry, properties and candidacidal action of this bovine α-hemoglobin-derived peptide. *Biochim. Biophys. Acta* **2015**, *1848*, 3140–3149. [CrossRef]
10. Radzishevsky, I.S.; Rotem, S.; Bourdetsky, D.; Navon-Venezia, S.; Carmeli, Y.; Mor, A. Improved antimicrobial peptides based on acyl-lysine oligomers. *Nat. Biotechnol.* **2007**, *25*, 657–659. [CrossRef]
11. Zhou, H.; Dou, J.; Wang, J.; Chen, L.; Wang, H.; Zhou, W.; Li, Y.; Zhou, C. The antibacterial activity of BF-30 in vitro and in infected burned rats is through interference with cytoplasmic membrane integrity. *Peptides* **2011**, *32*, 1131–1138. [CrossRef]
12. Xie, J.P.; Yue, J.; Xiong, Y.L.; Wang, W.Y.; Yu, S.Q.; Wang, H.H. In vitro activities of small peptides from snake venom against clinical isolates of drug-resistant *Mycobacterium tuberculosis*. *Int. J. Antimicrob. Agents* **2003**, *22*, 172–174. [CrossRef]
13. Zimmerman, L.M.; Vogel, L.A.; Bowden, R.M. Understanding the vertebrate immune system: Insights from the reptilian perspective. *J. Exp. Biol.* **2010**, *213*, 661–671. [CrossRef]
14. Rios, F.M.; Zimmerman, L.M. Immunology of Reptiles. In *eLS*; John Wiley & Sons, Ltd.: Chichester, UK, 2015.
15. Grego, K.F.; Alves, J.A.S.; Albuquerque, L.C.R.; Fernandes, W. Referências hematológicas para a jararaca de rabo branco (*Bothrops leucurus*) recém capturadas da natureza. *Arq. Bras. Med. Vet. Zootec.* **2006**, *58*, 1240–1243. [CrossRef]
16. Carvalho, M.P.N.; Queiroz-Hazarbassanov, N.G.T.; Massoco, C.O.; Sant'Anna, S.S.; Lourenço, M.M.; Levin, G.; Sogayar, M.C.; Grego, K.F.; Catão-Dias, J.L. Functional characterization of neotropical snakes peripheral blood leukocytes subsets: Linking flow cytometry cell features, microscopy images and serum corticosterone levels. *Dev. Comp. Immunol.* **2017**, *74*, 144–153. [CrossRef]
17. Farag, M.A.; El Ridi, R. Proliferative Responses of Snake Lymphocytes to Concanavalin A. *Dev. Comp. Immunol.* **1986**, *10*, 561–569. [CrossRef]
18. Saad, A.H. Sex-Associated Differences in the Mitogenic Responsiveness of Snake Blood Lymphocytes. *Dev. Comp. Immunol.* **1989**, *13*, 225–229. [CrossRef]
19. Uetz, P.; Freed, P.; Aguilar, R.; Hošek, J. (Eds.) The Reptile Database. 2022. Available online: https://www.reptile-database.org (accessed on 1 December 2022).
20. Vogel, C.W.; Muller-Eberhard, J. The Cobra Complement System: I. The Alternative Pathway of Activation. *Dev. Comp. Immunol.* **1985**, *9*, 311–325. [CrossRef]
21. Graham, S.P.; Fielman, K.T.; Mendonça, M.T. Thermal performance and acclimatization of a component of snake (*Agkistrodon piscivorus*) innate immunity. *J. Exp. Zool.* **2017**, *327*, 351–357. [CrossRef]
22. Bals, R.; Wilson, J.M. Cathelicidins—A family of multifunctional antimicrobial peptides. *Cell. Mol. Life Sci.* **2003**, *60*, 711–720. [CrossRef]
23. Graham, S.P.; Earley, R.L.; Guyer, C.; Mendonça, M.T. Innate immune performance and steroid hormone profiles of pregnant versus nonpregnant cottonmouth snakes (*Agkistrodon piscivorus*). *Gen. Comp. Endocr.* **2011**, *174*, 348–353. [CrossRef]
24. French, S.S.; Neuman-Lee, L.A. Improved ex vivo method for microbiocidal activity across vertebrate species. *Biol. Open* **2012**, *1*, 482–487. [CrossRef]
25. Figueiredo, A.C.; Nogueira, L.A.K.; Titon, S.C.M.; Gomes, F.R.; Carvalho, J.E. Immune and hormonal regulation of the *Boa constrictor* (Serpentes; Boidae) in response to feeding. *Comp. Biochem. Physiol. A* **2022**, *264*, 111–119. [CrossRef]
26. Brusch, G.A., IV; Mills, A.M.; Walman, R.M.; Masuda, G.; Byeon, A.; DeNardo, D.F.; Stahlschmidt, Z.R. Dehydration enhances cellular and humoral immunity in a mesic snake community. *J. Exp. Zool.* **2020**, *333*, 306–315. [CrossRef]
27. Brusch, G.A., IV; DeNardo, D.F. Egg desiccation leads to dehydration and enhanced innate immunity in python embryos. *Dev. Comp. Immunol.* **2019**, *90*, 147–151. [CrossRef]
28. Brusch, G.A., IV; DeNardo, D.F. When less means more: Dehydration improves innate immunity in rattlesnakes. *J. Exp. Biol.* **2017**, *220*, 2287–2295. [CrossRef]
29. Fabrıcio-Neto, A.; Madelaire, C.B.; Gomes, F.R.; Andrade, D.V. Exposure to fluctuating temperatures leads to reduced immunity and to stress response in rattlesnakes. *J. Exp. Biol.* **2019**, *222*, jeb208645. [CrossRef]
30. Baker, S.J.; Merchant, M.E. Antibacterial properties of plasma from the prairie rattlesnake (*Crotalus viridis*). *Dev. Comp. Immunol.* **2018**, *84*, 273–278. [CrossRef]

31. Brusch, G.A., IV; Christian, K.; Brown, G.P.; Shine, R.; DeNardo, D.F. Dehydration enhances innate immunity in a semiaquatic snake from the wet-dry tropics. *J. Exp. Zool.* **2019**, *331*, 245–252. [CrossRef]
32. Tripathi, M.K.; Singh, R. Differential Suppressive Effects of Testosterone on Immune Function in Fresh Water Snake, *Natrix piscator*: An In Vitro Study. *PLoS ONE* **2014**, *9*, e104431. [CrossRef]
33. Tripathi, M.K.; Singh, R.; Pati, A.K. Daily and Seasonal Rhythms in Immune Responses of Splenocytes in the Freshwater Snake, *Natrix piscator*. *PLoS ONE* **2015**, *10*, e0116588. [CrossRef]
34. Singh, A.; Singh, R.; Tripathi, M.K. Photoperiodic manipulation modulates the innate and cell mediated immune functions in the freshwater snake, *Natrix piscator*. *Sci. Rep.* **2020**, *10*, 14722. [CrossRef]
35. Luoma, R.L.; Butler, M.W.; Stahlschmidt, Z.R. Plasticity of immunity in response to eating. *J. Exp. Biol.* **2016**, *219*, 1965–1968. [CrossRef]
36. Lind, C.M.; Agugliaro, J.; Farrell, T.M. The metabolic response to an immune challenge in a viviparous snake, *Sistrurus miliarius*. *J. Exp. Biol.* **2020**, *223*, jeb225185. [CrossRef]
37. McCoy, C.M.; Lind, C.M.; Farrell, T.M. Environmental and physiological correlates of the severity of clinical signs of snake fungal disease in a population of pigmy rattlesnakes, *Sistrurus miliarius*. *Conserv. Physiol.* **2017**, *5*, cow077. [CrossRef]
38. Sparkman, A.M.; Palacios, M.G. A test of life-history theories of immune defence in two ecotypes of the garter snake, *Thamnophis elegans*. *J. Anim. Ecol.* **2009**, *78*, 1242–1248. [CrossRef]
39. Palacios, M.G.; Sparkman, A.M.; Bronikowski, A.M. Developmental plasticity of immune defence in two life-history ecotypes of the garter snake, *Thamnophis elegans*—A common-environment experiment. *J. Anim. Ecol.* **2011**, *80*, 431–437. [CrossRef]
40. Neuman-Lee, L.A.; Fokidis, H.B.; Spence, A.R.; van der Walt, M.; Smith, G.D.; Durham, S.; Smith, S.S. Food restriction and chronic stress alter energy use and affect immunity in an infrequent feeder. *Funct. Ecol.* **2015**, *29*, 1453–1462. [CrossRef]
41. Palacios, M.G.; Bronikowski, A.M. Immune variation during pregnancy suggests immune component-specific costs of reproduction in a viviparous snake with disparate life-history strategies. *J. Exp. Zool.* **2017**, *327*, 513–522. [CrossRef]
42. Palacios, M.G.; Gangloff, E.J.; Reding, D.M.; Bronikowski, A.M. Genetic background and thermal environment differentially influence the ontogeny of immune components during early life in an ectothermic vertebrate. *J. Anim. Ecol.* **2020**, *89*, 1883–1894. [CrossRef]
43. Spence, A.R.; French, S.S.; Hopkins, G.R.; Durso, A.M.; Hudson, S.B.; Smith, G.D.; Neuman-Lee, L.A. Long-term monitoring of two snake species reveals immune–endocrine interactions and the importance of ecological context. *J. Exp. Zool.* **2020**, *333*, 744–755. [CrossRef]
44. Combrink, L.L.; Bronikowski, A.M.; Miller, D.A.W.; Sparkman, A.M. Current and time-lagged effects of climate on innate immunity in two sympatric snake species. *Ecol. Evol.* **2021**, *11*, 3239–3250. [CrossRef]
45. Neuman-Lee, L.A.; van Wettere, A.J.; French, S.S. Interrelations among Multiple Metrics of Immune and Physiological Function in a Squamate, the Common Gartersnake (*Thamnophis sirtalis*). *Physiol. Biochem. Zool.* **2019**, *92*, 12–23. [CrossRef]
46. Kobolkuti, L.; Cadar, D.; Czirjak, G.; Niculae, M.; Kiss, T.; Sandru, C.; Spinu, M. The Effects of Environment and Physiological Cyclicity on the Immune System of *Viperinae*. *Sci. World J.* **2012**, *2012*, 574867. [CrossRef]
47. Stiles, B.G.; Sexton, F.W.; Weinstein, S.A. Antibacterial Effects of Different Snake Venoms: Purification and Characterization of Antibacterial Proteins from *Pseudechis australis* (Australian King Brown or Muga Snake) Venom. *Toxicon* **1991**, *29*, 1129–1141. [CrossRef]
48. Skarnes, R.C. L-Amino-acid Oxidase, a Bactericidal System. *Nature* **1970**, *225*, 1072–1073. [CrossRef]
49. Sulca-Lopez, M.A.; Remuzgo, C.; Cardenas, J.; Kiyota, S.; Cheng, E.; Bemquerer, M.P.; Machini, M.T. Venom of the Peruvian snake *Bothriopsis oligolepis*: Detection of antibacterial activity and involvement of proteolytic enzymes and C-type lectins in growth inhibition of *Staphylococcus aureus*. *Toxicon* **2017**, *134*, 30–40. [CrossRef]
50. Rheubert, J.L.; Meyer, M.F.; Strobel, R.M.; Pasternak, M.A.; Charvat, R.A. Predicting antibacterial activity from snake venom proteomes. *PLoS ONE* **2020**, *15*, e0226807. [CrossRef]
51. Arlinghaus, F.T.; Eble, J.A. C-type lectin-like proteins from snake venoms. *Toxicon* **2012**, *60*, 512–519. [CrossRef]
52. Murakami, M.T.; Zela, S.P.; Gava, L.M.; Michelan-Duarte, S.; Cintra, A.C.O.; Arni, R.K. Crystal structure of the platelet activator convulxin, a disulfide-linked a4b4 cyclic tetramer from the venom of *Crotalus durissus terrificus*. *Biochem. Biophys. Res. Commun.* **2003**, *310*, 478–482. [CrossRef]
53. Rádis-Baptista, G.; Moreno, F.B.M.B.; Nogueira, L.L.; Martins, A.M.C.; Toyama, D.O.; Toyama, M.H.; Cavada, B.S.; Azevedo, W.F., Jr.; Yamane, T. Crotacetin, a Novel Snake Venom C-Type Lectin Homolog of Convulxin, Exhibits an Unpredictable Antimicrobial Activity. *Cell Biochem. Biophys.* **2006**, *44*, 412–423. [CrossRef]
54. Castanheira, L.E.; Nunes, D.C.O.; Cardoso, T.M.; Santos, P.S.; Goulart, L.R.; Rodrigues, R.S.; Richardson, M.; Borges, M.H.; Yoneyama, K.A.G.; Rodrigues, V.M. Biochemical and functional characterization of a C-type lectin (BpLec) from *Bothrops pauloensis* snake venom. *Int. J. Biol. Macromol.* **2013**, *54*, 57–64. [CrossRef]
55. Nunes, E.S.; Souza, M.A.A.; Vaz, A.F.M.; Santana, G.M.S.; Gomes, F.S.; Coelho, L.C.B.B.; Paiva, P.M.G.; Silva, R.M.L.; Silva-Lucca, R.A.; Oliva, M.L.V.; et al. Purification of a lectin with antibacterial activity from *Bothrops leucurus* snake venom. *Comp. Biochem. Physiol. B* **2011**, *159*, 57–63. [CrossRef]
56. Klein, R.C.; Fabres-Klein, M.H.; de Oliveira, L.L.; Feio, R.N.; Malouin, F.; Ribon, A.O.B. A C-Type Lectin from *Bothrops jararacussu* Venom Disrupts Staphylococcal Biofilms. *PLoS ONE* **2015**, *10*, e0120514. [CrossRef]

57. Moura-da-Silva, A.M.; Theakston, R.D.G.; Crampton, J.M. Evolution of Disintegrin Cysteine-Rich and Mammalian Matrix-Degrading Metalloproteinases: Gene Duplication and Divergence of a Common Ancestor Rather than Convergent Evolution. *J. Mol. Evol.* **1996**, *43*, 263–269. [CrossRef]
58. Bazaa, A.; Juárez, P.; Marrakchi, N.; Lasfer, Z.B.; El Ayeb, M.; Harrison, R.A.; Calvete, J.J.; Sanz, L. Loss of Introns Along the Evolutionary Diversification Pathway of Snake Venom Disintegrins Evidenced by Sequence Analysis of Genomic DNA from *Macrovipera lebetina transmediterranea* and *Echis ocellatus*. *J. Mol. Evol.* **2007**, *64*, 261–271. [CrossRef]
59. Samy, R.P.; Gopalakrishnakone, P.; Chow, V.T.K.; Ho, B. Viper Metalloproteinase (*Agkistrodon halys* Pallas) with Antimicrobial Activity against Multi-Drug Resistant Human Pathogens. *J. Cell. Physiol.* **2008**, *216*, 54–68. [CrossRef]
60. Allane, D.; Oussedik-Oumehdi, H.; Harrat, Z.; Seve, M.; Laraba-Djebari, F. Isolation and characterization of an anti-leishmanial disintegrin from *Cerastes cerastes* venom. *J. Biochem. Mol. Toxicol.* **2018**, *32*, e22018. [CrossRef]
61. Serrano, S.M.T.; Maroun, R.C. Snake venom serine proteinases: Sequence homology vs. substrate specificity, a paradox to be solved. *Toxicon* **2005**, *45*, 1115–1132. [CrossRef]
62. Castro, H.C.; Zingali, R.B.; Albuquerque, M.G.; Pujol-Luz, M.; Rodrigues, C.R. Snake venom thrombin-like enzymes: From reptilase to now. *Cell. Mol. Life Sci.* **2004**, *61*, 843–856. [CrossRef]
63. Ali, S.A.; Stoeva, S.; Abbasi, A.; Alam, J.M.; Kayed, R.; Faigle, M.; Neumeister, B.; Voelter, W. Isolation, Structural, and Functional Characterization of an Apoptosis-Inducing L-Amino Acid Oxidase from Leaf-Nosed Viper (*Eristocophis macmahoni*) Snake Venom. *Arch. Biochem. Biophys.* **2000**, *384*, 216–226. [CrossRef]
64. Du, X.-Y.; Clemetson, K.J. Snake venom L-amino acid oxidases. *Toxicon* **2002**, *40*, 659–665. [CrossRef]
65. Takatsuka, H.; Sakurai, Y.; Yoshioka, A.; Kokubo, T.; Usami, Y.; Suzuki, M.; Matsui, T.; Titani, K.; Yagi, H.; Matsumoto, M.; et al. Molecular characterization of L-amino acid oxidase from *Agkistrodon halys blomhoffii* with special reference to platelet aggregation. *Biochim. Biophys. Acta* **2001**, *1544*, 267–277. [CrossRef]
66. Kasai, K.; Nakano, M.; Ohishi, M.; Nakamura, T.; Miura, T. Antimicrobial properties of L-amino acid oxidase: Biochemical features and biomedical applications. *Appl. Microbiol. Biotechnol.* **2021**, *105*, 4819–4832. [CrossRef]
67. Sun, M.-Z.; Guo, C.; Tian, Y.; Chen, D.; Greenaway, F.T.; Liu, S. Biochemical, functional and structural characterization of Akbu-LAAO: A novel snake venom L-amino acid oxidase from *Agkistrodon blomhoffii ussurensis*. *Biochimie* **2010**, *92*, 343–349. [CrossRef]
68. Zhang, H.; Yang, Q.; Sun, M.; Teng, M.; Niu, L. Hydrogen Peroxide produced by Two Amino Acid Oxidases Mediates Antibacterial Actions. *J. Microbiol.* **2004**, *42*, 336–339.
69. Muñoz, L.J.V.; Estrada-Gomez, S.; Núñez, V.; Sanz, L.; Calvete, J.J. Characterization and cDNA sequence of *Bothriechis schlegelii* L-aminoacid oxidase with antibacterial activity. *Int. J. Biol. Macromol.* **2014**, *69*, 200–207. [CrossRef]
70. Stábeli, R.G.; Marcussi, S.; Carlos, G.B.; Pietro, R.C.L.R.; Selistre-de-Araújo, H.S.; Giglio, J.R.; Oliveira, E.B.; Soares, A.M. Platelet aggregation and antibacterial effects of an L-amino acid oxidase purified from *Bothrops alternatus* snake venom. *Bioorg. Med. Chem.* **2004**, *12*, 2881–2886. [CrossRef]
71. Ciscotto, P.; Avila, R.A.M.; Coelho, E.A.F.; Oliveira, J.; Diniz, C.G.; Farías, L.M.; Carvalho, M.A.R.; Maria, W.S.; Sanchez, E.F.; Borges, A.; et al. Antigenic, microbicidal and antiparasitic properties of an L-amino acid oxidase isolated from *Bothrops jararaca* snake venom. *Toxicon* **2009**, *53*, 330–341. [CrossRef]
72. Torres, A.F.C.; Dantas, R.T.; Toyama, M.H.; Diz Filho, E.; Zara, F.J.; Queiroz, M.G.R.; Nogueira, N.A.P.; Oliveira, M.R.; Toyama, D.O.; Monteiro, H.S.A.; et al. Antibacterial and antiparasitic effects of *Bothrops marajoensis* venom and its fractions: Phospholipase A_2 and L-amino acid oxidase. *Toxicon* **2010**, *55*, 795–804. [CrossRef]
73. Okubo, B.M.; Silva, O.N.; Migliolo, L.; Gomes, D.G.; Porto, W.F.; Batista, C.L.; Ramos, C.S.; Holanda, H.H.S.; Dias, S.C.; Franco, O.L.; et al. Evaluation of an Antimicrobial L-Amino Acid Oxidase and Peptide Derivatives from *Bothropoides mattogrosensis* Pitviper Venom. *PLoS ONE* **2012**, *7*, e33639. [CrossRef]
74. Stábeli, R.G.; Sant'Ana, C.D.; Ribeiro, P.H.; Costa, T.R.; Ticli, F.K.; Pires, M.G.; Nomizo, A.; Albuquerque, S.; Malta-Neto, N.R.; Marins, M.; et al. Cytotoxic L-amino acid oxidase from *Bothrops moojeni*: Biochemical and functional characterization. *Int. J. Biol. Macromol.* **2007**, *41*, 132–140. [CrossRef]
75. Rodrigues, R.S.; Silva, J.F.; França, J.B.; Fonseca, F.P.P.; Otaviano, A.R.; Silva, F.H.; Hamaguchi, A.; Magro, A.J.; Braz, A.S.K.; Santos, J.I.; et al. Structural and functional properties of Bp-LAAO, a new L-amino acid oxidase isolated from *Bothrops pauloensis* snake venom. *Biochimie* **2009**, *91*, 490–501. [CrossRef]
76. Izidoro, L.F.M.; Ribeiro, M.C.; Souza, G.R.L.; Sant'Ana, C.D.; Hamaguchi, A.; Homsi-Brandeburgo, M.I.; Goulart, L.R.; Beleboni, R.O.; Nomizo, A.; Sampaio, S.V.; et al. Biochemical and functional characterization of an L-amino acid oxidase isolated from *Bothrops pirajai* snake venom. *Bioorg. Med. Chem.* **2006**, *14*, 7034–7043. [CrossRef]
77. Costa, T.R.; Menaldo, D.L.; Silva, C.P.; Sorrechia, R.; Albuquerque, S.; Pietro, R.C.L.R.; Ghisla, S.; Antunes, L.M.G.; Sampaio, S.V. Evaluating the microbicidal, antiparasitic and antitumor effects of CR-LAAO from *Calloselasma rhodostoma* venom. *Int. J. Biol. Macromol.* **2015**, *80*, 489–497. [CrossRef]
78. Toyama, M.H.; Toyama, D.O.; Passero, L.F.D.; Laurenti, M.D.; Corbett, C.E.; Tomokane, T.Y.; Fonseca, F.V.; Antunes, E.; Joazeiro, P.P.; Beriam, L.O.S.; et al. Isolation of a new L-amino acid oxidase from *Crotalus durissus cascavella* venom. *Toxicon* **2006**, *47*, 47–57. [CrossRef]
79. Vargas, L.J.; Quintana, J.C.; Pereañez, J.A.; Núñez, V.; Sanz, L.; Calvete, J. Cloning and characterization of an antibacterial L-amino acid oxidase from *Crotalus durissus cumanensis* venom. *Toxicon* **2013**, *64*, 1–11. [CrossRef]

80. Zhong, S.-R.; Jin, Y.; Wu, J.-B.; Jia, Y.-H.; Xu, G.-L.; Wang, G.-C.; Xiong, Y.-L.; Lu, Q.-M. Purification and characterization of a new L-amino acid oxidase from *Daboia russellii siamensis* venom. *Toxicon* **2009**, *54*, 763–771. [CrossRef]
81. Samel, M.; Tonismagi, K.; Ronnholm, G.; Vija, H.; Siigur, J.; Kalkkinen, N.; Siigur, E. L-Amino acid oxidase from *Naja naja oxiana* venom. *Comp. Biochem. Physiol. B* **2008**, *149*, 572–580. [CrossRef]
82. Lee, M.L.; Tan, N.H.; Fung, S.Y.; Sekaran, S.D. Antibacterial action of a heat-stable form of L-amino acid oxidase isolated from king cobra (*Ophiophagus hannah*) venom. *Comp. Biochem. Physiol. C* **2011**, *153*, 237–242. [CrossRef]
83. Phua, C.S.; Vejayan, J.; Ambu, S.; Ponnudurai, G.; Gorajana, A. Purification and antibacterial activities of an L-amino acid oxidase from king cobra (*Ophiophagus hannah*) venom. *J. Venom. Anim. Toxins Incl. Trop. Dis.* **2012**, *18*, 198–207. [CrossRef]
84. Zhang, Y.-J.; Wang, J.-H.; Lee, W.-H.; Wang, Q.; Liu, H.; Zheng, Y.-T.; Zhang, Y. Molecular characterization of *Trimeresurus stejnegeri* venom L-amino acid oxidase with potential anti-HIV activity. *Biochem. Biophys. Res. Commun.* **2003**, *309*, 598–604. [CrossRef]
85. Tõnismagi, K.; Samel, M.; Trummal, K.; Ronnholm, G.; Siigur, J.; Kalkkinen, N.; Siigur, E. L-Amino acid oxidase from *Vipera lebetina* venom: Isolation, characterization, effects on platelets and bacteria. *Toxicon* **2006**, *48*, 227–237. [CrossRef]
86. Arias, S.P.; Rey-Suárez, P.; Pereáñez, J.A.; Acosta, C.; Rojas, M.; Santos, L.D.; Ferreira, R.S., Jr.; Núñez, V. Isolation and Functional Characterization of an Acidic Myotoxic Phospholipase A_2 from Colombian *Bothrops asper* Venom. *Toxins* **2017**, *9*, 342. [CrossRef]
87. Lomonte, B. Lys49 myotoxins, secreted phospholipase A_2-like proteins of viperid venoms: A comprehensive review. *Toxicon* **2023**, *224*, 107024. [CrossRef]
88. Vargas, L.J.; Londoño, M.; Quintana, J.C.; Rua, C.; Segura, C.; Lomonte, B.; Núñez, V. An acidic phospholipase A_2 with antibacterial activity from *Porthidium nasutum* snake venom. *Comp. Biochem. Physiol. B Biochem. Mol. Biol.* **2012**, *161*, 341–347. [CrossRef]
89. Costa, T.R.; Menaldo, D.L.; Oliveira, C.Z.; Santos-Filho, N.A.; Teixeira, S.S.; Nomizo, A.; Fuly, A.L.; Monteiro, M.C.; Souza, B.M.; Palma, M.S.; et al. Myotoxic phospholipases A_2 isolated from *Bothrops brazili* snake venom and synthetic peptides derived from their C-terminal region: Cytotoxic effect on microorganism and tumor cells. *Peptides* **2008**, *29*, 1645–1656. [CrossRef]
90. Páramo, L.; Lomonte, B.; Pizarro-Cerdá, J.; Bengoechea, J.A.; Gorvel, J.-P.; Moreno, E. Bactericidal activity of Lys49 and Asp49 myotoxic phospholipases A_2 from *Bothrops asper* snake venom Synthetic Lys49 myotoxin II-(115−129)-peptide identifies its bactericidal region. *Eur. J. Biochem.* **1998**, *253*, 452–461. [CrossRef]
91. Muller, V.D.M.; Russo, R.R.; Cintra, A.C.O.; Sartim, M.A.; Alves-Paiva, R.M.; Figueiredo, L.T.M.; Sampaio, S.V.; Aquino, V.H. Crotoxin and phospholipases A_2 from *Crotalus durissus terrificus* showed antiviral activity against dengue and yellow fever viruses. *Toxicon* **2012**, *59*, 507–515. [CrossRef]
92. Muller, V.D.; Soares, R.O.; dos Santos-Junior, N.N.; Trabuco, A.C.; Cintra, A.C.; Figueiredo, L.T.; Caliri, A.; Sampaio, S.V.; Aquino, V.H. Phospholipase A_2 Isolated from the Venom of *Crotalus durissus terrificus* Inactivates Dengue virus and Other Enveloped Viruses by Disrupting the Viral Envelope. *PLoS ONE* **2014**, *9*, e112351. [CrossRef]
93. Brenes, H.; Loría, G.D.; Lomonte, B. Potent virucidal activity against Flaviviridae of a group IIA phospholipase A_2 isolated from the venom of *Bothrops asper*. *Biologicals* **2020**, *63*, 48–52. [CrossRef]
94. Fenard, D.; Lambeau, G.; Valentin, E.; Lefebvre, J.-C.; Lazdunski, M.; Doglio, A. Secreted phospholipases A_2, a new class of HIV inhibitors that block virus entry into host cells. *J. Clin. Investig.* **1999**, *104*, 611–618. [CrossRef]
95. Shimizu, J.F.; Pereira, C.M.; Bittar, C.; Batista, M.N.; Campos, G.R.F.; da Silva, S.; Cintra, A.C.O.; Zothner, C.; Harris, M.; Sampaio, S.V.; et al. Multiple effects of toxins isolated from *Crotalus durissus terrificus* on the hepatitis C virus life cycle. *PLoS ONE* **2017**, *12*, e0187857. [CrossRef]
96. Roberto, P.G.; Kashima, S.; Marcussi, S.; Pereira, J.O.; Astolfi-Filho, S.; Nomizo, A.; Giglio, J.R.; Fontes, M.R.M.; Soares, A.M.; França, S.C. Cloning and Identification of a Complete cDNA Coding for a Bactericidal and Antitumoral Acidic Phospholipase A_2 from *Bothrops jararacussu* Venom. *Protein J.* **2004**, *23*, 273–285. [CrossRef]
97. Santamaría, C.; Larios, S.; Ângulo, Y.; Pizarro-Cerda, J.; Gorvel, J.-P.; Moreno, E.; Lomonte, B. Antimicrobial activity of myotoxic phospholipases A_2 from crotalid snake venoms and synthetic peptide variants derived from their C-terminal region. *Toxicon* **2005**, *45*, 807–815. [CrossRef]
98. Aragão, E.A.; Chioato, L.; Ward, R.J. Permeabilization of *E. coli* K12 inner and outer membranes by bothropstoxin-I, A LYS49 phospholipase A_2 from *Bothrops jararacussu*. *Toxicon* **2008**, *51*, 538–546. [CrossRef]
99. Barbosa, P.S.F.; Martins, A.M.C.; Havt, A.; Toyama, D.O.; Evangelista, J.S.A.M.; Ferreira, D.P.P.; Joazeiro, P.P.; Beriam, L.O.S.; Toyama, M.H.; Fonteles, M.C.; et al. Renal and antibacterial effects induced by myotoxin I and II isolated from *Bothrops jararacussu* venom. *Toxicon* **2005**, *46*, 376–386. [CrossRef]
100. Corrêa, E.A.; Kayano, A.M.; Diniz-Sousa, R.; Setúbal, S.S.; Zanchi, F.B.; Zuliani, J.P.; Matos, N.B.; Almeida, J.R.; Resende, L.M.; Marangoni, S.; et al. Isolation, structural and functional characterization of a new Lys49 phospholipase A_2 homologue from *Bothrops neuwiedi urutu* with bactericidal potential. *Toxicon* **2016**, *115*, 13–21. [CrossRef]
101. Xu, C.; Ma, D.; Yu, H.; Li, Z.; Liang, J.; Lin, G.; Zhang, Y.; Lai, R. A bactericidal homodimeric phospholipases A_2 from *Bungarus fasciatus* venom. *Peptides* **2007**, *28*, 969–973. [CrossRef]
102. Wen, Y.-L.; Wu, B.-J.; Kao, P.-H.; Fu, Y.-S.; Chang, L.-S. Antibacterial and membrane-damaging activities of β-bungarotoxin B chain. *J. Pept. Sci.* **2013**, *19*, 1–8. [CrossRef]
103. Samy, R.P.; Kandasamy, M.; Gopalakrishnakone, P.; Stiles, B.G.; Rowan, E.G.; Becker, D.; Shanmugam, M.K.; Sethi, G.; Chow, V.T.K. Wound Healing Activity and Mechanisms of Action of an Antibacterial Protein from the Venom of the Eastern Diamondback Rattlesnake (*Crotalus adamanteus*). *PLoS ONE* **2014**, *9*, e80199. [CrossRef]

104. Toyama, M.H.; Toyama, D.O.; Joazeiro, P.P.; Carneiro, E.M.; Beriam, L.O.S.; Marangoni, L.S.; Boschero, A.C. Biological and Structural Characterization of a New PLA_2 from the *Crotalus durissus collilineatus* Venom. *Protein J.* **2005**, *24*, 103–112. [CrossRef]
105. Diz Filho, E.B.S.; Marangoni, S.; Toyama, D.O.; Fagundes, F.H.R.; Oliveira, S.C.B.; Fonseca, F.V.; Calgarotto, A.K.; Joazeiro, P.P.; Toyama, M.H. Enzymatic and structural characterization of new PLA_2 isoform isolated from white venom of *Crotalus durissus ruruima*. *Toxicon* **2009**, *53*, 104–114. [CrossRef]
106. Samy, R.P.; Pachiappan, A.; Gopalakrishnakone, P.; Thwin, M.M.; Hian, Y.E.; Chow, V.T.K.; Bow, H.; Weng, J.T. In vitro antimicrobial activity of natural toxins and animal venoms tested against *Burkholderia pseudomallei*. *BMC Infect. Dis.* **2006**, *6*, 100.
107. Samy, R.P.; Gopalakrishnakone, P.; Thwin, M.M.; Chow, T.K.V.; Bow, H.; Yap, E.H.; Thong, T.W.J. Antibacterial activity of snake, scorpion and bee venoms: A comparison with purified venom phospholipase A_2 enzymes. *J. Appl. Microbiol.* **2007**, *102*, 650–659. [CrossRef]
108. Almeida, J.R.; Lancellotti, M.; Soares, A.M.; Calderon, L.A.; Ramírez, D.; González, W.; Marangoni, S.; da Silva, S.L. CoaTx-II, a new dimeric Lys49 phospholipase A2 from *Crotalus oreganus abyssus* snake venom with bactericidal potential: Insights into its structure and biological roles. *Toxicon* **2016**, *120*, 147–158. [CrossRef]
109. Sudarshan, S.; Dhananjaya, B.L. Antibacterial Potential of a Basic Phospholipase A_2 (VRV_PL_V) of *Daboia russellii pulchella* (Russell's Viper) Venom. *Biochemistry (Mosc.)* **2014**, *79*, 1237–1244. [CrossRef]
110. Sudharshan, S.; Dhananjaya, B.L. Antibacterial potential of a basic phospholipase A_2 (VRV-PL-VIIIa) from *Daboia russelii pulchella* (Russell's viper) venom. *J. Venom. Anim. Toxins Incl. Trop. Dis.* **2015**, *21*, 17. [CrossRef]
111. Samy, R.P.; Stiles, B.G.; Chinnathambi, A.; Zayed, M.E.; Alharbi, S.A.; Franco, O.L.; Rowan, E.G.; Kumar, A.P.; Lim, L.H.K.; Sethi, G. Viperatoxin-II: A novel viper venom protein as an effective bactericidal Agent. *FEBS Open Bio* **2015**, *5*, 928–941. [CrossRef]
112. Samy, R.P.; Gopalakrishnakone, P.; Bow, H.; Puspharaj, P.N.; Chow, V.T.K. Identification and characterization of a phospholipase A_2 from the venom of the Saw-scaled viper: Novel bactericidal and membrane damaging activities. *Biochimie* **2010**, *92*, 1854–1866. [CrossRef]
113. Diniz-Sousa, R.; Caldeira, C.A.S.; Kayano, A.M.; Paloschi, M.V.; Pimenta, D.C.; Simões-Silva, R.; Ferreira, A.S.; Zanchi, F.B.; Matos, N.B.; Grabner, F.P.; et al. Identification of the Molecular Determinants of the Antibacterial Activity of LmutTX, a Lys49 Phospholipase A2 Homologue Isolated from *Lachesis muta muta* Snake Venom (Linnaeus, 1766). *Basic Clin. Pharmacol. Toxicol.* **2018**, *122*, 413–423. [CrossRef]
114. Accary, C.; Mantash, A.; Mallem, Y.; Fajloun, Z.; Elkak, A. Separation and Biological Activities of Phospholipase A2 (Mb-PLA2) from the Venom of *Montivipera bornmuelleri*, a Lebanese Viper. *J. Liq. Chromatogr. Relat. Technol.* **2015**, *38*, 833–839. [CrossRef]
115. Sudarshan, S.; Dhananjaya, B.L. The Antimicrobial Activity of an Acidic Phospholipase A_2 (NN-XIa-PLA_2) from the Venom of *Naja naja naja* (Indian Cobra). *Appl. Biochem. Biotechnol.* **2015**, *176*, 2027–2038. [CrossRef]
116. Sudarshan, S.; Dhananjaya, B.L. Antibacterial activity of an acidic phospholipase A2 (NN-XIb-PLA2) from the venom of *Naja naja* (Indian cobra). *SpringerPlus* **2016**, *5*, 112. [CrossRef]
117. Adade, C.M.; Carvalho, A.L.O.; Tomaz, M.A.; Costa, T.F.R.; Godinho, J.L.; Melo, P.A.; Lima, A.P.C.A.; Rodrigues, J.C.F.; Zingali, R.B.; Souto-Padrón, T. Crovirin, a Snake Venom Cysteine-Rich Secretory Protein (CRISP) with Promising Activity against Trypanosomes and *Leishmania*. *PLoS Negl. Trop. Dis.* **2014**, *8*, e3252. [CrossRef]
118. Castillo, J.C.Q.; Vargas, L.J.; Segura, C.; Gutiérrez, J.M.; Pérez, J.C.A. In Vitro Antiplasmodial Activity of Phospholipases A_2 and a Phospholipase Homologue Isolated from the Venom of the Snake *Bothrops asper*. *Toxins* **2012**, *4*, 1500–1516. [CrossRef]
119. Mello, C.P.; Lima, D.B.; Menezes, R.R.P.P.B.; Bandeira, I.C.J.; Tessarolo, L.D.; Sampaio, T.L.; Falcao, C.B.; Radis-Baptista, G.; Martins, A.M.C. Evaluation of the antichagasic activity of batroxicidin, a cathelicidin-related antimicrobial peptide found in *Bothrops atrox* venom gland. *Toxicon* **2017**, *130*, 56–62. [CrossRef]
120. Tempone, A.G.; Andrade, H.F., Jr.; Spencer, P.J.; Lourenço, C.O.; Rogero, J.R.; Nascimento, N. *Bothrops moojeni* Venom Kills *Leishmania* spp. with Hydrogen Peroxide Generated by Its L-Amino Acid Oxidase. *Biochem. Biophys. Res. Commun.* **2001**, *280*, 620–624. [CrossRef]
121. Wang, Y.; Hong, J.; Liu, X.; Yang, H.; Liu, R.; Wu, J.; Wang, A.; Lin, D.; Lai, R. Snake Cathelicidin from *Bungarus fasciatus* Is a Potent Peptide Antibiotics. *PLoS ONE* **2008**, *3*, e3217. [CrossRef]
122. Quintana, J.C.; Chacón, A.M.; Vargas, L.; Segura, C.; Gutiérrez, J.M.; Alarcón, J.C. Antiplasmodial effect of the venom of *Crotalus durissus cumanensis*, crotoxin complex and Crotoxin B. *Acta Trop.* **2012**, *124*, 126–132. [CrossRef]
123. Yamane, E.S.; Bizerra, F.C.; Oliveira, E.B.; Moreira, J.T.; Rajabi, M.; Nunes, G.L.C.; Souza, A.O.; Silva, I.D.C.G.; Yamane, T.; Karpel, R.L.; et al. Unraveling the antifungal activity of a South American rattlesnake toxin Crotamine. *Biochimie* **2013**, *95*, 231–240. [CrossRef]
124. El Chamy Maluf, S.; Dal Mas, C.; Oliveira, E.B.; Melo, P.M.S.; Carmona, A.K.; Gazarini, M.L.; Hayashi, M.A.F. Inhibition of malaria parasite *Plasmodium falciparum* development by crotamine, a cell penetrating peptide from the snake venom. *Peptides* **2016**, *78*, 11–16. [CrossRef]
125. Bandeira, I.C.J.; Bandeira-Lima, D.; Mello, C.P.; Pereira, T.P.; de Menezes, R.R.P.P.B.; Sampaio, T.L.; Falcão, C.B.; Rádis-Baptista, G.; Martins, A.M.C. Antichagasic effect of crotalicidin, a cathelicidin-like vipericidin, found in *Crotalus durissus terrificus* rattlesnake's venom gland. *Parasitology* **2018**, *145*, 1059–1064. [CrossRef]
126. Wei, L.; Gao, J.; Zhang, S.; Wu, S.; Xie, Z.; Ling, G.; Kuang, Y.-Q.; Yang, Y.; Yu, H.; Wang, Y. Identification and Characterization of the First Cathelicidin from Sea Snakes with Potent Antimicrobial and Anti-inflammatory Activity and Special Mechanism. *J. Biol. Chem.* **2015**, *290*, 16633–16652. [CrossRef]

127. Guillaume, C.; Deregnaucourt, C.; Clavey, V.; Schrévela, J. Anti-Plasmodium properties of group IA, IB, IIA and III secreted phospholipases A_2 are serum-dependent. *Toxicon* **2004**, *43*, 311–318. [CrossRef]
128. Wang, A.; Zhang, F.; Guo, Z.; Chen, Y.; Zhang, M.; Yu, H.; Wang, Y. Characterization of a Cathelicidin from the Colubrinae Snake, *Sinonatrix annularis*. *Zoolog. Sci.* **2019**, *36*, 68–76. [CrossRef]
129. Nair, D.G.; Fry, B.G.; Alewood, P.; Kumar, P.P.; Kini, R.M. Antimicrobial activity of omwaprin, a new member of the waprin family of snake venom proteins. *Biochem. J.* **2007**, *402*, 93–104. [CrossRef]
130. Torres, A.M.; Wong, H.Y.; Desai, M.; Moochhala, S.; Kuchel, P.W.; Kini, R.M. Identification of a Novel Family of Proteins in Snake Venoms. Purification and Structural Characterization of Nawaprin from *Naja nigricollis* Snake Venom. *J. Biol. Chem.* **2003**, *278*, 40097–40104. [CrossRef]
131. Chen, L.-W.; Kao, P.-H.; Fu, Y.-S.; Lin, S.-R.; Chang, L.-S. Membrane-damaging activity of Taiwan cobra cardiotoxin 3 is responsible for its bactericidal activity. *Toxicon* **2011**, *58*, 46–53. [CrossRef]
132. Chen, L.-W.; Kao, P.-H.; Fu, Y.-S.; Hu, W.-P.; Chang, L.-S. Bactericidal effect of *Naja nigricollis* toxin is related to its membrane-damaging activity. *Peptides* **2011**, *32*, 1755–1763. [CrossRef]
133. Kao, P.-H.; Lin, S.-R.; Hu, W.-P.; Chang, L.-S. *Naja naja atra* and *Naja nigricollis* cardiotoxins induce fusion of *Escherichia coli* and *Staphylococcus aureus* membrane-mimicking liposomes. *Toxicon* **2012**, *60*, 367–377. [CrossRef]
134. Martin, E.; Ganz, T.; Lehrer, R.I. Defensins and other endogenous peptide antibiotics of vertebrates. *Leukoc. Biol.* **1995**, *58*, 128–136. [CrossRef]
135. Nguyen, L.T.; Haney, E.F.; Vogel, H.J. The expanding scope of antimicrobial peptide structures and their modes of action. *Trends Biotechnol.* **2011**, *29*, 464–472. [CrossRef]
136. Gomes, V.M.; Carvalho, A.O.; Cunha, M.; Keller, M.N.; Bloch, C., Jr.; Deolindo, P.; Alves, E.W. Purification and characterization of a novel peptide with antifungal activity from *Bothrops jararaca* venom. *Toxicon* **2005**, *45*, 817–827. [CrossRef]
137. Oguiura, N.; Boni-Mitake, M.; Rádis-Baptista, G. New view on crotamine, a small basic polypeptide myotoxin from South American rattlesnake venom. *Toxicon* **2005**, *46*, 363–370. [CrossRef]
138. Coronado, M.A.; Gabdulkhakov, A.; Georgieva, D.; Sankaran, B.; Murakami, M.T.; Arni, R.K.; Betzel, C. Structure of the polypeptide crotamine from the Brazilian rattlesnake *Crotalus durissus terrificus*. *Acta Cryst.* **2013**, *D69*, 1958–1964.
139. Costa, B.A.; Sanches, L.; Gomide, A.B.; Bizerra, F.; Dal Mas, C.; Oliveira, E.B.; Perez, K.R.; Itri, R.; Oguiura, N.; Hayashi, M.A.F. Interaction of the Rattlesnake Toxin Crotamine with Model Membranes. *J. Phys. Chem. B* **2014**, *118*, 5471–5479. [CrossRef]
140. Yount, N.Y.; Kupferwasser, D.; Spisni, A.; Dutz, S.M.; Ramjan, Z.H.; Sharma, S.; Waring, A.J.; Yeaman, M.R. Selective reciprocity in antimicrobial activity versus cytotoxicity of hBD-2 and crotamine. *Proc. Natl. Acad. Sci. USA* **2009**, *106*, 14972–14977. [CrossRef]
141. Oguiura, N.; Boni-Mitake, M.; Affonso, R.; Zhang, G. In vitro antibacterial and hemolytic activities of crotamine, a small basic myotoxin from rattlesnake *Crotalus durissus*. *J. Antibiot. (Tokyo)* **2011**, *64*, 327–331. [CrossRef]
142. Oguiura, N.; Corrêa, P.G.; Rosmino, I.L.; de Souza, A.O.; Pasqualoto, K.F.M. Antimicrobial Activity of Snake β-Defensins and Derived Peptides. *Toxins* **2022**, *14*, 1. [CrossRef]
143. Scheetz, T.; Bartlett, J.A.; Walters, J.D.; Schutte, B.C.; Casavant, T.L.; McCray, P.B., Jr. Genomics-based approaches to gene discovery in innate immunity. *Immunol. Rev.* **2002**, *190*, 137–145. [CrossRef]
144. Schutte, B.C.; Mitros, J.P.; Bartlett, J.A.; Walters, J.D.; Jia, H.P.; Welsh, M.J.; Casavant, T.L.; McCray, P.B., Jr. Discovery of five conserved β-defensin gene clusters using a computational search strategy. *Proc. Natl. Acad. Sci. USA* **2002**, *99*, 2129–2133. [CrossRef]
145. Oliveira, Y.S.; Corrêa, P.G.; Oguiura, N. Beta-defensin genes of the Colubridae snakes *Phalotris mertensi*, *Thamnodynastes hypoconia*, and *T. strigatus*. *Toxicon* **2018**, *146*, 124–128. [CrossRef]
146. Corrêa, P.G.; Oguiura, N. Phylogenetic analysis of β-defensin-like genes of *Bothrops*, *Crotalus* and *Lachesis* snakes. *Toxicon* **2013**, *69*, 65–74. [CrossRef]
147. Campbell, J.A.; Lamar, W.W. *The Venomous Reptiles of the Western Hemisphere*, 2nd ed.; Publishing C, Cornell University Press: Ithaca, NY, USA, 2004; 976p.
148. Kaerse, M.; Wilson, A.; Stones-Havas, S.; Cheung, M.; Sturrock, S.; Buxton, S.; Cooper, A.; Markowitz, S.; Duran, C.; Thierer, T.; et al. Geneious Basic: An integrated and extendable desktop software platform for the organization and analysis of sequence data. *Bioinformatics* **2012**, *28*, 1647–1649. [CrossRef]
149. Rádis-Baptista, G.; Kubo, T.; Oguiura, N.; Svartman, M.; Almeida, T.M.B.; Batistic, R.F.; Oliveira, E.B.; Vianna-Morgante, A.M.; Yamane, T. Structure and chromosomal localization of the gene for crotamine, a toxin from the South American rattlesnake, *Crotalus durissus terrificus*. *Toxicon* **2003**, *42*, 747–752. [CrossRef]
150. Edgar, R.C. MUSCLE: A multiple sequence alignment method with reduced time and space complexity. *BMC Bioinform.* **2004**, *5*, 113. [CrossRef]
151. Hall, T.A. BioEdit: A user-friendly biological sequence alignment editor and analysis program for windows 95/98/NT. *Nucleic Acids Symp. Ser.* **1999**, *41*, 95–98.
152. Rádis-Baptista, G.; Kubo, T.; Oguiura, N.; Silva, A.R.B.P.; Hayashi, M.A.F.; Oliveira, E.B.; Yamane, T. Identification of crotasin, a crotamine-related gene of *Crotalus durissus terrificus*. *Toxicon* **2004**, *43*, 751–759. [CrossRef]
153. Jobb, G.; von Haeseler, A.; Strimmer, K. TREEFINDER: A powerful graphical analysis environment for molecular phylogenetics. *BMC Evol. Biol.* **2004**, *4*, 18. [CrossRef]

154. Strimmer, K.; Rambaut, A. Inferring confidence sets of possibly misspecified gene trees. *Proc. R. Soc. Lond. B* **2002**, *269*, 137–142. [CrossRef]
155. Zanetti, M. Cathelicidins, multifunctional peptides of the innate immunity. *J. Leukoc. Biol.* **2004**, *75*, 39–48. [CrossRef]
156. Zhao, H.; Gan, T.-X.; Liu, X.-D.; Jin, Y.; Lee, W.-H.; Shen, J.-H.; Zhang, Y. Identification and characterization of novel reptile cathelicidins from elapid snakes. *Peptides* **2008**, *29*, 1685–1691. [CrossRef]
157. Zanetti, M.; Gennaro, R.; Romeo, D. Cathelicidins: A novel protein family with a common proregion and a variable C-terminal antimicrobial domain. *FEBS Lett.* **1995**, *374*, 1–5. [CrossRef]
158. Kosciuczuk, E.M.; Lisowski, P.; Jarczak, J.; Strzałkowska, N.; Jozwik, A.; Horbanczuk, J.; Krzyzewski, J.; Zwierzchowski, L.; Bagnicka, E. Cathelicidins: Family of antimicrobial peptides. A review. *Mol. Biol. Rep.* **2012**, *39*, 10957–10970. [CrossRef]
159. Dalla Valle, L.; Benato, F.; Paccanaro, M.C.; Alibardi, L. Bioinformatic and molecular characterization of cathelicidin-like peptides isolated from the green lizard *Anolis carolinensis* (Reptilia: Lepidosauria: Iguanidae). *Ital. J. Zool.* **2013**, *80*, 177–186. [CrossRef]
160. Zanetti, M.; Gennaro, R.; Skerlavaj, B.; Tomasinsig, L.; Circo, R. Cathelicidin Peptides as Candidates for a Novel Class of Antimicrobials. *Curr. Pharm. Des.* **2002**, *8*, 779–793. [CrossRef]
161. Du, H.; Samuel, R.L.; Massiah, M.A.; Gillmor, S.D. The structure and behavior of the NA-CATH antimicrobial peptide with liposomes. *Biochim. Biophys. Acta* **2015**, *1848*, 2394–2405. [CrossRef]
162. Samuel, R.; Gillmor, S. Membrane phase characteristics control NA-CATH activity. *Biochim. Biophys. Acta* **2016**, *1858*, 1974–1982. [CrossRef]
163. Azim, S.; McDowell, D.; Cartagena, A.; Rodriguez, R.; Laughlin, T.F.; Ahmada, Z. Venom peptides cathelicidin and lycotoxin cause strong inhibition of *Escherichia coli* ATP synthase. *Int. J. Biol. Macromol.* **2016**, *87*, 246–251. [CrossRef]
164. de Barros, E.; Gonçalves, R.M.; Cardoso, M.H.; Santos, N.C.; Franco, O.L.; Cândido, E.S. Snake Venom Cathelicidins as Natural Antimicrobial Peptides. *Front. Pharmacol.* **2019**, *10*, 1415. [CrossRef]
165. Falcao, C.B.; La Torre, B.G.; Pérez-Peinado, C.; Barron, A.E.; Andreu, D.; Rádis-Baptista, G. Vipericidins: A novel family of cathelicidin-related peptides from the venom gland of South American pit vipers. *Amino Acids* **2014**, *46*, 2561–2571. [CrossRef]
166. Oliveira-Júnior, N.G.; Freire, M.; Almeida, J.A.; Rezende, T.M.B.; Franco, O.L. Antimicrobial and proinflammatory effects of two vipericidins. *Cytokine* **2018**, *111*, 309–316. [CrossRef]
167. Wang, Y.; Zhang, Z.; Chen, L.; Guang, H.; Li, Z.; Yang, H.; Li, J.; You, D.; Yu, H.; Lai, R. Cathelicidin-BF, a Snake Cathelicidin-Derived Antimicrobial Peptide, Could Be an Excellent Therapeutic Agent for *Acne Vulgaris*. *PLoS ONE* **2011**, *6*, e22120. [CrossRef]
168. Hao, Q.; Wang, H.; Wang, J.; Dou, J.; Zhang, M.; Zhou, W.; Zhou, C. Effective antimicrobial activity of Cbf-K16 and Cbf-A7A13 against NDM-1-carrying *Escherichia coli* by DNA binding after penetrating the cytoplasmic membrane in vitro. *J. Pept. Sci.* **2013**, *19*, 173–180. [CrossRef]
169. Wang, J.; Li, B.; Li, Y.; Dou, J.; Hao, Q.; Tian, Y.; Wang, H.; Zhou, C. BF-30 effectively inhibits ciprofloxacin-resistant bacteria in vitro and in a rat model of vaginosis. *Arch. Pharm. Res.* **2014**, *37*, 927–936. [CrossRef]
170. Li, L.; Wang, Q.; Li, H.; Yuan, M.; Yuan, M. Preparation, Characterization, In Vitro Release and Degradation of Cathelicidin-BF-30-PLGA Microspheres. *PLoS ONE* **2014**, *9*, e100809. [CrossRef]
171. Xia, X.; Zhang, L.; Wang, Y. The antimicrobial peptide cathelicidin-BF could be a potential therapeutic for *Salmonella typhimurium* infection. *Microbiol. Res.* **2015**, *171*, 45–51. [CrossRef]
172. Liu, C.; Qi, J.; Shan, B.; Gao, R.; Gao, F.; Xie, H.; Yuan, M.; Liu, H.; Jin, S.; Wu, F.; et al. Pretreatment with cathelicidin-BF ameliorates *Pseudomonas aeruginosa* pneumonia in mice by enhancing NETosis and the autophagy of recruited neutrophils and macrophages. *Int. Immunopharmacol.* **2018**, *65*, 382–391. [CrossRef]
173. Tajbakhsh, M.; Akhavan, M.M.; Fallah, F.; Karimi, A. A Recombinant Snake Cathelicidin Derivative Peptide: Antibiofilm Properties and Expression in *Escherichia coli*. *Biomolecules* **2018**, *8*, 118. [CrossRef]
174. Carlile, S.R.; Shiels, J.; Kerrigan, L.; Delaney, R.; Megaw, J.; Gilmore, B.F.; Weldon, S.; Dalton, J.P.; Taggart, C.C. Sea snake cathelicidin (Hc-cath) exerts a protective effect in mouse models of lung inflammation and infection. *Sci. Rep.* **2019**, *9*, 6071. [CrossRef]
175. Blower, R.J.; Barksdale, S.M.; van Hoek, M.L. Snake Cathelicidin NA-CATH and Smaller Helical Antimicrobial Peptides Are Effective against *Burkholderia thailandensis*. *PLoS Negl. Trop. Dis.* **2015**, *9*, e0003862. [CrossRef]
176. Blower, R.J.; Popov, S.G.; van Hoek, M.L. Cathelicidin peptide rescues *G. mellonella* infected with *B. anthracis*. *Virulence* **2018**, *9*, 287–293. [CrossRef]
177. Amer, L.S.; Bishop, B.M.; van Hoek, M.L. Antimicrobial and antibiofilm activity of cathelicidins and short, synthetic peptides against *Francisella*. *Biochem. Biophys. Res. Commun.* **2010**, *396*, 246–251. [CrossRef]
178. Latour, F.A.; Amer, L.S.; Papanstasiou, E.A.; Bishop, B.M.; van Hoek, M.L. Antimicrobial activity of the *Naja atra* cathelicidin and related small peptides. *Biochem. Biophys. Res. Commun.* **2010**, *396*, 825–830. [CrossRef]
179. Li, S.-A.; Xiang, Y.; Wang, Y.-J.; Liu, J.; Lee, W.-H.; Zhang, Y. Naturally Occurring Antimicrobial Peptide OH-CATH30 Selectively Regulates the Innate Immune Response to Protect against Sepsis. *J. Med. Chem.* **2013**, *56*, 9136–9145. [CrossRef]
180. Zhao, F.; Lan, X.-Q.; Du, Y.; Chen, P.-Y.; Zhao, J.; Zhao, F.; Lee, W.-H.; Zhang, Y. King cobra peptide OH-CATH30 as a potential candidate drug through clinic drug-resistant isolates. *Zool. Res.* **2018**, *39*, 87–96.
181. Creane, S.E.; Carlile, S.R.; Downey, D.; Weldon, S.; Dalton, J.P.; Taggart, C.C. The Impact of Lung Proteases on Snake-Derived Antimicrobial Antimicrobial Peptides. *Biomolecules* **2021**, *11*, 1106. [CrossRef]

182. Kim, D.; Soundrarajan, N.; Lee, J.; Cho, H.-S.; Choi, M.; Cha, S.-Y.; Ahn, B.; Jeon, H.; Le, M.T.; Song, H.; et al. Genomewide analysis of the antimicrobial peptides in *Python bivittatus* and characterization of cathelicidins with potent antimicrobial activity and low cytotoxicity. *Antimicrob. Agents Chemother.* **2017**, *61*, e00530-17. [CrossRef]
183. Cai, S.; Qiao, X.; Feng, L.; Shi, N.; Wang, H.; Yang, H.; Guo, Z.; Wang, M.; Chen, Y.; Wang, Y.; et al. *Python* Cathelicidin CATHPb1 Protects against Multidrug-resistant Staphylococcal Infections by Antimicrobial-Immunomodulatory Duality. *J. Med. Chem.* **2018**, *61*, 2075–2086. [CrossRef]
184. Zhang, H.; Xia, X.; Han, F.; Jiang, Q.; Rong, Y.; Song, D.; Wang, Y. Cathelicidin-BF, a Novel Antimicrobial Peptide from *Bungarus fasciatus*, Attenuates Disease in a Dextran Sulfate Sodium Model of Colitis. *Mol. Pharm.* **2015**, *12*, 1648–1661. [CrossRef]
185. Yi, H.; Yu, C.; Zhang, H.; Song, D.; Jiang, D.; Du, H.; Wang, Y. Cathelicidin-BF suppresses intestinal inflammation by inhibiting the nuclear factor-κB signaling pathway and enhancing the phagocytosis of immune cells via STAT-1 in weanling piglets. *Int. Immunopharmacol.* **2015**, *28*, 61–69. [CrossRef]
186. Zhang, H.; Zhang, B.; Zhang, X.; Wang, X.; Wu, K.; Guan, Q. Effects of cathelicidin-derived peptide from reptiles on lipopolysaccharide-induced intestinal inflammation in weaned piglets. *Vet. Immunol. Immunopathol.* **2017**, *192*, 41–53. [CrossRef]
187. Li, S.-A.; Lee, W.-H.; Zhang, Y. Efficacy of OH-CATH30 and Its Analogs against drug-resistant Bacteria In Vitro and in Mouse Models. *Antimicrob. Agents Chemother.* **2012**, *56*, 3309–3317. [CrossRef]
188. Nei, M.; Rooney, A.P. Concerted and birth-and-death evolution of multigene families. *Annu. Rev. Genet.* **2005**, *39*, 121–152. [CrossRef]
189. Pizzolato-Cezar, L.R.; Okuda-Shinagawa, M.; Machini, M.T. Combinatory Therapy Antimicrobial Peptide-Antibiotic to Minimize the Ongoing Rise of Resistance. *Front. Microbiol.* **2019**, *10*, 1703. [CrossRef]
190. Mendes, B.; Almeida, J.R.; Vale, N.; Gomes, P.; Gadelha, F.R.; Silva, S.L.; Miguel, D.C. Potential use of 13-mer peptides based on phospholipase and oligoarginine as leishmanicidal agents. *Comp. Biochem. Physiol. C* **2019**, *226*, 108612. [CrossRef]
191. Almeida, J.R.; Mendes, B.; Lancellotti, M.; Franchi, G.C., Jr.; Passos, Ó.; Ramos, M.J.; Fernandes, P.A.; Alves, C.; Vale, N.; Gomes, P.; et al. Lessons from a Single Amino Acid Substitution: Anticancer and Antibacterial Properties of Two Phospholipase A2-Derived Peptides. *Curr. Issues Mol. Biol.* **2022**, *44*, 46–62. [CrossRef]
192. Lomonte, B.; Pizarro-Cerda, J.; Angulo, Y.; Gorvel, J.P.; Moreno, E. Tyr-Trp-substituted peptide 115-129 of a Lys49 phospholipase A_2 expresses enhanced membrane-damaging activities and reproduces its in vivo myotoxic eject. *Biochim. Biophys. Acta* **1999**, *1461*, 19–26. [CrossRef]
193. Santos-Filho, N.A.; Lorenzon, E.N.; Ramos, M.A.S.; Santos, C.T.; Piccoli, J.P.; Bauab, T.M.; Fusco-Almeida, A.M.; Cilli, E.M. Synthesis and characterization of an antibacterial and non-toxic dimeric peptide derived from the C-terminal region of Bothropstoxin-I. *Toxicon* **2015**, *103*, 160–168. [CrossRef]
194. Santos-Filho, N.A.; Fernandes, R.S.; Sgardioli, B.F.; Ramos, M.A.S.; Piccoli, J.P.; Camargo, I.L.B.C.; Bauab, T.M.; Cilli, E.M. Antibacterial Activity of the Non-Cytotoxic Peptide (p-BthTX-I)2 and Its Serum Degradation Product against Multidrug-resistant Bacteria. *Molecules* **2017**, *22*, 1898. [CrossRef]
195. Peña-Carrillo, M.; Pinos-Tamayo, E.A.; Mendes, B.; Domínguez-Borbor, C.; Proaño-Bolaños, C.; Miguel, D.C.; Almeida, J.R. Dissection of phospholipases A2 reveals multifaceted peptides targeting cancer cells, Leishmania and bacteria. *Bioorg. Chem.* **2021**, *114*, 105041. [CrossRef]
196. Chen, W.; Yang, B.; Zhou, H.; Sun, L.; Dou, J.; Qian, H.; Huang, W.; Mei, Y.; Han, J. Structure–activity relationships of a snake cathelicidin-related peptide, BF-15. *Peptides* **2011**, *32*, 2497–2503. [CrossRef]
197. Li, B.; Kang, W.; Liu, H.; Wang, Y.; Yu, C.; Zhu, X.; Dou, J.; Cai, H.; Zhou, C. The antimicrobial activity of Cbf-K16 against MRSA was enhanced by b-lactamantibiotics through cell wall non-integrity. *Arch. Pharm. Res.* **2016**, *39*, 978–988. [CrossRef]
198. Jin, L.; Bai, X.; Luan, N.; Yao, H.; Zhang, Z.; Liu, W.; Chen, Y.; Yan, X.; Rong, M.; Lai, R.; et al. A Designed Tryptophan- and Lysine/Arginine-Rich Antimicrobial Peptide with Therapeutic Potential for Clinical Antibiotic-Resistant *Candida albicans* Vaginitis. *J. Med. Chem.* **2016**, *59*, 1791–1799. [CrossRef]
199. Ma, L.; Wang, Y.; Wang, M.; Tian, Y.; Kang, W.; Liu, H.; Wang, H.; Dou, J.; Zhou, C. Effective antimicrobial activity of Cbf-14, derived from a cathelin-like domain, against penicillin-resistant bacteria. *Biomaterials* **2016**, *87*, 32–45. [CrossRef]
200. Tajbakhsh, M.; Karimi, A.; Tohidpour, A.; Abbasi, N.; Fallah, F.; Akhavan, M.M. The antimicrobial potential of a new derivative of cathelicidin from *Bungarus fasciatus* against methicillin-resistant *Staphylococcus aureus*. *J. Microbiol.* **2018**, *56*, 128–137. [CrossRef]
201. Falcao, C.B.; Perez-Peinado, C.; Torre, B.G.; Mayol, X.; Zamora-Carreras, H.; Jimenez, M.Á.; Rádis-Baptista, G.; Andreu, D. Structural Dissection of Crotalicidin, a Rattlesnake Venom Cathelicidin, Retrieves a Fragment with Antimicrobial and Antitumor Activity. *J. Med. Chem.* **2015**, *58*, 8553–8563. [CrossRef]
202. Dal Mas, C.; Pinheiro, D.A.; Campeiro, J.D.; Mattei, B.; Oliveira, V.; Oliveira, E.B.; Miranda, A.; Perez, K.R.; Hayashi, M.A.F. Biophysical and biological properties of small linear peptides derived from crotamine, a cationic antimicrobial/antitumoral toxin with cell penetrating and cargo delivery abilities. *Biochim. Biophys. Acta Biomembr.* **2017**, *1859*, 2340–2349. [CrossRef]
203. Cavalcante, C.S.P.; Aguiar, F.L.L.; Fontenelle, R.O.S.; Menezes, R.R.P.P.B.; Martins, A.M.C.; Falcao, C.B.; Andreu, D.; Radis-Baptista, G. Insights into the candidacidal mechanism of Ctn[15–34]—A carboxyl-terminal, crotalicidin-derived peptide related to cathelicidins. *J. Med. Microbiol.* **2018**, *67*, 129–138. [CrossRef]
204. Ponnappan, N.; Budagavi, D.P.; Chugh, A. CyLoP-1: Membrane-active peptide with cell-penetrating and antimicrobial properties. *Biochim. Biophys. Acta* **2017**, *1859*, 167–176. [CrossRef]

205. Pérez-Peinado, C.; Dias, S.A.; Domingues, M.M.; Benfield, A.H.; Freire, J.M.; Rádis-Baptista, G.; Gaspar, D.; Castanho, M.A.R.B.; Craik, D.J.; Henriques, S.T.; et al. Mechanisms of bacterial membrane permeabilization by crotalicidin (Ctn) and its fragment Ctn(15–34), antimicrobial peptides from rattlesnake venom. *J. Biol. Chem.* **2018**, *293*, 1536–1549. [CrossRef]
206. Priya, A.; Aditya, A.; Budagavi, D.P.; Chugh, A. Tachyplesin and CyLoP-1 as efficient anti-mycobacterial peptides: A novel finding. *Biochim. Biophys. Acta Biomembr.* **2022**, *1864*, 183895. [CrossRef]
207. Almeida, J.R.; Mendes, B.; Lancellotti, M.; Marangoni, S.; Vale, N.; Passos, Ó.; Ramos, M.J.; Fernandes, P.A.; Gomes, P.; da Silva, S.L. A novel synthetic peptide inspired on Lys49 phospholipase A2 from *Crotalus oreganus abyssus* snake venom active against multidrug-resistant clinical isolates. *Eur. J. Med. Chem.* **2018**, *149*, 248–256. [CrossRef]
208. Dean, S.N.; Bishop, B.M.; van Hoek, M.L. Natural and synthetic cathelicidin peptides with anti-microbial and anti-biofilm activity against *Staphylococcus aureus*. *BMC Microbiol.* **2011**, *11*, 114. [CrossRef]
209. Juba, M.; Porter, D.; Dean, S.; Gillmor, S.; Bishop, B. Characterization and Performance of Short Cationic Antimicrobial Peptide Isomers. *Biopolymers (Pept. Sci.)* **2013**, *100*, 387–401. [CrossRef]
210. Sala, A.; Cabassi, C.S.; Santospirito, D.; Polverini, E.; Flisi, S.; Cavirani, S.; Taddei, S. Novel *Naja atra* cardiotoxin 1 (CTX-1) derived antimicrobial peptides with broad spectrum activity. *PLoS ONE* **2018**, *13*, e0190778. [CrossRef]
211. Zhang, Y.; Zhao, H.; Yua, G.-Y.; Liu, X.-D.; Shen, J.-H.; Lee, W.-H.; Zhang, Y. Structure–function relationship of king cobra cathelicidin. *Peptides* **2010**, *31*, 1488–1493. [CrossRef]
212. De Aguiar, F.L.L.; Cavalcante, C.S.d.P.; Fontenelle, R.O.S.; Falcao, C.B.; Andreu, D.; Radis-Baptista, G. The antiproliferative peptide Ctn[15–34] is active against multidrug-resistant yeasts *Candida albicans* and *Cryptococcus neoformans*. *J. Appl. Microbiol.* **2019**, *128*, 414–425. [CrossRef]
213. Perez-Peinado, C.; Valle, J.; Freire, J.M.; Andreu, D. Tumor Cell Attack by Crotalicidin (Ctn) and Its Fragment Ctn[15–34]: Insights into Their Dual Membranolytic and Intracellular Targeting Mechanism. *ACS Chem. Biol.* **2020**, *15*, 2945–2957. [CrossRef]
214. El-Aziz, T.M.A.; Soares, A.G.; Stockand, J.D. Snake Venoms in Drug Discovery: Valuable Therapeutic Tools for Life Saving. *Toxins* **2019**, *11*, 564. [CrossRef]
215. Alam, M.I.; Ojha, R.; Alam, M.A.; Quasimi, H.; Alam, O. Therapeutic potential of snake venoms as antimicrobial agents. *Front. Drug Chem. Clin. Res.* **2019**, *2*, 1–9. [CrossRef]
216. Boldrini-França, J.; Cologna, C.T.; Pucca, M.B.; Bordon, K.C.F.; Amorim, F.G.; Anjolette, F.A.P.; Cordeiro, F.A.; Wiezel, G.A.; Cerni, F.A.; Pinheiro-Junior, E.L.; et al. Minor snake venom proteins: Structure, function and potential applications. *Biochim. Biophys. Acta Gen. Subjects* **2017**, *1861*, 824–838. [CrossRef]
217. Fry, B.G.; Koludarov, I.; Jackson, T.N.W.; Holford, M.; Terrat, Y.; Casewell, N.R.; Undheim, E.A.B.; Vetter, I.; Ali, S.A.; Low, D.H.W.; et al. Seeing the Woods for the Trees: Understanding Venom Evolution. In *Venoms to Drugs: Venom as a Source for the Development of Human Therapeutics*; King, G.F., Ed.; Royal Society of Chemistry: London, UK, 2015; pp. 1–36.
218. Rambault, A. FigTree for Windows v. 1.4.4. Available online: http://tree.bio.ed.ac.uk/software/figtree/ (accessed on 12 March 2020).
219. Wheeler, D.L.; Barrett, T.; Benson, D.A.; Bryant, S.H.; Canese, K.; Chetvernin, V.; Church, D.M.; DiCuccio, M.; Edgar, R.; Federhen, S.; et al. Database resources of the national center for biotechnology information. *Nucleic Acids Res.* **2007**, *36*, D13–D21. [CrossRef]
220. Altschul, S.F.; Gish, W.; Miller, W.; Myers, E.W.; Lipman, D.J. Basic local alignment search tool. *J. Mol. Biol.* **1990**, *215*, 403–410. [CrossRef]
221. Morgulis, A.; Coulouris, G.; Raytselis, Y.; Madden, T.L.; Agarwala, R.; Schäffer, A.A. Database indexing for production MegaBLAST searches. *Bioinformatics* **2008**, *24*, 1757–1764. [CrossRef]
222. Camacho, C.; Coulouris, G.; Avagyan, V.; Ma, N.; Papadopoulos, J.; Bealer, K.; Madden, T.L. BLAST+: Architecture and applications. *BMC Bioinform.* **2009**, *10*, 421. [CrossRef]
223. R Core Team. R: A Language and Environment for Statistical Computing [Internet]. Vienna, Austria, 2016. Available online: https://www.R-project.org/ (accessed on 12 June 2022).
224. Pagès, H.; Aboyoun, P.; Gentleman, R.; DebRoy, S. Biostrings: Efficient Manipulation of Biological Strings; R Package Version 2.64.21. Bioconductor. Available online: https://rdrr.io/bioc/Biostrings/ (accessed on 23 September 2022).
225. Osorio, D.; Rondón-Villarreal, P.; Torres, R. Peptides: A package for data mining of antimicrobial peptides. *Small* **2015**, *12*, 44–444. [CrossRef]
226. Katoh, K.; Rozewicki, J.; Yamada, K.D. MAFFT online service: Multiple sequence alignment, interactive sequence choice, and visualization. *Brief. Bioinform.* **2019**, *20*, 1160–1166. [CrossRef]

 animals

Review

Dinosaurs: Comparative Cytogenomics of Their Reptile Cousins and Avian Descendants

Darren K. Griffin [1,*], Denis M. Larkin [2], Rebecca E. O'Connor [1] and Michael N. Romanov [1]

[1] School of Biosciences, University of Kent, Canterbury CT2 7NJ, UK
[2] Department of Comparative Biomedical Sciences, Royal Veterinary College, University of London, London NW1 0TU, UK
* Correspondence: d.k.griffin@kent.ac.uk

Simple Summary: Dinosaurs have been in scientific and popular culture since early fossil discoveries, but increased interest, particularly in their genomes, is expanding. Birds are reptiles, specifically theropod dinosaurs, meaning that if we compare the genomes of related reptile relations, we can get an idea of what the extinct dinosaur genomes looked like. In all animals/plants/fungi, we think of genome organization in terms of chromosomes. Genes sit on chromosomes and each cell of each individual of each species has its own unique organization. Every gene is in exactly the same spot on each chromosome, organized like continents and islands, with the genes as the cities/towns/villages. All reptiles apart from crocodilians have both big and small chromosomes in their genomes but birds particularly so, like the Philippines or Polynesia. Birds have ~80 chromosomes (far more than most organisms) and this is very consistent in most species. Recent studies suggest that this pattern was probably established ~255 million years ago as it is also mostly present in some turtles. In other words, most dinosaurs probably had chromosomes (genome organization) like chickens or emus. In this paper, we present ideas of how this may have contributed to dinosaurs being so diverse in appearance and function.

Abstract: Reptiles known as dinosaurs pervade scientific and popular culture, while interest in their genomics has increased since the 1990s. Birds (part of the crown group Reptilia) are living theropod dinosaurs. Chromosome-level genome assemblies cannot be made from long-extinct biological material, but dinosaur genome organization can be inferred through comparative genomics of related extant species. Most reptiles apart from crocodilians have both macro- and microchromosomes; comparative genomics involving molecular cytogenetics and bioinformatics has established chromosomal relationships between many species. The capacity of dinosaurs to survive multiple extinction events is now well established, and birds now have more species in comparison with any other terrestrial vertebrate. This may be due, in part, to their karyotypic features, including a distinctive karyotype of around $n = 40$ (~10 macro and 30 microchromosomes). Similarity in genome organization in distantly related species suggests that the common avian ancestor had a similar karyotype to e.g., the chicken/emu/zebra finch. The close karyotypic similarity to the soft-shelled turtle ($n = 33$) suggests that this basic pattern was mostly established before the Testudine–Archosaur divergence, ~255 MYA. That is, dinosaurs most likely had similar karyotypes and their extensive phenotypic variation may have been mediated by increased random chromosome segregation and genetic recombination, which is inherently higher in karyotypes with more and smaller chromosomes.

Keywords: dinosaurs; birds; reptiles; chromosome; karyotype; cytogenomics; comparative genomics; genome evolution

Citation: Griffin, D.K.; Larkin, D.M.; O'Connor, R.E.; Romanov, M.N. Dinosaurs: Comparative Cytogenomics of Their Reptile Cousins and Avian Descendants. *Animals* **2023**, *13*, 106. https://doi.org/10.3390/ani13010106

Academic Editor: Ettore Olmo

Received: 15 November 2022
Revised: 22 December 2022
Accepted: 23 December 2022
Published: 27 December 2022

1. Introduction

The question of the origin of reptiles, birds and their relationship to extinct dinosaurs has challenged many generations of biologists; it also continues to interest the lay public.

In recent years, this interest has increased due to new paleontological findings and developments in the field of genomics (e.g., [1,2]). In light of recent paleontological findings, the hypothesis that dinosaurs were completely eradicated by the most recent mass extinction event [3,4] has been pervasive in the scientific literature, as well as fiction, film, television, popular culture and the media. However, this scientific dogma has undergone a fundamental revision in recent times; dinosaurs are now thought to be reptile survivors of the most recent extinction event through their evolution into modern birds (e.g., [1,2]). In other words, birds are both reptiles and dinosaurs.

A karyotype represents a map of the genome of interest and every genome sequence assembly would benefit from an accurate cytogenomic map [5]. However, while we can do this directly in extant species by sampling live material, the chromosomal composition of extinct dinosaurs can only be derived by inference. This conclusion can be reached by examining the whole genome chromosome-level assemblies (CLA) of extant species [5]. With information about several species' CLAs at our disposal, comparative genomics is much more practical in silico [6]. While the auxiliary method of cross-species fluorescence in situ hybridization (zoo-FISH) can uncover further chromosome rearrangements that are difficult to detect using conventional karyotyping (e.g., [7–10]), comparative genomics enables us to outline the genome structure of less well-studied species (e.g., [8,11,12]) and reveal the chromosome rearrangements that led to each species' distinct karyotype (e.g., [10]) using a reference species as a benchmark, e.g., chickens. The relevance and genomic correlates of such chromosome constituents as evolutionary breakpoint regions (EBRs) and homologous synteny blocks (HSBs) that are features of chromosome evolution [6], as well as the mechanisms behind chromosomal breakage and fusion, can all be addressed with the use of CLAs. Given the prevalence of genomics in modern scientific enquiry, cytogenetics (or, more precisely, cytogenomics) is not only a descriptive discipline but also offers a conceptual framework for the organization of any genome. It also provides an original framework for delineating genome–phenome relationships.

Aligning genome assemblies with the respective sets of chromosomes for the majority of species is a challenging task requiring different technologies. The difficulty in developing a genomic roadmap in birds is that the small microchromosomes belie the accurate identification with even contemporary methodologies. Herewith, cytogenomics specialists can examine CLAs in a wide range of bird species, offering new information about the genome structure of the extinct dinosaurs, the ancestors of modern birds and phylogenetic "cousins" of other extant reptiles.

2. Reptilia: Their Phylogeny and Karyotypes

The crown group Reptilia [13] incorporates extinct and existing clades of reptiles, dinosaurs and birds (Figure 1). In particular, it encompasses the diapsid reptiles including the Lepidosauria (tuatara, lizards and snakes) and the Archosauria (extinct dinosaurs, pterosaurs, crocodilians and birds); the latter having originated ~250 million years ago (MYA) [14]. The divergence of synapsids (mammals and their extinct ancestors) in one branch, and anapsids (turtles) and diapsids (other reptiles and birds) in the other, occurred about 310–350 MYA. Evolutionarily, birds represent a monophyletic group of homoeothermic reptiles and are believed to have arisen from theropod dinosaurs about 150 MYA (e.g., [14–16]). *Archaeopteryx* discovered from the late Jurassic (~150 MYA) is recognized as one of the earliest birds. Fossils of most orders of modern birds appear in the early part of the Cenozoic era (65–0 MYA). According to mitochondrial DNA comparisons with extant reptiles, birds are most closely linked to crocodilians, and the divergence between the two lineages is thought to has happened between 210 and 250 MYA (reviewed in [17]). The order Testudines (turtles, tortoises and terrapins) were separated from the Lepidosauria and the Archosauria in the traditional phylogeny because they were thought to be the only survivors of a presumed early anapsid reptile group. The results from molecular phylogeny data estimated from the nucleotide sequences of complete mitochondrial genomes and nuclear genes suggest that turtles should be grouped within the Archelosauria along with crocodilians and birds, while squamates (scaled reptiles including snakes and lizards) are classified into a different clade of Lepidosauria (e.g., [17–21]; Figure 1).

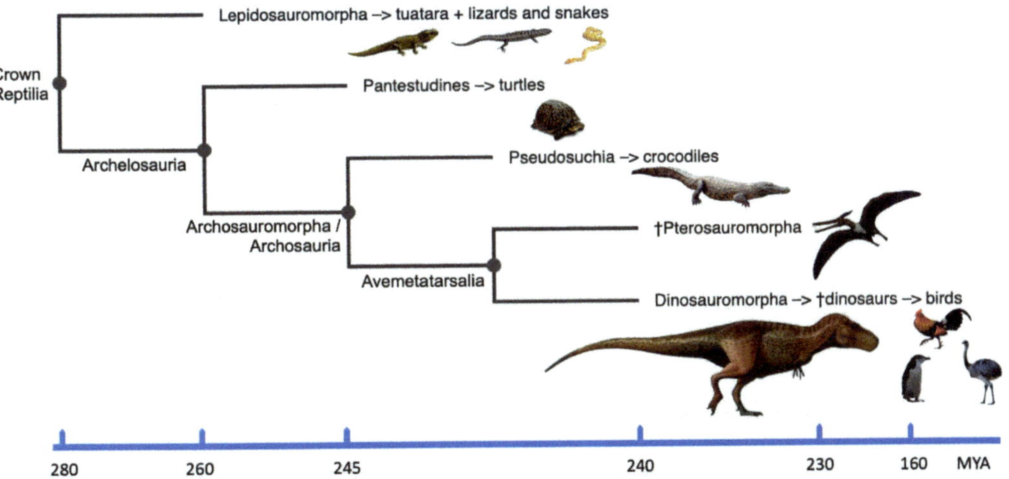

Figure 1. A simplified cladogram of the crown group Reptilia for major evolutionary groups including dinosaurs, birds and reptiles, based on [22–24] and plotted using the Phylo.io webtool [25]. † extinct groups. Time scale is not linear.

Similar to birds, two main chromosomal components of the karyotypes of snakes, turtles, lizards and tuatara (but not crocodilians) are the macro- and microchromosomes. Snakes have a limited spectrum of karyotypic variation. The diploid number $2n = 36$, including 8 pairs of macro- and 10 pairs of microchromosomes, is the most prevalent karyotype in snakes (reviewed in [17]). Lizards also have a low karyotypic variation, mostly with 32–44 chromosomes (e.g., [26,27]) and the extremes being 16 [28] and 62 chromosomes [29]. The diploid number in the lizard *Anolis monticola* is 48 chromosomes, including 24 macro- and 24 microchromosomes. The fission of chromosomes has been demonstrated in conjunction with lower diploid numbers [30]. The karyotype of the indigenous New Zealand lizard genus *Sphenodon* (tuatara) has not changed for at least one million years. It has 36 chromosomes, with 14 pairs of macrochromosomes and 4 pairs of microchromosomes. The similarity of the karyotypes of *Sphenodon* and most Testudines (turtles) points to an ancestral karyotype with a complement of 14 pairs of macrochromosomes and varying numbers of microchromosome pairs [31]. The chromosome number of most crocodilians has long been known [32]; the American alligator (*Alligator mississippiensis*) karyotype ($2n$) consists of 32 macrochromosomes, but notably no microchromosomes (in contrast to other reptiles including birds, e.g., [33]). This peculiar feature, unique among reptiles, suggests a derived karyotype arising as a result of wholesale microchromosomal fusion, probably of single origin (given the small number of monophyletic species in which it is observed). Why crocodilians underwent this change and other reptiles did not is unclear.

Using cDNA clones of functional reptile genes and zoo-FISH, Matsuda et al. [17] created comparative cytogenetic maps of the Japanese four-striped rat snake (*Elaphe quadrivirgata*) and the Chinese soft-shelled turtle (*Pelodiscus sinensis*). The six biggest chromosomes were found to be near-identical between the chicken and turtle, indicating that chromosome homology was well conserved between the two species. However, compared to the turtle, the snake's homology to the chicken chromosomes is lower. The chicken Z chromosome shares conserved synteny with the turtle 6q and the snake 2p chromosomes. These findings imply that conserved sequence blocks have survived during the evolution of Testudines and Archosauria in the genomes of turtles and birds. The lineage of snakes has a karyotype with a number of large-sized macrochromosomes and fewer microchromosomes due to a greater frequency of interchromosomal rearrangements that happened between the macrochromosomes and also between macro- and microchromosomes [17]. The suggested that the molecular phylogenetic links between the three genera are supported by the higher conserved synteny in the comparison between the chicken and turtle than in the comparison between the chicken and snake [18,20].

In the 2000s, bacterial artificial chromosome (BAC) libraries became available for the genomes of five reptilian species, American alligator (*Alligator mississippiensis*), garter snake (*Thamnophis sirtalis*), tuatara (*Sphenodon punctatus*), painted turtle (*Chrysemys picta*) and gila monster (*Heloderma suspectum*), which represent all five major lineages of extant reptiles [33,34]. The green anole lizard (*Anolis carolinensis*) was the first reptilian target species for which the genome sequence and CLA were produced [35], with the painted turtle [36], American alligator [37], garter snake [38] and a variety of other reptile species having followed. These advances, along with the progress in avian genomics, make it possible to study the evolutionary relationships and genome history of higher vertebrates (reptiles, birds and mammals) in a broader context [39]. Comparative mapping of birds and reptiles sheds additional light on the amniotes' evolutionary history [17].

3. Defining Dinosaurs

According to *Britannica* [40], dinosaurs are described as "*Triceratops*, contemporary birds, their most recent common ancestor and all of their descendants." However, for biologists, it could be simpler to picture dinosaurs as reptiles with hind limbs held erect beneath the trunk, similar to how mammals' hind limbs are held. This sets dinosaurs apart from the majority of other reptiles, including lizards and crocodilians, whose legs are often placed to the side. The related evolutionary clades of dinosaurs, birds and reptiles within the crown group Reptilia [13] are shown in Figure 1. Dinosaurs can therefore be straightforwardly discerned from other animals if its easily identifiable sidelong sister branch of pterosaurs is taken out. With this in mind, dinosaurs are survivors of many extinction events including the most recent Cretaceous–Paleogene (K–Pg) [4]. Data combined from molecular cytogenetics and bioinformatics help demonstrate that their adaptability and capacity to survive extinction events may be due, at least in part, to their karyotypic features.

4. Dinosaurian Forefathers and Avian Heirs

The amniote lineage divided into the reptile/bird lineage (diapsids) and the synapsids, which eventually evolved into mammals (and others), ~325 MYA. Over 17,500 diapsid species exist on the planet, the majority of which are birds (~11,000 species). Turtles (Testudines) diverged first (~255 MYA), followed by crocodilians (~ 252 MYA), pterosaurs (~245 MYA) and then true dinosaurs (including birds) ~240 MYA [41,42]. All of these organisms, including dinosaurs and birds, share a common ancestor (Figure 1) that lived 275 MYA. Dinosaur species remained few in number for the following 30 million years, but during the Jurassic period, their numbers, geographic range and body sizes all increased [43]. The subsequent 135 million years of dinosaur evolution were remarkable because they were the dominant vertebrates on Earth and manifested an extraordinary diversity of species [1]. Amazingly, the dinosaurs survived the catastrophic extinction events of the Carnian–Norian and end-Triassic eras (228 and 201 MYA, respectively). There are currently more than 1000 known species of fossil, with around 30 new species (excluding birds) added each year [44].

Usually, the wide diversity and species abundance of dinosaurs is attributed to the extinction of competing species, which allowed the dinosaurs to prosper. However, it has also been suggested that these remarkable levels of abundance and diversity were a result of dinosaur-specific genetic adaptations, which let them outlive other species in hostile habitats. Examples include unusual bone development rates and highly adapted respiration systems [45], such as unidirectional respiration [46]. Avian species may have evolved successfully due to these types of adaptations; evidence for this may be found in the organization and structure of their genomes.

Multiple bird genome sequencing projects have corrected the important dates of avian diversification, thanks to a revised avian phylogeny based on genome assemblies [47,48]. When the Neognathae (Galloanserae/Neoaves) and the Palaeognathae (Ratites/Tinamous) split apart, this was the time of the first bird evolutionary divergence occurring around 100 MYA. The second divergence occurred when the Galloanserae (Galliformes and Anseriformes) and the Neoaves split 80 MYA, with the divergence of the Galliformes (landfowl, such as chicken, turkey, quail and pheasant) and the Anseriformes (waterfowl, i.e., geese, ducks and swans) occurring around 66 MYA. A further significant split of the Neoaves into the Columbea (including pigeons) and the Passerea (including songbirds) was earlier in evolutionary time (67–69 MYA). Around the time of these two major divergences and after the K–Pg mass extinction event [3,4], a total of 36 neoavian lineages evolved due to diversification in a very brief evolutionary period of 10–15 million years, as shown by Jarvis et al. [48] and Prum et al. [49]. Thus, comparative studies using genomics have revised our understanding of the evolution of dinosaurs, providing fascinating insights into the diversification and the evolution of phenotype [47,48], and prompting further research of the dinosaur karyotype.

5. Characterizing a Hypothetical Dinosaur Genome Organization

With no intact DNA available from dinosaur fossils, researchers can infer information about extinct dinosaur karyotypes by studying enough avian and reptile CLAs. Romanov et al. [50] were able to determine the most likely ancestral karyotype of all birds by aligning (near) chromosome-level assemblies from six extant birds and an outgroup of the *Anolis* lizard. This research strategy revealed that the common avian ancestor had a karyotype comparable to that of a chicken or ratite bird [1,50], being a bipedal, terrestrial, tiny Jurassic dinosaur with some flight capacity [1,51]. The next step was to retrace the most likely sequence of rearrangement occurrences that resulted in the avian species' characteristic karyotypes (e.g., [10]). The zebra finch (*Taeniopygia guttata*) and budgerigar (*Melopsittacus undulatus*) were likely subject to the most intra- and interchromosomal changes, while the reconstructed ancestral genome makeup was actually closest to the common chicken karyotype among the birds explored [1,50]. Damas et al. [52] used the method DESCHRAMBLER on fragmented genome assemblies to rebuild the ancestral avian karyotype. A thorough examination of the structure of primitive avian chromosomes was conducted around 14 significant nodes in the evolution of birds. These findings elucidated the varying rates of rearrangement that took place throughout bird evolution. Additionally, it enabled the identification of patterns in the distribution of EBRs along the micro- and macrochromosomes.

A similar method was used by O'Connor et al. [53] to reproduce the diapsids' most likely ancestral karyotype. A universally hybridizing BAC FISH probe set was created for this purpose [10], which was capable of directly hybridizing across species that diverged hundreds of millions of years ago [54]. The BAC probes used in zoo-FISH investigations produced distinctive signals on the chromosomes of anole lizard (*Anolis carolinensis*) and further on those of the red-eared slider (*Trachemys scripta*) and spiny soft-shelled turtle (*Apalone spinifera*). Based on these zoo-FISH examinations, the chromosome rearrangement events might then be anchored from the viewpoint of an ancestral archelosaur (bird–turtle). The chromosomal modifications from the diapsid ancestor through the archelosaur ancestor [55] and the theropod lineage, and to birds, including chickens, were thus recreated by merging molecular cytogenetics with bioinformatics data [1].

In addition to detecting macro- and microchromosomal homologues, the hybridization of BACs to *Trachemys scripta* ($2n = 50$) and *Anolis carolinensis* ($2n = 36$) metaphases also revealed the ancestral diapsid karyotype (275 MYA) with $2n = 36$–46 and with the ratio of macro- to microchromosomes being approximately 1:1 [1,35,56]. The majority of the key characteristics linked to a typical bird karyotype were already set in the archelosaur progenitor 255 MYA [1,57], which experienced rapid transformation in the preceding 20 million years. We know this because the majority of the *Apalone spinifera* ($2n = 66$) and chicken (i.e., ancestral avian) chromosomes (numbered 1-28 + Z) are perfectly syntenic [1]. Studies using chicken chromosome painting on the chromosomes of the painted turtle (*Chrysemys picta*) [58], red-eared slider (*Trachemys scripta*; both $2n = 50$) [9] and Chinese soft-shelled turtle (*Pelodiscus sinensis*; $2n = 66$) [17] further support the hypothesis that macrochromosomes of birds and turtles are syntenic. Given this information, the only parsimonious explanation is that birds and *Pelodiscus sinensis* share a common ancestor in terms of their karyotypic structure, as the number of independent convergent events to achieve the same pattern would be statistically extremely unlikely.

To achieve the common avian karyotype pattern from this (~255 MYA) common archelosaur ancestor, to that present in the majority of the main groups of birds, including the Ratites, Galliformes, Anseriformes, Columbea, Passeriformes and others, only about seven fissions would be required. At the rate of chromosomal change occurring at the time, a complete bird-like karyotype would have most likely formed prior to the emergence of the earliest dinosaurs and pterosaurs ~240 MYA. That is, if the same fission rate that had been present for the preceding 20 million years was maintained for another 15 million years, the early dinosaurs probably had bird-like karyotypes [1,59].

The data available therefore strongly imply that not only in most birds, but also with a high degree of certainty, in many, if not most, extinct dinosaurs, the avian chromosomal pattern was maintained mostly unchanged [60]. Figure 2 illustrates this.

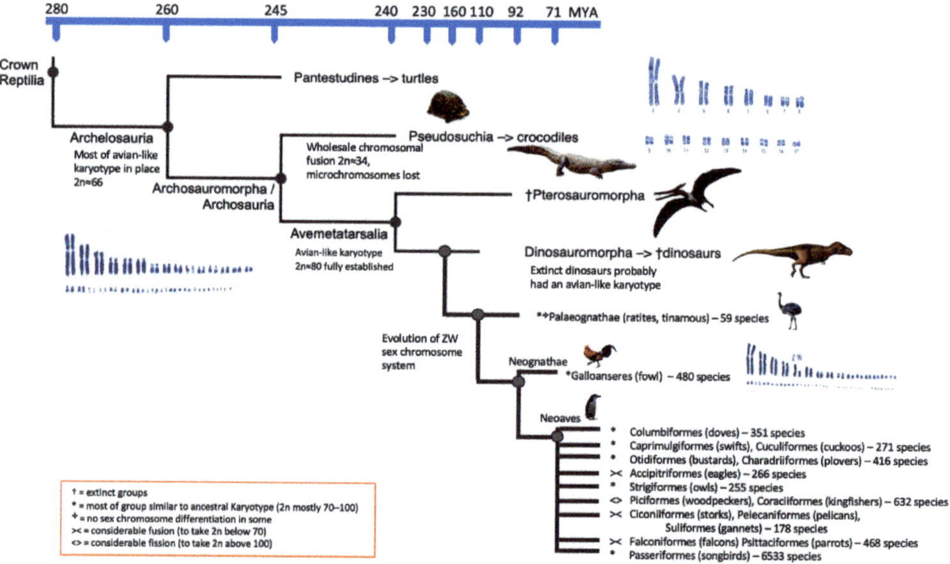

Figure 2. Cladogram of the major evolutionary reptilian groups including dinosaurs and several groups of birds. Likely karyotypic changes given, time scale is not linear.

6. Further Insights into Karyotype Evolution

It had already been suggested that the genome of avian ancestors dating back to more than 80 MYA already had microchromosomes [61,62]. O'Connor et al. [53] asserted that this karyotype organization existed far earlier. They also disputed the idea that the fragmented genome organization (i.e., a karyotype with $2n \approx 80$ chromosomes) accompanied the genome size decrease in birds that has occasionally been linked to the evolution of flight. In other words, a certain correlation was previously thought to exist between genomes with fewer chromosomes (and no microchromosomes) and greater genome sizes (2.5–3 Gb), e.g., in mammals and crocodilians [37,63]. However, O'Connor et al. [62] hypothesized that the bird-like karyotype evolved first, followed by a decrease in genome size and then by the evolution of flight.

In theory, there are two potential reasons why a near-identical karyotype pattern has persisted for ~255 million years: either there is minimal opportunity for change, or the arrangement has been so successful in driving evolution that there is no need to alter. For the former, interchromosomal rearrangement, which is frequently observed in mammals but almost never observed in avian species, is facilitated by repeated elements. This implies that the lack of recombination hotspots [64,65], repeat structures [66–68] or endogenous retroviruses [50,69,70] in the avian karyotypes limit the options for interchromosomal rearrangement. Additionally, purifying selection acting on some of the smallest microchromosomes was demonstrated by Damas et al. [52]. However, a karyotype with little variation over 255 million years also suggests that it is an evolutionary success. The significant degree of phenotypic variation that we observe in dinosaurs (including birds) may be caused by the high rates of chromosome recombination and large number of chromosomes, particularly microchromosomes, [53]. That is, this variation is likely mediated by random chromosome segregation and increased genetic recombination. Although the presence of numerous chromosomes is by no means the sole means via which variation can be created, it may help to explain the apparent paradox of the dinosaurs' enormous phenotypic diversity but low karyotypic diversity. Phenotypic variation is the driving force behind evolution. O'Connor et al. [53] acknowledged the possibility/likelihood that some dinosaurs underwent a significant amount of interchromosomal alteration. Modern instances include multiple fusions in parrots [7,71], falcons [53,72] and many fissions in kingfishers [73]; see Figure 2. It may never be known which particular extinct dinosaur groups accomplished this, if any.

7. What Else Can Be Learned from Cytogenomics?

The main mechanism for chromosomal change in the evolution of the dinosaur genome was likely chromosome inversion with few or no interchromosomal rearrangements. Contiguous ancestral regions (CARs), which are most likely to reflect the chromosomes of the diapsid ancestor, were described by Griffin et al. [1] using the ancestral genome reconstruction program Multiple Genome Rearrangement and Analysis (MGRA) [74]. Although this number was likely underestimated, 49 inversions along the route from the diapsid progenitor to the present chicken were found [1]. Even in chickens, the rate of intrachromosomal alteration may have accelerated in modern times [50]. However, several bird clades, especially the songbirds, the group with the greatest number of species, showed an even larger degree of rearrangement [47,70,75]. The possibility that periods of rapid speciation may have also coincided with higher rates of chromosomal inversion in other dinosaur groups seems plausible [50,62,75].

Around 400 HSBs flanked by EBRs that define the evolution of the dinosaur genome were discovered by O'Connor et al. [62]. The EBRs frequently exist in gene-dense regions with genes involved in lineage-specific biology, transposable elements and other repetitive sequences, according to prior genomic studies in other species (mainly mammals) [76–80]. In contrast, HSBs have a greater number of regulatory and developmental genes [67,77]. Chromosome breaks that damage important genes or do not offer a selective benefit are more likely to be repaired in populations even if found in regions prone to breakage, such as open chromatin regions or recombination hotspots [70].

Using gene ontology (GO) methods, substantial enrichments in the HSB regions were determined with respect to the genes responsible for the development of sensory organs, amino acid transmembrane transport and signaling, plus synapse/neurotransmitter transport, nucleoside metabolism, cell morphogenesis and cytoskeleton [62]. The dinosaur findings reported by O'Connor et al. [62] corroborate the concept that HSBs are enriched for GO terms associated with evolutionary constant phenotypic traits [79]. One such feature is the *Hox* code and its relevance to facilitating both the species diversity and evolutionary success of tetrapods [81]

On the other hand, EBRs are frequently suggested as active spots in genome evolution [82]. In avian EBRs, we observe enrichment for GO terms pertaining to certain adaptation traits, such as forebrain development in budgerigar EBRs (relevant to vocal learning) [70]. There are other significant enrichments of EBRs with genes and single GO terms related to chromatin modification, chromosomal architecture and proteasome/signalosome structure [62]. Chromosomal rearrangements arise, mechanistically, as a result of genetic recombination, DNA repair and/or replication mechanisms occurring following double-strand breaks or replication fork breakage/stalling. Non-Allelic Homologous Recombination (NAHR), Non-Homologous End-Joining (NHEJ), Fork Stalling and Template Switching (FoSTeS) and Microhomology-Mediated Break-Induced Replication (MMBIR) are all mechanisms implicated in this process. The importance of high-resolution analysis in determining the DNA sequence around EBRs has been highlighted [83] as a means of unravelling which specific mechanisms are involved. Transposable elements are also a source of diverse cis-regulatory sequences, constituting a large part of eukaryotic genomes [84]. As our understanding of the biological influence of genomic transposable elements increases, their relevance in chromosomal change, especially in birds is becoming apparent. Flighted birds appear to have smaller genomes yet more transposable elements than their flightless counterparts and all display genomic instability in the genomic regions that are enriched for these sequences [85]. Integrating cytogenomics, single-molecule technologies and genome assemblies in which the repetitive elements have been properly defined is therefore essential to understand how and why the genomes of birds (and their forebears, the extinct dinosaurs) have evolved.

Two recent studies [86,87] introduced a new concept in the context of genomic rearrangements affecting regulation in constantly evolving systems—that of topologically associating domains (TADs). The TADs are conserved inter-species; they buffer evolutionary rearrangements and conserve long-range interactions. Surprisingly, they nonetheless often span EBRs in close proximity to genes with species-specific expression (e.g., in immunological cells). They thus generate novel enhancer-promoter interactions exclusive to the species of interest. In other words, the TAD boundaries are disrupted by EBRs and enable sequence-conserved enhancer elements (from various locations in the genome) inter-species to create unique regulatory modules [86]. All animal genomes are thought to be sequestered into TADs, and they also insulate gene promoters from enhancers. Evolutionary chromosome rearrangements disrupt TAD structure and thereby generate novel regulatory interactions between promoters and enhancers that were historically physically separated. In turn, this could lead to new genomic expression patterns. These could cause deleterious phenotypes but could, nonetheless, create patterns and phenotypes that are evolutionarily advantageous. The EBRs therefore may influence TAD structure in the context of the evolution of gene regulation and of phenotypes of the various different species that arise [87]. No doubt attention will turn to TADs in the study of reptilian chromosome evolution in the coming years.

8. Conclusions

The finding that the avian-like karyotype probably predates the appearance of dinosaurs adds to the paleontological research showing that feathers and pneumatized skeletons initially appeared in more ancient dinosaur or archosaurian forebears [58,82]. For 200 million years, dinosaurs dominated the animal kingdom, with substantial radiations following two great extinction events. Plasticity of the dinosaur clade (including modern birds) in terms of remarkable variation and number of species [88] is noticeable in spite of the near eradication after the K–Pg extinction event [4].

In comparison with other established methodologies, the cytogenomic examination of the possible dinosaur karyotype shines new light on genome evolution, with insights regarding phenotype and an alternate avenue of inquiry [89]. In this regard, this is much more than just a curation effort or a conjectural one. The recent studies outlined and discussed here have shown a peculiar paradox of the dinosaur genome structure that is quite possibly the cause of such phenotypic evolutionary change yet being strikingly karyotypically unchanged in the course of evolution.

Author Contributions: Conceptualization, D.K.G. and D.M.L.; writing—original draft preparation, D.K.G., D.M.L., R.E.O. and M.N.R.; writing—review and editing, D.K.G., D.M.L., R.E.O. and M.N.R.; visualization, M.N.R.; supervision, D.K.G.; project administration, D.K.G. and D.M.L.; funding acquisition, D.K.G. and D.M.L. All authors have read and agreed to the published version of the manuscript.

Funding: This research was funded by the Biotechnology and Biological Sciences Research Council (BB/K008226/1 and BB/J010170/1 to D.M.L, and BB/K008161/1 to D.K.G.).

Institutional Review Board Statement: Not applicable.

Informed Consent Statement: Not applicable.

Data Availability Statement: Not applicable.

Acknowledgments: The authors would like to thank Rafael Kretschmer for his input on Figure 2.

Conflicts of Interest: The authors declare no conflict of interest.

References

1. Brusatte, S.L.; Lloyd, G.T.; Wang, S.C.; Norell, M.A. Gradual Assembly of Avian Body Plan Culminated in Rapid Rates of Evolution across the Dinosaur-Bird Transition. *Curr. Biol.* **2014**, *24*, 2386–2392. [CrossRef] [PubMed]
2. Moon, Y.S. The Link between Birds and Dinosaurs: Aves Evolved from Dinosaurs. *Korean J. Poult. Sci.* **2022**, *49*, 167–180. [CrossRef]
3. Schulte, P.; Alegret, L.; Arenillas, I.; Arz, J.A.; Barton, P.J.; Bown, P.R.; Bralower, T.J.; Christeson, G.L.; Claeys, P.; Cockell, C.S.; et al. The Chicxulub Asteroid Impact and Mass Extinction at the Cretaceous-Paleogene Boundary. *Science* **2010**, *327*, 1214–1218. [CrossRef] [PubMed]
4. Berv, J.S.; Field, D.J. Genomic Signature of an Avian Lilliput Effect across the K-Pg Extinction. *Syst. Biol.* **2018**, *67*, 1–13. [CrossRef] [PubMed]
5. Lewin, H.A.; Larkin, D.M.; Pontius, J.; O'Brien, S.J. Every genome sequence needs a good map. *Genome Res.* **2009**, *19*, 1925–1928. [CrossRef]
6. Larkin, D.M.; Everts-van der Wind, A.; Rebeiz, M.; Schweitzer, P.A.; Bachman, S.; Green, C.; Wright, C.L.; Campos, E.J.; Benson, L.D.; Edwards, J.; et al. A Cattle–Human Comparative Map Built with Cattle BAC-Ends and Human Genome Sequence. *Genome Res.* **2003**, *13*, 1966–1972. [CrossRef] [PubMed]
7. Nanda, I.; Karl, E.; Griffin, D.K.; Schartl, M.; Schmid, M. Chromosome repatterning in three representative parrots (Psittaciformes) inferred from comparative chromosome painting. *Cytogenet. Genome Res.* **2007**, *117*, 43–53. [CrossRef]
8. Modi, W.S.; Romanov, M.; Green, E.D.; Ryder, O. Molecular Cytogenetics of the California Condor: Evolutionary and Conservation Implications. *Cytogenet. Genome Res.* **2009**, *127*, 26–32. [CrossRef]
9. Kasai, F.; O'Brien, P.C.M.; Martin, S.; Ferguson-Smith, M.A. Extensive Homology of Chicken Macrochromosomes in the Karyotypes of Trachemys scripta elegans and Crocodylus niloticus Revealed by Chromosome Painting despite Long Divergence Times. *Cytogenet. Genome Res.* **2012**, *136*, 303–307. [CrossRef]
10. Lithgow, P.E.; O'Connor, R.; Smith, D.; Fonseka, G.; Rathje, C.; Frodsham, R.; O'Brien, P.C.; Ferguson-Smith, M.A.; Skinner, B.M.; Griffin, D.K.; et al. Novel Tools for Characterising Inter- and Intra-Chromosomal Rearrangements in Avian Microchromosomes. In Proceedings of the 2014 Meeting on Avian Model Systems, Cold Spring Harbor, NY, USA, 5–8 March 2014; Cold Spring Harbor Laboratory: Cold Spring Harbor, NY, USA, 2014; p. 56.
11. Romanov, M.N.; Koriabine, M.; Nefedov, M.; de Jong, P.J.; Ryder, O.A. Construction of a California condor BAC library and first-generation chicken–condor comparative physical map as an endangered species conservation genomics resource. *Genomics* **2006**, *88*, 711–718. [CrossRef]
12. Romanov, M.N.; Dodgson, J.B.; Gonser, R.A.; Tuttle, E.M. Comparative BAC-based mapping in the white-throated sparrow, a novel behavioral genomics model, using interspecies overgo hybridization. *BMC Res. Notes* **2011**, *4*, 211. [CrossRef]
13. Pritchard, A.C.; Nesbitt, S.J. A bird-like skull in a Triassic diapsid reptile increases heterogeneity of the morphological and phylogenetic radiation of Diapsida. *R. Soc. Open Sci.* **2017**, *4*, 170499. [CrossRef]
14. Pereira, S.L.; Baker, A.J. A Mitogenomic Timescale for Birds Detects Variable Phylogenetic Rates of Molecular Evolution and Refutes the Standard Molecular Clock. *Mol. Biol. Evol.* **2006**, *23*, 1731–1740. [CrossRef]
15. Kumar, S.; Hedges, S.B. A molecular timescale for vertebrate evolution. *Nature* **1998**, *392*, 917–920. [CrossRef]

16. Schmid, M.; Nanda, I.; Hoehn, H.; Schartl, M.; Haaf, T.; Buerstedde, J.M.; Arakawa, H.; Caldwell, R.B.; Weigend, S.; Burt, D.W.; et al. Second report on chicken genes and chromosomes 2005. *Cytogenet. Genome Res.* **2005**, *109*, 415–479. [CrossRef]
17. Matsuda, Y.; Nishida-Umehara, C.; Tarui, H.; Kuroiwa, A.; Yamada, K.; Isobe, T.; Ando, J.; Fujiwara, A.; Hirao, Y.; Nishimura, O.; et al. Highly conserved linkage homology between birds and turtles: Bird and turtle chromosomes are precise counterparts of each other. *Chromosome Res.* **2005**, *13*, 601–615. [CrossRef]
18. Zardoya, R.; Meyer, A. Complete mitochondrial genome suggests diapsid affinities of turtles. *Proc. Natl. Acad. Sci. USA* **1998**, *95*, 14226–14231. [CrossRef]
19. Hedges, S.B.; Poling, L.L. A Molecular Phylogeny of Reptiles. *Science* **1999**, *283*, 998–1001. [CrossRef]
20. Cao, Y.; Sorenson, M.D.; Kumazawa, Y.; Mindell, D.P.; Hasegawa, M. Phylogenetic position of turtles among amniotes: Evidence from mitochondrial and nuclear genes. *Gene* **2000**, *259*, 139–148. [CrossRef]
21. Cotton, J.A.; Page, R.D.M. Going nuclear: Gene family evolution and vertebrate phylogeny reconciled. *Proc. Biol. Sci.* **2002**, *269*, 1555–1561. [CrossRef]
22. Nesbitt, S.J. The Early Evolution of Archosaurs: Relationships and the Origin of Major Clades. *Bull. Am. Mus. Nat. Hist.* **2011**, *352*, 1–292. [CrossRef]
23. Nesbitt, S.J.; Butler, R.J.; Ezcurra, M.D.; Barrett, P.M.; Stocker, M.R.; Angielczyk, K.D.; Smith, R.M.H.; Sidor, C.A.; Niedźwiedzki, G.; Sennikov, A.G.; et al. The earliest bird-line archosaurs and the assembly of the dinosaur body plan. *Nature* **2017**, *544*, 484–487. [CrossRef]
24. Lee, M.S.Y. Turtle origins: Insights from phylogenetic retrofitting and molecular scaffolds. *J. Evol. Biol.* **2013**, *26*, 2729–2738. [CrossRef]
25. Robinson, O.; Dylus, D.; Dessimoz, C. *Phylo.io*: Interactive Viewing and Comparison of Large Phylogenetic Trees on the Web. *Mol. Biol. Evol.* **2016**, *33*, 2163–2166. [CrossRef]
26. Lamborot, M. A New Derived and Highly Polymorphic Chromosomal Race of Liolaemus Monticola (Iguanidae) from the 'Norte Chico' of Chile. *Chromosome Res.* **1998**, *6*, 247–254. [CrossRef]
27. dos Santos, R.M.; Bertolotto, C.E.; Pellegrino, K.C.; Rodrigues, M.T.; Yonenaga-Yassuda, Y. Chromosomal studies on sphaerodactyl lizards of genera Gonatodes and Coleodactylus (Squamata, Gekkonidae) using differential staining and fragile sites analyses. *Cytogenet. Genome Res.* **2003**, *103*, 128–134. [CrossRef]
28. Schmid, M.; Feichtinger, W.; Nanda, I.; Schakowski, R.; Visbal Garcia, R.; Manzanilla Puppo, J.; Fernández Badillo, A. An Extraordinarily Low Diploid Chromosome Number in the Reptile Gonatodes taniae (Squamata, Gekkonidae). *J. Hered.* **1994**, *85*, 255–260. [CrossRef]
29. Pellegrino, K.C.; Rodrigues, M.T.; Yonenaga-Yassuda, Y. Chromosomal polymorphisms due to supernumerary chromosomes and pericentric inversions in the eyelidless microteiid lizard Nothobachia ablephara (Squamata, Gymnophthalmidae). *Chromosome Res.* **1999**, *7*, 247–254. [CrossRef]
30. Webster, T.P.; Hall, W.P.; Williams, E.E. Fission in the Evolution of a Lizard Karyotype. *Science* **1972**, *177*, 611–613. [CrossRef]
31. Norris, T.B.; Rickards, G.K.; Daugherty, C.H. Chromosomes of tuatara, Sphenodon, a chromosome heteromorphism and an archaic reptilian karyotype. *Cytogenet. Genome Res.* **2004**, *105*, 93–99. [CrossRef]
32. Cohen, M.M.; Gans, C. The chromosomes of the order Crocodilia. *Cytogenetics* **1970**, *9*, 81–105. [CrossRef]
33. Shedlock, A.M.; Botka, C.W.; Zhao, S.; Shetty, J.; Zhang, T.; Liu, J.S.; Deschavanne, P.J.; Edwards, S.V. Phylogenomics of nonavian reptiles and the structure of the ancestral amniote genome. *Proc. Natl. Acad. Sci. USA* **2007**, *104*, 2767–2772. [CrossRef]
34. Wang, Z.; Miyake, T.; Edwards, S.V.; Amemiya, C.T. Tuatara (Sphenodon) Genomics: BAC Library Construction, Sequence Survey, and Application to the DMRT Gene Family. *J. Hered.* **2006**, *97*, 541–548. [CrossRef]
35. Alföldi, J.; Di Palma, F.; Grabherr, M.; Williams, C.; Kong, L.; Mauceli, E.; Russell, P.; Lowe, C.B.; Glor, R.E.; Jaffe, J.D.; et al. The genome of the green anole lizard and a comparative analysis with birds and mammals. *Nature* **2011**, *477*, 587–591. [CrossRef]
36. Shaffer, H.B.; Minx, P.; Warren, D.E.; Shedlock, A.M.; Thomson, R.C.; Valenzuela, N.; Abramyan, J.; Amemiya, C.T.; Badenhorst, D.; Biggar, K.K.; et al. The western painted turtle genome, a model for the evolution of extreme physiological adaptations in a slowly evolving lineage. *Genome Biol.* **2013**, *14*, R28. [CrossRef]
37. St John, J.A.; Braun, E.L.; Isberg, S.R.; Miles, L.G.; Chong, A.Y.; Gongora, J.; Dalzell, P.; Moran, C.; Bed'hom, B.; Abzhanov, A.; et al. Sequencing three crocodilian genomes to illuminate the evolution of archosaurs and amniotes. *Genome Biol.* **2012**, *13*, 415. [CrossRef]
38. Perry, B.W.; Card, D.C.; McGlothlin, J.W.; Pasquesi, G.I.M.; Adams, R.H.; Schield, D.R.; Hales, N.R.; Corbin, A.B.; Demuth, J.P.; Hoffmann, F.G.; et al. Molecular Adaptations for Sensing and Securing Prey and Insight into Amniote Genome Diversity from the Garter Snake Genome. *Genome Biol. Evol.* **2018**, *10*, 2110–2129. [CrossRef]
39. Modi, W.S.; Crews, D. Sex chromosomes and sex determination in reptiles. *Curr. Opin. Genet. Dev.* **2005**, *15*, 660–665. [CrossRef]
40. Dinosaur ancestors. Encyclopedia Britannica. Available online: https://www.britannica.com/animal/dinosaur/Dinosaur-ancestors (accessed on 24 October 2022).
41. Shedlock, A.M.; Edwards, S.V. Amniotes (Amniota). In *The Timetree of Life*; Hedges, S.B., Kumar, S., Eds.; Oxford University Press: Oxford, UK, 2009; pp. 375–379.
42. Hedges, S.B.; Marin, J.; Suleski, M.; Paymer, M.; Kumar, S. Tree of Life Reveals Clock-Like Speciation and Diversification. *Mol. Biol. Evol.* **2015**, *32*, 835–845. [CrossRef]
43. Benton, M.J.; Forth, J.; Langer, M.C. Models for the Rise of the Dinosaurs. *Curr. Biol.* **2014**, *24*, R87–R95. [CrossRef]

44. Weishampel, D.O. *The Dinosauria*, 2nd ed.; University of California Press: Berkeley, CA, USA, 2004.
45. O'Connor, P.M.; Claessens, L.P.A.M. Basic avian pulmonary design and flow-through ventilation in non-avian theropod dinosaurs. *Nature* **2005**, *436*, 253–256. [CrossRef]
46. Farmer, C.G.; Sanders, K. Unidirectional Airflow in the Lungs of Alligators. *Science* **2010**, *327*, 338–340. [CrossRef]
47. Zhang, G.; Li, C.; Li, Q.; Li, B.; Larkin, D.M.; Lee, C.; Storz, J.F.; Antunes, A.; Greenwold, M.J.; Meredith, R.W.; et al. Comparative genomics reveals insights into avian genome evolution and adaptation. *Science* **2014**, *346*, 1311–1320. [CrossRef]
48. Jarvis, E.D.; Mirarab, S.; Aberer, A.J.; Li, B.; Houde, P.; Li, C.; Ho, S.Y.W.; Faircloth, B.C.; Nabholz, B.; Howard, J.T.; et al. Whole-genome analyses resolve early branches in the tree of life of modern birds. *Science* **2014**, *346*, 1320–1331. [CrossRef]
49. Prum, R.O.; Berv, J.S.; Dornburg, A.; Field, D.J.; Townsend, J.P.; Lemmon, E.M.; Lemmon, A.R. A comprehensive phylogeny of birds (Aves) using targeted next-generation DNA sequencing. *Nature* **2015**, *526*, 569–573. [CrossRef]
50. Romanov, M.N.; Farré, M.; Lithgow, P.E.; Fowler, K.E.; Skinner, B.M.; O'Connor, R.; Fonseka, G.; Backström, N.; Matsuda, Y.; Nishida, C.; et al. Reconstruction of gross avian genome structure, organization and evolution suggests that the chicken lineage most closely resembles the dinosaur avian ancestor. *BMC Genom.* **2014**, *15*, 1060. [CrossRef]
51. Witmer, L.M. The debate on avian ancestry: Phylogeny, function, and fossils. In *Mesozoic Birds: Above the Heads of Dinosaurs*; Chiappe, L.M., Witmer, L.M., Eds.; University of California Press: Berkeley, CA, USA, 2002; pp. 3–30.
52. Damas, J.; Kim, J.; Farré, M.; Griffin, D.K.; Larkin, D.M. Reconstruction of avian ancestral karyotypes reveals differences in the evolutionary history of macro- and microchromosomes. *Genome Biol.* **2018**, *19*, 155. [CrossRef]
53. O'Connor, R.E.; Romanov, M.N.; Kiazim, L.G.; Barrett, P.M.; Farré, M.; Damas, J.; Ferguson-Smith, M.; Valenzuela, N.; Larkin, D.M.; Griffin, D.K. Reconstruction of the diapsid ancestral genome permits chromosome evolution tracing in avian and non-avian dinosaurs. *Nat. Commun.* **2018**, *9*, 1883. [CrossRef]
54. Damas, J.; O'Connor, R.; Farré, M.; Lenis, V.P.E.; Martell, H.J.; Mandawala, A.; Fowler, K.; Joseph, S.; Swain, M.T.; Griffin, D.K.; et al. Upgrading short-read animal genome assemblies to chromosome level using comparative genomics and a universal probe set. *Genome Res.* **2017**, *27*, 875–884. [CrossRef]
55. Benton, M.J.; Donoghue, P.C.J.; Asher, R.J.; Friedman, M.; Near, T.J.; Vinther, J. Constraints on the timescale of animal evolutionary history. *Palaeontol. Electron.* **2015**, *18*, 1–106. [CrossRef]
56. Beçak, W.; Beçak, M.L.; Nazareth, H.R.S.; Ohno, S. Close karyological kinship between the reptilian suborder serpentes and the class aves. *Chromosoma* **1964**, *15*, 606–617. [CrossRef] [PubMed]
57. Uno, Y.; Nishida, C.; Tarui, H.; Ishishita, S.; Takagi, C.; Nishimura, O.; Ishijima, J.; Ota, H.; Kosaka, A.; Matsubara, K.; et al. Inference of the Protokaryotypes of Amniotes and Tetrapods and the Evolutionary Processes of Microchromosomes from Comparative Gene Mapping. *PLoS ONE* **2012**, *7*, e53027. [CrossRef] [PubMed]
58. Badenhorst, D.; Hillier, L.W.; Literman, R.; Montiel, E.E.; Radhakrishnan, S.; Shen, Y.; Minx, P.; Janes, D.E.; Warren, W.C.; Edwards, S.V.; et al. Physical Mapping and Refinement of the Painted Turtle Genome (*Chrysemys picta*) Inform Amniote Genome Evolution and Challenge Turtle-Bird Chromosomal Conservation. *Genome Biol. Evol.* **2015**, *7*, 2038–2050. [CrossRef]
59. Baron, M.G.; Norman, D.B.; Barrett, P.M. A new hypothesis of dinosaur relationships and early dinosaur evolution. *Nature* **2017**, *543*, 501–506. [CrossRef] [PubMed]
60. O'Connor, R.E.; Kiazim, L.; Skinner, B.; Fonseka, G.; Joseph, S.; Jennings, R.; Larkin, D.M.; Griffin, D.K. Patterns of microchromosome organization remain highly conserved throughout avian evolution. *Chromosoma* **2019**, *128*, 21–29. [CrossRef]
61. Cracraft, J.; Houde, P.; Ho, S.Y.W.; Mindell, D.P.; Fjeldså, J.; Lindow, B.; Edwards, S.V.; Rahbek, C.; Mirarab, S.; Warnow, T.; et al. Response to Comment on "Whole-genome analyses resolve early branches in the tree of life of modern birds". *Science* **2015**, *349*, 1460. [CrossRef] [PubMed]
62. Burt, D.W. Origin and evolution of avian microchromosomes. *Cytogenet. Genome Res.* **2002**, *96*, 97–112. [CrossRef]
63. Kapusta, A.; Suh, A.; Feschotte, C. Dynamics of genome size evolution in birds and mammals. *Proc. Natl. Acad. Sci. USA* **2017**, *114*, E1460–E1469. [CrossRef]
64. Kawakami, T.; Smeds, L.; Backström, N.; Husby, A.; Qvarnström, A.; Mugal, C.F.; Olason, P.; Ellegren, H. A high-density linkage map enables a second-generation collared flycatcher genome assembly and reveals the patterns of avian recombination rate variation and chromosomal evolution. *Mol. Ecol.* **2014**, *23*, 4035–4058. [CrossRef]
65. Smeds, L.; Mugal, C.F.; Qvarnström, A.; Ellegren, H. High-Resolution Mapping of Crossover and Non-crossover Recombination Events by Whole-Genome Re-sequencing of an Avian Pedigree. *PLoS Genet.* **2016**, *12*, e1006044. [CrossRef]
66. Mason, A.S.; Fulton, J.E.; Hocking, P.M.; Burt, D.W. A new look at the LTR retrotransposon content of the chicken genome. *BMC Genom.* **2016**, *17*, 688. [CrossRef]
67. Warren, W.C.; Hillier, L.W.; Tomlinson, C.; Minx, P.; Kremitzki, M.; Graves, T.; Markovic, C.; Bouk, N.; Pruitt, K.D.; Thibaud-Nissen, F.; et al. A New Chicken Genome Assembly Provides Insight into Avian Genome Structure. *G3 Genes Genomes Genet.* **2017**, *7*, 109–117. [CrossRef]
68. Gao, B.; Wang, S.; Wang, Y.; Shen, D.; Xue, S.; Chen, C.; Cui, H.; Song, C. Low diversity, activity, and density of transposable elements in five avian genomes. *Funct. Integr. Genomics* **2017**, *17*, 427–439. [CrossRef]
69. Cui, J.; Zhao, W.; Huang, Z.; Jarvis, E.D.; Gilbert, M.T.P.; Walker, P.J.; Holmes, E.C.; Zhang, G. Low frequency of paleoviral infiltration across the avian phylogeny. *Genome Biol.* **2014**, *15*, 539. [CrossRef]
70. Farré, M.; Narayan, J.; Slavov, G.T.; Damas, J.; Auvil, L.; Li, C.; Jarvis, E.D.; Burt, D.W.; Griffin, D.K.; Larkin, D.M. Novel Insights into Chromosome Evolution in Birds, Archosaurs, and Reptiles. *Genome Biol. Evol.* **2016**, *8*, 2442–2451. [CrossRef]

71. O'Connor, R.E.; Farré, M.; Joseph, S.; Damas, J.; Kiazim, L.; Jennings, R.; Bennett, S.; Slack, E.A.; Allanson, E.; Larkin, D.M.; et al. Chromosome-level assembly reveals extensive rearrangement in saker falcon and budgerigar, but not ostrich, genomes. *Genome Biol.* **2018**, *19*, 171. [CrossRef]
72. Joseph, S.; O'Connor, R.E.; Al Mutery, A.F.; Watson, M.; Larkin, D.M.; Griffin, D.K. Chromosome Level Genome Assembly and Comparative Genomics between Three Falcon Species Reveals an Unusual Pattern of Genome Organisation. *Diversity* **2018**, *10*, 113. [CrossRef]
73. Christidis, L. *Animal Cytogenetics, Vol. 4: Chordata 3 B: Aves*; Gebrüder Borntraeger: Berlin, Germany, 1990.
74. Avdeyev, P.; Jiang, S.; Aganezov, S.; Hu, F.; Alekseyev, M.A. Reconstruction of Ancestral Genomes in Presence of Gene Gain and Loss. *J. Comput. Biol.* **2016**, *23*, 150–164. [CrossRef]
75. Skinner, B.M.; Griffin, D.K. Intrachromosomal rearrangements in avian genome evolution: Evidence for regions prone to breakpoints. *Heredity* **2012**, *108*, 37–41. [CrossRef]
76. Nadeau, J.H.; Taylor, B.A. Lengths of chromosomal segments conserved since divergence of man and mouse. *Proc. Natl. Acad. Sci. USA* **1984**, *81*, 814–818. [CrossRef]
77. Pevzner, P.; Tesler, G. Human and mouse genomic sequences reveal extensive breakpoint reuse in mammalian evolution. *Proc. Natl. Acad. Sci. USA* **2003**, *100*, 7672–7677. [CrossRef] [PubMed]
78. International Chicken Genome Sequencing Consortium. Sequence and comparative analysis of the chicken genome provide unique perspectives on vertebrate evolution. *Nature* **2004**, *432*, 695–716. [CrossRef] [PubMed]
79. Larkin, D.M.; Pape, G.; Donthu, R.; Auvil, L.; Welge, M.; Lewin, H.A. Breakpoint regions and homologous synteny blocks in chromosomes have different evolutionary histories. *Genome Res.* **2009**, *19*, 770–777. [CrossRef]
80. Rao, M.; Morisson, M.; Faraut, T.; Bardes, S.; Fève, K.; Labarthe, E.; Fillon, V.; Huang, Y.; Li, N.; Vignal, A. A duck RH panel and its potential for assisting NGS genome assembly. *BMC Genomics* **2012**, *13*, 513. [CrossRef] [PubMed]
81. Böhmer, C.; Rauhut, O.W.M.; Wörheide, G. Correlation between *Hox* code and vertebral morphology in archosaurs. *Proc. R. Soc. B: Boil. Sci.* **2015**, *282*, 20150077. [CrossRef] [PubMed]
82. Sankoff, D. The where and wherefore of evolutionary breakpoints. *J. Biol.* **2009**, *8*, 66. [CrossRef]
83. Burssed, B.; Zamariolli, M.; Bellucco, F.T.; Melaragno, M.I. Mechanisms of structural chromosomal rearrangement formation. *Mol. Cytogenet.* **2022**, *15*, 23. [CrossRef]
84. Sundaram, V.; Wysocka, J. Transposable elements as a potent source of diverse *cis*-regulatory sequences in mammalian genomes. *Philos. Trans. R. Soc. B Biol. Sci.* **2020**, *375*, 20190347. [CrossRef]
85. Kapusta, A.; Suh, A. Evolution of bird genomes—A transposon's-eye view. *Ann. N. Y. Acad. Sci.* **2017**, *1389*, 164–185. [CrossRef]
86. Gilbertson, S.E.; Walter, H.C.; Gardner, K.; Wren, S.N.; Vahedi, G.; Weinmann, A.S. Topologically associating domains are disrupted by evolutionary genome rearrangements forming species-specific enhancer connections in mice and humans. *Cell Rep.* **2022**, *39*, 110769. [CrossRef] [PubMed]
87. Preger-Ben Noon, E.; Frankel, N. Can changes in 3D genome architecture create new regulatory landscapes that contribute to phenotypic evolution? *Essays Biochem.* **2022**, *66*, 745–752. [CrossRef] [PubMed]
88. Barrowclough, G.F.; Cracraft, J.; Klicka, J.; Zink, R.M. How Many Kinds of Birds Are There and Why Does It Matter? *PLoS ONE* **2016**, *11*, e0166307. [CrossRef] [PubMed]
89. Zhou, Z. The origin and early evolution of birds: Discoveries, disputes, and perspectives from fossil evidence. *Naturwissenschaften* **2004**, *91*, 455–471. [CrossRef] [PubMed]

MDPI
St. Alban-Anlage 66
4052 Basel
Switzerland
Tel. +41 61 683 77 34
Fax +41 61 302 89 18
www.mdpi.com

Animals Editorial Office
E-mail: animals@mdpi.com
www.mdpi.com/journal/animals

www.ingramcontent.com/pod-product-compliance
Lightning Source LLC
LaVergne TN
LVHW080504180726
843515LV00016B/1385
* 9 7 8 3 0 3 6 5 8 1 7 0 5 *